ENVIRONMENT, POPULATION AND DEVELOPMENT

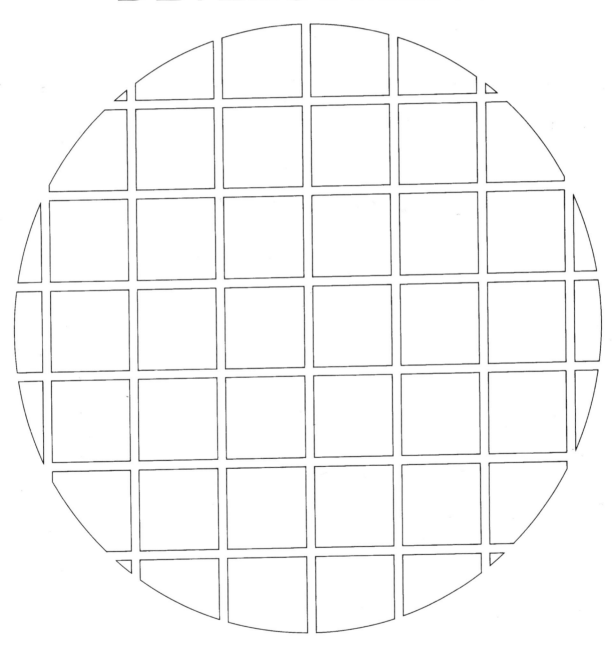

This book is the second in a series published by Hodder & Stoughton
in association with The Open University.

Environment and Society
edited by Philip Sarre and Alan Reddish

Environment, Population and Development
edited by Philip Sarre and John Blunden

Energy, Resources and Environment
edited by John Blunden and Alan Reddish

Global Environmental Issues
edited by Roger Blackmore and Alan Reddish

The final form of the text is the joint responsibility of chapter authors,
book editors and course team commentators.

ENVIRONMENT, POPULATION AND DEVELOPMENT

EDITED BY
PHILIP SARRE AND JOHN BLUNDEN
FOR AN OPEN UNIVERSITY COURSE TEAM

Hodder & Stoughton
A MEMBER OF THE HODDER HEADLINE GROUP

IN ASSOCIATION WITH

The Open University

This book has been printed on paper made from pulp which was bleached without chlorine; other, environmentally-friendly, oxidation agents were used. The paper mill concerned emits very low levels of effluent, and uses raw material from sources renewable through reforestation.

British Library Cataloguing-in-Publication Data

Environment, population and development.
 –(Environment)
 I. Sarre, Dr Philip   II. Blunden, Dr John
 333.7

  ISBN 0–340–66354–5

First published in the United Kingdom 1991. Second edition 1996.

Edited and designed by The Open University.

Index compiled by Sue Robertson.

Typeset by Wearset, Boldon, Tyne and Wear.

Printed in the United Kingdom for Hodder & Stoughton Educational, a division of Hodder Headline Plc., 338 Euston Road, London NW1 3BH, by Butler & Tanner Ltd, Frome, Somerset.

This text forms part of an Open University second level course, U206 *Environment*. If you would like a copy of Studying with the Open University, please write to the Central Enquiry Service, PO Box 200, The Open University, Milton Keynes, MK7 6YZ.

Contents

This book is the second in a series of four which presents an interdisciplinary and integrated explanation of environmental issues. Throughout it is stressed that environmental issues are complex and that to understand them one must combine scientific evidence and theory, analysis of social processes, knowledge of technological possibilities and awareness of underlying value positions. The series considers a range of environmental issues, some local, many transnational and some global. In doing so, it aims to widen readers' *awareness* of environmental issues, to deepen their ability to *analyse* them and to equip them to *evaluate* policies to influence them. It stresses that solutions to particular problems should be complementary with, and ideally contributory to, the solution of international and global problems, including global warming.

The first book, *Environment and Society*, explores three aspects of the context within which contemporary environmental issues occur, and teaches basic concepts and skills to be applied to understand environmental problems. The first aspect is the ecological analysis of life on Earth, stressing the role of living things in modifying the environment through geological time, the variety of life-forms and their mutual dependence. The second aspect is the impact of human societies on environments over millennia and the greatly augmented power, for good or ill, of technological society. The third aspect is that of values, considered both from a dispassionate philosophical viewpoint and through the passionate convictions of environmentalist pressure groups. The perspectives explored in the first book remain highly relevant, but later books in the series become more deeply involved in the detail of particular issues which the course team has recognised as of strategic importance in the interaction of society and environment.

This book focuses on population growth and economic development, which have both contributed to environmental change and affected each other. The book starts by considering the growth of the human population, which has doubled since 1950 to a current figure of about 5.5 billion, and is likely to double again before it ceases to grow. This growth raises questions about both past and future: how has it been possible to achieve economic development in the past so that larger populations could be supported at higher standards of living and – perhaps more crucially – how might this be done in future without excessive damage to the environment? The book explores these questions in the case of agriculture, trade and urbanisation. In so doing it establishes a paradoxical relationship: population growth has historically been stabilised only in countries which experienced economic and technological development, but development greatly increases the impacts of a given population on the environment.

The analyses in the chapters of the book suggest that the relationships between environment, population and development are rather more complex than they often seem in mass media discussions, particularly because of the paradoxical relationship between population growth and economic development. Chapter 1 confirms that population growth has only been controlled in societies which have experienced rapid economic development – but shows that this does not occur automatically, nor does development lead to more rational environmental management in all cases, as is spelled out for agriculture in Chapters 2, 3 and 4. Demand for food and raw materials has led to dramatic agricultural change, but rather than promoting environmentally friendly forms of intensification, the move has been towards capital-intensive mechanical and chemical inputs which have both over-produced certain commodities and created new problems of land degradation and water pollution. Analysis of agriculture suggests that world economic and political systems are compounding both population and environmental problems. Chapter 5 confirms that British and European agricultural policy has reflected political expediency rather than balancing agricultural output, environmental and wider social goals. However, the collapse of the argument for production at all costs leaves the British countryside open to a range of pressures from new non-agricultural uses, many of which threaten further impacts on the quality of life. The centrality of economic and political processes is confirmed by Chapter 6's analysis of the role played by trade in development, and in shaping the unequal relations between developed and less developed countries. Nowhere are the relations more intense or the inequalities more extreme than in cities, which are considered in Chapter 7. Past and present cities can be seen both as the leading edge of civilised achievement and as the most polluted and least sustainable environments. The final question about how optimistic or pessimistic we should be about future cities depends on whether we can achieve forms of development and world trade which can yield better living standards, stabilise population growth and control environmental impacts.

The subsequent books in the series take up two sets of issues arising from this one. *Energy, Resources and Environment* analyses the materials and energy technologies which have been applied in past economic growth, with a critical look at the negative environmental impacts of current systems of supply and demand. Later parts of the book go on to evaluate technical and policy options which are now available, showing that there is room for improvement but that it is not being pursued very energetically. However, a case study of nuclear waste disposal suggests that environmental politics is progressing beyond minority reaction to become a force to be reckoned with. *Global Environmental Issues* uses new and emerging scientific knowledge of global issues, notably atmospheric change (both ozone depletion and greenhouse warming), biodiversity and management of the oceans, to emphasise the need for the adoption of new policies. Finally, the concept and reality of sustainable development is examined to see whether more equitable and less environmentally damaging approaches to the future are possible.

If such a concept can be delivered, it will inevitably have a lot to say about agriculture, urbanisation and alternative forms of economic development. In effect this will mean coming back to the topics in this book, and in fact the book has been written to try to open up options for the future as well as to explain the present and past.

1 Introduction

As you read this chapter, look out for answers to the following key questions:
- What is the rate of population growth in different parts of the world?
- How can these differences be explained?
- To what extent are different economic systems and their attendant environment capable of sustaining current population trends?

Human population growth is not new: at a global level there has been a continuous increase since the late seventeenth and early eighteenth centuries. What is of recent concern is the sustained and accelerating high rates of growth which have been recorded in many countries of the so-called third world (see Box 1.1) during the twentieth century. By the year 2000 the world's population will have reached 6.3 billion (that is, 6300 million or 6.3×10^9) people and will be increasing at about 1.6% per annum. But what do figures like these mean? In terms of the immensity of the numbers involved it is perhaps easier to think of there being an extra 100 million people to feed and shelter each year. This is a population about twice the size of that living in England in 1991. Population growth is not, however, simply an issue of quantities. Not all of the millions of babies born each year place the same demands on the Earth and its resources. In many respects the 80 million extra people added to the populations of the less developed countries result in a much smaller net impact on their environments than the 20 million extra people living in the wealthier nations. This is so simply because of the differentials in living standards and thus of consumption of the world's resources: ecologists, for example, have estimated that on average a North American baby is 50 times as demanding on the world's resources as an Indian child. Thus the question is whether there is a population problem, a resource problem or an allocation problem.

Some ecologists and conservationists (such as some members of the British Green Party) have argued that faced with a finite planet, a fragile physical environment and a limited resource base, human populations cannot be allowed to continue to grow but must stabilise or be reduced. By contrast some population specialists (demographers) have argued that certain population trends are unavoidable, that rates of world population growth have already peaked and that by the twenty-second century the world's population will once again have stabilised of its own accord, albeit at a much higher level than before. Clearly the debate is a critical one. What attitudes and actions should be adopted concerning the implications of population growth in relationship to environmental change? This chapter examines these issues by answering three broad questions (set out in the margin).

Section 2 outlines the broad pattern of population growth in different parts of the world. Section 3 looks at several attempts to explain population change: are different rates of population change the result of environmental constraints or problems of food supply? More recent explanations stress the need to account for fertility and mortality independently, and these are the concern of Section 4. Section 5 brings together earlier analyses to argue that population change is closely linked to the nature and extent of economic development and that population policies must take account of economic and social factors if they are to slow current growth rates in a humane fashion.

Box 1.1   The third world: concept and reality

The concept of a third world emerged during the cold war period when many of the countries of the world were divided into two opposed blocs – the first world, consisting of the United States of America and its allies (the liberal-democratic or advanced capitalist countries), and the second world of the USSR and its allies in eastern Europe (the state socialist or command economies). In the early years *political non-alignment* was the characteristic emphasised as typical of the third world, but as time has passed, emphasis has shifted to the relatively *low level of economic development* of this group, under such terms as 'developing', 'underdeveloped' or 'less developed' countries.

The very origin of the concept – as the 'leftovers' after the first and second worlds – suggests that this might not be a very clearly defined or homogeneous group. This is also probable given the number of countries involved and their varied histories and geographies. When the United Nations was founded, it had about thirty members which would now be regarded as members of the third world, but this number has quadrupled as former colonies have gained political independence. Today, even omitting China, the third world covers about half the land surface of the Earth and contains somewhat over half the world's population (over half of whom live in chronic poverty), but produces only about one-fifth of the world's goods and services.

Since the 1980s it has become common to recognise the great variety of the third world countries by dividing it into sub-groups. For example, the *Encyclopaedia of the Third World* recognises four such groups:

(a)   *Petroleum exporters*: following the rise in oil prices in the 1970s a number of third world countries rose rapidly in terms of income per capita; indeed countries with a sparse population like Kuwait, Libya and Saudi Arabia rose to levels associated with first world countries. However, falling oil prices in the 1980s have caused setbacks, especially in oil exporters with large populations, like Mexico, Indonesia and Nigeria.

(b)   *Advanced developing countries*, otherwise known as newly industrialising countries: a small number of countries (notably Brazil, South Korea, Taiwan, Hong Kong, Singapore and more recently Thailand) achieved rapid rates of industrial and economic growth in the 1970s and have begun to close the income gap between themselves and the less dynamic economies of the first and second worlds.

(c)   *Middle developing countries*: these are another residual category consisting of the third world countries (like Egypt or Peru) which have been neither strikingly successful nor strikingly unsuccessful, with average annual incomes per capita of around a thousand dollars.

(d)   *Least developed countries*: sometimes referred to as the fourth world, these 35 countries contain over one third of the world's population (half being in India) but produce only 3% of its wealth. If China is included in spite of its regime, this group would comprise half the world's population. The average annual income of people in these countries is only a few hundred dollars. To make matters worse, members of this group often suffer serious environmental problems from desertification in countries of the Sahel, like Niger and Chad, to frequent and catastrophic flooding in Bangladesh.

Other more complex classifications exist, notably that of the World Bank, but enough has been said to emphasise that the concept of the third world covers a very complex and varied reality. However, that reality should not conceal the fact that most countries in Latin America, Africa and Asia have incomes per capita which are at best a quarter of that of even a relatively poor first world country like the UK. At worst the populations of these countries have hardly enough to maintain a bare subsistence.

Further information about the origins and variability of the third world can be obtained from A. Thomas *et al.* (eds) *Third World Atlas* (Buckingham, Open University Press; 1994, second edition).

2   The growth of the human population

For most of human history there have only been a few million people living on the Earth. This is not really surprising since they had to depend on hunting and gathering activities to meet their food needs. As human beings gradually developed more advanced methods of meeting their basic needs for food, clothing and shelter, population numbers began to rise. At first this seems to have happened in a rather uncertain fashion with the general trend being upwards, but with crises such as wars, plagues and famines periodically leading to geographically selective reductions in population numbers. Two thousand years ago there were perhaps 300 million people on Earth. It took at least the next 1500 years for the population to double. This relatively slow rate of growth contrasts starkly with what was to follow. By the late eighteenth and early nineteenth centuries some parts of the globe were being introduced to innovative technologies, new economic systems were being adopted and a radically different social order was emerging. In parallel with the changes in the structure and organisation of society came changes in demographic regimes. The result was that between 1750 and 1900 world population size doubled to reach 1.7 billion. Temporal extension and geographical expansion of forces promoting population increase meant that, by the twentieth century, population growth rates had accelerated still further to the point that within the thirty-year period, 1950 to 1980, world population nearly doubled again to 4.8 billion. By 1992 the world population had risen to an estimated 5.45 billion persons. Projection of these very rapid post-war population growth rates produces alarming results, since the inevitable consequences of the world's population doubling every 30 to 35 years are not only a mushrooming of the numbers of people but an explosion in the human demands placed upon the global ecosystem. Long-term extrapolations suggesting, for example, that in 600 years' time there will be standing room only on the Earth are never likely to be matched by real population trends, but they do emphasise the finite capacity of the world. Fortunately projections of this kind are based on statistical extrapolation without adequate reference to the demographic processes that underpin population change.

> *Activity 1*
>
> Draw a graph of the growth of world population over time using the information above. You can check yours against the one at the end of the chapter.

2.1   Geographical variation in growth rates

Even the most superficial examination of population growth rates for different parts of the world shows that there are very significant geographical variations in the rate and character of population change, illustrating the need to base statements about population trends upon a more detailed understanding of the processes which account for population change.

Sometimes the impact of population growth rates is illustrated by considering how long it will take a population to double its present size: doubling times are shown in Table 1.1. The table indicates that the future size of a nation's population is critically determined by its population growth rate rather than by its absolute size. For example, although India's current (1992) population is 880 million to China's 1.181 billion, it has been estimated that by the year 2050 India will reach the staggering total of 1.68 billion (while China will be 1.45 billion), because of India's higher population growth rate of 2.1% per annum. Figure 1.1 shows some striking contrasts between countries in terms of population increase. The first feature which deserves comment is that the global pattern of population doubling times does not mirror the oft-referred to north–south divide in wealth. Although the most rapidly expanding populations are found in Africa and the Islamic world, there is significant variation in the rate of population growth within the developing countries. Nine African states have population doubling times of 20 years or less, but by contrast in South America and in parts of Asia, the doubling times are 41 years or more, putting them on a par with the countries of North America. The part of the world where

Table 1.1

Annual growth rate (%)	Doubling time (years)
1.0	70
2.0	35
3.0	24
4.0	17

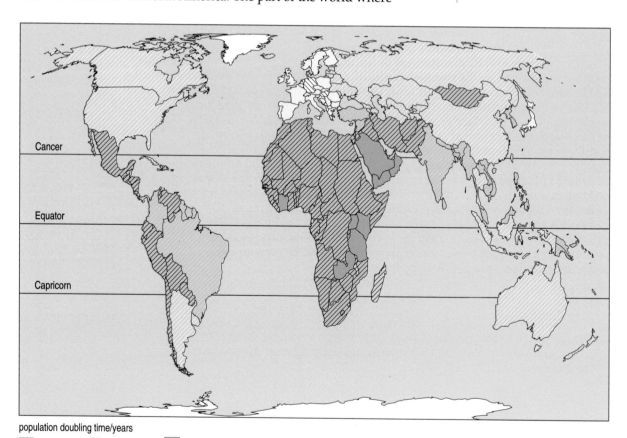

population doubling time/years

over 100 | 21–30 | no data
41–100 | less than 21
31–40 | population decline

▲ *Figure 1.1   Population 'doubling times' (years at current rate of growth).*
Source: Findlay, A. (1994) 'Population doubling times', in Unwin, T. (ed.) *Atlas of World Development*, Chichester, Wiley.

population growth is leading to the most stagnant **demographic regimes** is
western Europe. Here there are many older industrial nations which are
either declining in population terms or where doubling times are over 100
years.

Why do population growth rates vary so greatly between countries?
There are, of course, economic, social, cultural and political forces which
interact with demographic processes to produce the variations evident in
Figure 1.1, but a more immediate answer to the question is that there exist
very substantial geographical variations in mortality and **fertility rates**
between nations and in particular between different regions sharing similar
cultures. Table 1.2 presents some data on regional fertility patterns for the
early 1980s. It is necessary to consider in more detail how fertility is
measured before evaluating the patterns which the table reveals.

The **crude birth rate (CBR)** is the simplest measure of fertility,
measuring the number of births in a year per thousand persons of the mid-
year population of a region or country. Since not all members of a
population (males, children, older females) have the potential for child-
bearing and since these elements of a population may vary in their
numerical importance between one place and another and through time,
the CBR is only of value in comparing fertility levels. More useful are **age-
specific birth rates** which compare the number of births in a given year to
the number of women in specific age cohorts. A particularly sensitive
measure which population geographers and demographers like to use in

Table 1.2 Fertility rates by world region for the early 1980s

	Total population (millions)	Crude birth rate (per 000)	Total fertility rate	GNP per capita (US$)
North Africa	124	41	5.9	1 240
West Africa	161	49	6.8	660
East Africa	150	47	6.6	330
Middle Africa	60	45	6.2	420
Southern Africa	36	36	5.2	2 490
South-west Asia	110	37	5.5	4 110
Middle South Asia	1036	37	5.3	260
South-east Asia	393	33	4.5	720
East Asia	1243	20	2.6	1 360
Middle America	102	34	4.9	1 970
Caribbean	31	26	3.4	–
Tropical South America	220	32	4.3	2 120
Temperate South America	44	24	2.7	2 440
North America	262	15	1.8	13 000
North and west Europe	236	12	1.7	11 500
Eastern and southern Europe	253	15	2.0	5 270 [1]
USSR	274	20	2.5	5 940
Oceania [2]	24	21	2.5	8 700

Notes: [1] Southern Europe only.

[2] Australasia, Polynesia, Micronesia and Melanesia.

Source: Coward. J. (1986) 'Fertility patterns in the modern world', in Pacione, M. (ed.) *Population Geography: progress and prospects*. London, Croom Helm, p. 63.

making comparisons of fertility is the **total fertility rate (TFR)**. This measures the average number of children that would be born to a woman passing through the child-bearing cohorts. Although its calculation is more complex than the CBR, its analysis is more useful since variations in the TFR do not reflect the demographic composition of a population, but rather they mirror more fundamental patterns influencing fertility: these include knowledge of and prevailing attitudes towards contraception, and attitudes relating to family size, structure and formation. Some of these determining influences on fertility are discussed later in the chapter, but more immediately it is important to become aware of the variations in fertility levels which exist across the globe.

Activity 2

Compare the CBR, TFR, and GNP per capita measures shown in Table 1.2 and describe the geographical associations which you perceive to exist. Then read the following discussion.

Table 1.2 suggests the existence of a dichotomy between the more and less developed countries. North America and northern Europe have CBRs at or below 15 per thousand and TFRs which indicate that women have on average less than two children during their lifetime; in Africa CBRs of over 40 per thousand are recorded and TFRs suggest that on average women have six or more children.

Given such stark contrasts it is difficult not to jump to the conclusion that economic environments and demographic behaviour are closely linked, but closer examination of the table shows that wide variations do exist in fertility patterns both in the richer and the poorer countries of the world. For example, south-west Asia – which includes many of the oil-rich states – boasts quite high per capita incomes but also sustains high TFRs. Conversely the regions described in Table 1.2 as east Asia and temperate South America have only modest levels of GNP per capita, but have fertility levels only slightly higher than those of the most developed countries. Low fertility rates have emerged not only in some of the newly industrialising countries of the less developed world, but also in socialist states. Thus, for example, Singapore, Hong Kong and Cuba all now have TFRs of less than two. Demographers place particular significance on the TFR threshold of 2.1 since below this value a population is failing to replace itself by natural processes. This is to say, there will be more deaths than births in the long run, and population decrease rather than population growth can be anticipated.

In summary, Table 1.2 does confirm that major differences in fertility levels underpin the geographical divide observed in national population growth rates. There are also major variations in fertility rates between developing countries which suggests that social, political and cultural factors as well as economic ones have a role in influencing population change.

Consider the following illustrations of demographic change in both developed and less developed countries. In 1991 the population census of Great Britain recorded the presence of 55 million persons, a figure almost exactly the same as the population in 1981. In the previous intercensal decade, population growth had been only 0.57%, while between 1961 and 1971 population growth was at 5.25%. Even this level of growth was low compared with that achieved in the nineteenth century when rates of 10 to 18% were recorded between censuses. Most of the prosperous industrial societies of western Europe and North America are currently experiencing very low levels of population growth, and in some cases numbers have actually been in decline. Denmark is currently ranked as the eighth wealthiest country in the world in terms of income per head of population, yet its population is expected to fall by 3.9% by the year 2000. Social and demographic forces have already produced substantial reductions in household size from 2.96 persons in 1960 to 2.39 in 1980. In 1983 only 27.3% of Danish households were families comprising a married couple and children, while 30.5% were single persons living alone.

Q   From the material presented in the preceding paragraph, what conclusions would you reach about the relationships between population trends and economic growth in western Europe?

A   It is difficult and misleading to make direct causal statements. Nevertheless, indirect relationships are clearly important. In western Europe demographic change occurred during a period of economic growth and social reorganisation. Only in the wealthiest European countries has the social structure evolved to the point where stable demographic structures have also emerged.

In stark contrast with the examples discussed above, which have been selected from the more developed countries, the countries of Africa south of the Sahara had an average population growth rate of over 3% in the 1980s compared with a rate of only 2.6% between 1961 and 1973. In this part of the world population growth is therefore high and accelerating. At current growth rates of 3.3% per annum the 500 million people of these countries will double their numbers by the year 2010. An extreme case is the state of Kenya whose population growth rate is expected to rise to 4.4% per annum during the 1990s, requiring that the country's economy achieve an incredible pace of expansion even to maintain the current living standards of the population. In many countries wealth simply cannot be produced fast enough to keep pace with demographic growth. In the 1980s the per capita incomes of sub-Saharan countries fell

▲   Contrasting conditions for child care: organised day care in Denmark, and a family in Mathare Valley shanty town near Nairobi, Kenya. Third world families face more difficult conditions even where family size is small.

by over 2% per annum, taking the standard of living back to that of over thirty years ago.

It would be very easy to conclude that rapid population growth has caused underdevelopment and human suffering, but in practice there is much evidence to show once again that direct causal statements should be avoided. For example, while the level of maternal mortality in sub-Saharan Africa was tragically high at more than 500 maternal deaths per 100 000 live births, China (with a larger population and with a not dissimilar average income per head) had a comparable figure of only 44 deaths per 100 000 births. Contrasts like these make it difficult to provide simple explanations for population change, as will be shown in Section 3.

2.2   Summary

It would appear that population growth as a phenomenon is not new, but the accelerating and very rapid rates of population increase in the least developed countries of the world in the latter part of the twentieth century have been a cause for concern. The uneven patterns of economic and demographic growth are also highlighted by the trend in many of the wealthiest nations towards stable or gently declining populations. Examination of the data would refute both the view that slow population growth causes rapid economic growth and the common misconception that rapid population growth causes underdevelopment. This is not to say, however, that economic and demographic processes are not affected by the common underlying influences of social organisation and structure.

3   Explanation of population change

Given the importance of population growth, many theorists have attempted to explain what causes or prevents it. However, some of the difficulties of providing a simple explanation have already been referred to in interpreting Figure 1.1 and Table 1.2. This section assesses a range of explanations which have been proposed, moving from those which emphasise environmental constraints, through those which relate population to food supply to one which emphasises the role of industrialisation.

3.1   Environmental constraints and demographic trends

Human populations are, of course, part of the global ecosystem. Like other species the size of the human population is affected by the forces which determine birth, death and migration rates. The physical environment presents society with a series of opportunities and constraints which have affected whether, at any time or place, births have exceeded deaths and in-migration exceeded out-migration. Human interaction with the environment has been very strongly affected, particularly in the twentieth century, by technological and sociological factors. There has, however, in some quarters been a tendency to over-emphasise the human ability to mould the environment and to forget how inseparable people are from the environment and how complex are the energy and material flows which link them to the global ecosystem. Environmental limits on human populations are imposed by the distribution of weather regimes, soil types, land forms and other physical features. As human geographers have frequently pointed out, most limits are specific to technological circumstances and cultural perceptions prevailing at specific points in time. It is important to ask, however, whether there are certain physical elements which have a common or universal effect on all human beings. Figure 1.2, for example, suggests that there are absolute limits to human endurance of certain climatic conditions. Examinations of a world population map readily shows the significance of environmental limits of this kind, with very little human settlement in those parts of the globe with very low temperatures such as the Antarctic or extreme aridity such as the Rub al Khali. Technological advance may have made it technically possible for people to survive under severe conditions over limited time-periods, but the fundamental constraints remain. Figure 1.2 also shows that there are considerable margins around what may be defined as the **human comfort zone** where human beings can exist.

An awareness of climatology makes one realise that weather conditions at any particular place on the Earth's surface are highly variable through time, with temperature and humidity levels changing on a daily and seasonal basis as well as being different from one year to the next. The

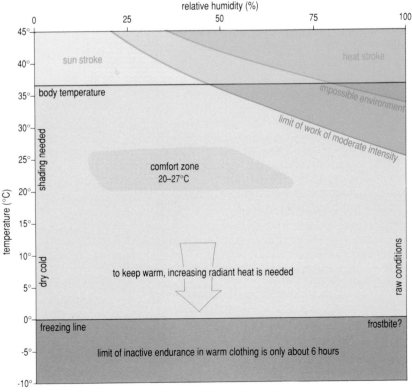

▲ Figure 1.2   The human relevance of climatic ranges.

resulting uncertainty produces geographical patterns which do not usually reflect absolute spatial limits to specific conditions but highly variable spatial weather patterns which often take people unawares in the form of so-called 'natural hazards'. These, regrettably, are still frequent contributory causes of human deaths through circumstances such as famine or floods.

People are affected not only by the climate of a particular environment, but are widely dependent on the capacity of the physical environment to produce the nutrients they need to support human life and to absorb the waste products resulting from human activities. For most of the course of human history food supply to human populations has occurred on a very local scale. The potential of an area to supply the critical food needs of its human population (like its animal populations) was an important influence on population size. Where there was too little food for all the individuals in a population, starvation would result in more deaths than normal thus leading to a reduction in population size to a level closer to sustainable levels in terms of the food supply. Of course, the relationships within this self-regulating system are much more complex than have been described here. For example, it has been established from research on the effects of twentieth-century famines such as the great Chinese famine of 1960–61, that poor nutrition levels reduce human fertility even before they lead to increased mortality. Thus failures in the food supply not only reduce future food demands through increased deaths, but also through reduced numbers of births (see Figure 1.3). As will be discussed later, the causes of famine are not merely environmental ones, but, whatever the causes, the consequences in circumstances of limited human planning and organisation are starvation and population decline.

Q   Describe the absolute level of crude birth and death rates shown in Figure 1.3 and attempt to relate changes in the birth and death rates to trends in the total population size.

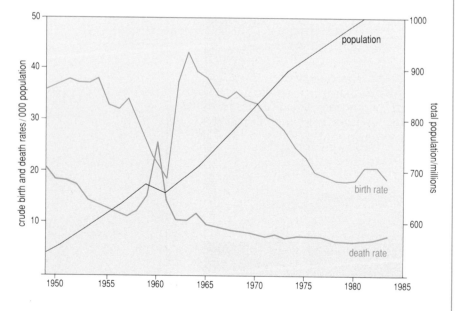

▲ *Figure 1.3   Demographic trends in China.*

A   Crude death rates in China fell from around 20 per 1000 in the late
     1940s to below 10 per 1000 in the mid 1960s. This general trend was
     interrupted by the crisis years of 1959, 1960 and 1961 when drought
     and famine greatly increased the death rate. In the same years the
     adverse circumstances caused the crude birth rate to drop below 20 per
     1000, compared with a norm of 35 to 38 per 1000 in the early 1950s. The
     other very marked feature of the figure is the dramatic and sustained
     decline of crude birth rates in the 1970s to a level of around 20 births
     per 1000 persons. The effect of the considerable excess of births over
     deaths is evident in the rapid growth in the total population during the
     1950s and 1960s. As the vital rates begin to converge in the 1970s, so too
     the rate of population growth slackens, resulting in a slower rate of
     increase in the size of the total population.

It has been suggested above that the physical environment may impose
certain constraints on humans' use of the Earth. The world distribution of
population density (the number of people living in a specified unit of land)
is shown in Figure 1.4 (over). Consider the spatial inequalities which exist
in population distribution: China and India account for less than 9% of the
Earth's land surface, yet together they are home to more than 36% of
humankind. They, along with much of western Europe, Japan, Java, the
Nile delta and the north-eastern United States, have population densities of
over 100 persons per square kilometre over much of their land area. By
contrast Figure 1.5 plots the world distribution of biological productivity
(which is expressed in terms of the potential number of grams of carbon
which could be produced by a square metre of land in a year; this can be
taken as an indicator of the ecological resource potential of different parts
of the Earth). Comparison of the patterns of Figures 1.4 and 1.5 makes it
evident that there is little association between areas of high population
concentration and high potential ecological productivity.

Q   How is it possible that the distributions of population and biological
     productivity are so dissimilar?

A   Society has developed the capacity through technology and through
     social and economic organisation to relax some of the constraints
     imposed by the physical environment. As a result the parts of the world
     yielding the highest output of cereals and livestock are not necessarily
     the same as the areas with the highest potential biological productivity.
     (The next three chapters look at this subject.) In addition it should be
     remembered that human settlement patterns also related to non-food
     production activities.

A wider investigation of the development of the human use of the Earth
would show that as economic systems have become larger in scale and as
social structures have become more complex, so also the potential for
altering the natural environment has grown, as has the potential for
destroying it. It has not simply been technological advances which have
increased the environmental impact of human society, but also the more
powerful and more subtle influences of expanding economic systems, in
forms such as the economic imperialism of western nations with regard to
the less developed countries, which have had profound effects on the
context of subsequent economic, demographic and environmental trends.

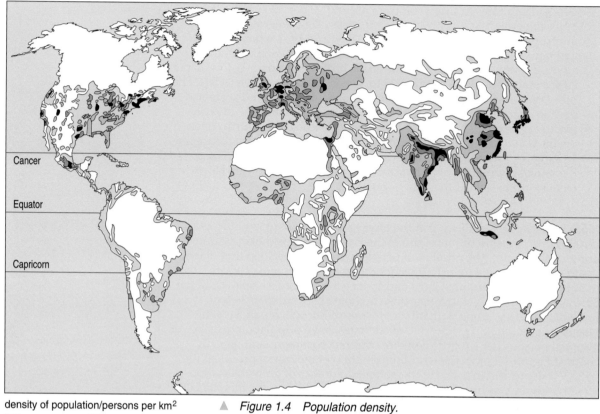

density of population/persons per km²

■ over 200   ▨ 3–25
▨ 100–200   ▢ 0–3
▨ 25–100

▲ *Figure 1.4   Population density.*

3.2   *Population and food resources*

Having recognised the complexity of relationships between society and environment, it is now useful to examine some of the basic ideas which have been suggested to explain how population change affects the environment and vice versa. First of all it is interesting to consider the relationships which might exist between population density and food production systems within subsistence agriculture. Table 1.3 attempts to relate the intensity of agricultural production to population densities in 29 tropical communities.

Q   What conclusions would you draw from examining the survey results?

A   The evidence of the table would seem to suggest a positive correlation between population density and the intensity of land use in certain types of tropical food production. In the cases chosen for study by the researchers, forest fallow production was to be found mainly at population densities of under four persons per square kilometre. Indeed no other system was recorded amongst the communities studied in the most sparsely populated areas. Inversely, at population densities of over 64 persons per square kilometre short fallow or annual cropping was the norm.

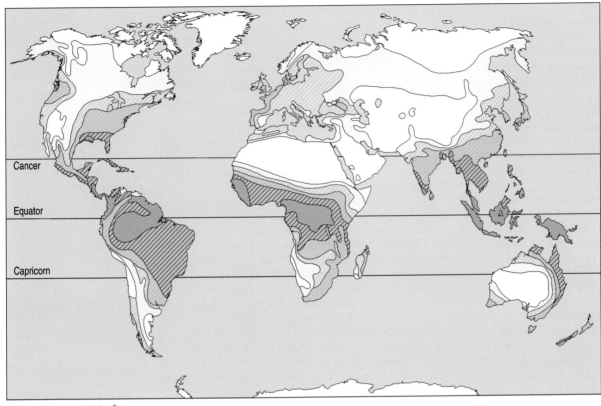

biological productivity/ gC/m²/yr

▨ over 800	▨ 200–399
▨ 600–799	☐ 100–199
▨ 400–599	☐ 0–99

▲ Figure 1.5   World distribution of biological productivity (expressed in grams of carbon).

Table 1.3 Population density and agricultural intensity: evidence from 29 tropical agricultural communities

Population density (persons per square km)	System of supply for vegetable food			
	Forest fallow (FF)	Bush fallow (BF)	Short fallow (SF)	Annual cropping (AC)
less than 4	3	0	0	0
4–16	1	4	1	0
16–64	1	8	0	0
over 64	0	2	6	3

Notes: following Boserup the following definitions are used:

FF: one or two crops followed by 15–25 years fallow

BF: two or more crops followed by 8–10 years fallow

SF: one or two crops followed by one or two years fallow

AC: one crop each year with only a few months fallow.

Source: Turner, B. L., Hanham. R. O. and Portararo. A. V. (1977) Population pressure and agricultural intensity. *Annals of the Association of American Geographers*. Vol. 67, pp. 386–7.

Some writers such as Boserup (1980) have taken this evidence to argue that rising population densities may actually act as a spur to innovation and development. This argument is implausible, however, in the contemporary world since relatively few communities operate closed systems producing food only to meet their own requirements. As soon as one moves from predominantly subsistence agriculture to food production systems governed by other methods of production, the society–land–food relationships become very complex, and the link between population density and agricultural productivity becomes obscure. More important become the social and economic forces which control access to markets and to means of production.

In the world today most people cannot freely organise either their own labour or the land which they work to meet the food and non-food needs of themselves and their immediate geographical community. For example, much of the world's cultivable land is concentrated into large landholdings so rural labourers in areas of high population densities do not have access to land and are not free to organise production methods simply to meet their own food needs. A United Nations study of land use in Central America showed that 10% of land-owners controlled 80% of all farmland. Those farmers who had four hectares or less cultivated 72% of their land, while farmers with over 35 hectares cultivated only 14%. In other words, those with the greatest potential for making improvements had the least incentive to do so, while those in the greatest need were least well positioned for reasons of capital and land shortages to increase production. An ever-increasing proportion of the world's agricultural labour force is employed in producing food for other people and in particular for multinational corporations who sell this production to the world's richest countries. For example, the United Nations estimates that multinational companies control production of approximately 90% of forest products, 90% of coffee, 85% of world cocoa, 85% of tea and 60% of sugar. Agricultural labour is therefore caught up increasingly in patterns of world crop production and the technologies which they use are not linked to local population densities or local food needs.

Even where subsistence agricultural systems persist, conditions of inadequate food production are much more likely in the short run to give rise to migration than to agricultural innovation. Boserup (1980) herself admits that rapid changes in population–resource ratios can lead to the adoption of agricultural technologies which may have environmentally harmful effects. She also recognises that in certain physical environments continual intensification of production may not be possible with rising population pressures. The semi-arid areas of Africa which experienced prolonged droughts in the 1970s and early 1980s would seem to be one example of an environment where population–land relationships have not been able to evolve in the way that might be expected given the relationship implied in Table 1.3, that is, local agricultural production has not been able to reorganise rapidly in response to population demands. See Box 1.2 (on pp. 18–19).

Unfortunately Ethiopia is not an isolated example. Rapid population growth and inadequate agricultural development led no less than 27 of the 39 countries of sub-Saharan Africa to face chronic food shortages in the 1970s and 1980s. Broadening the analysis to the world scale there were 70 countries out of a total of 126 in which food production did not keep pace with population growth between 1970 and 1980. Although global rates of food output grew at 2.3% per annum or above between 1960 and 1980, when population growth is allowed for, per capita growth was only of the order of 0.5%. Even at this lower rate of growth there should have been enough food for the world's people, but trends were geographically

Table 1.4   Indicators of food security in selected countries

	Food production per capita index (1979–81 = 100)	Food import dependency ratio index (1969–71 = 100)	Daily per capita calorie supply as percentage of requirements
	1991	1988–90	1988–90
Ethiopia	86	855	71
Afghanistan	71	193	76
Mozambique	77	300	77
Angola	79	366	80
Rwanda	84	322	80
Somalia	78	134	81
Sudan	80	156	83
Burundi	91	165	85
Haiti	84	364	94

Source: United Nations Development Programme (1994) *Human Development Report*, Oxford, Oxford University Press.

uneven. In the 1980s in sub-Saharan Africa, some 240 million people (about 30% of the total population) were undernourished. At the same time, in South Asia, approximately one third of all babies were born under-weight. Table 1.4 provides a range of indicators of food security in a number of developing countries which clearly show that by 1991 a significant number of countries in the developing world experienced poorer food production figures per capita than they did at the end of the 1970s. Furthermore, food import dependency ratios had worsened and daily calorie intake per capita had deteriorated. Given statistics such as these, and the frequent reporting of famine conditions in many parts of the developing world, it is easy to see why some commentators have returned to considering the ideas of Thomas Malthus, an English demographer of the eighteenth century.

Malthus wrote two influential essays, in 1798 and 1803, in which he laid out his basic argument that the capacity of human populations for natural growth was geometric (as you showed in the graph you drew for Activity 1), while the potential for expanding food production was highly constrained and ultimately was limited by the amount of cultivable land. As a result population growth would always seem in the long run to outpace food production. When this happened Malthus suggested population increase would be brought to a halt by what have come to be termed **Malthusian checks**, namely war, vice and human misery as in the case of famine. In a capitalist society, in which labour is sold as a factor of production, uncontrolled population growth may lead to rising food costs (due to growing demand) and falling wage levels (as the labour supply expands). Ensuing conditions of poverty and malnutrition ultimately lead to population growth being halted by rising mortality levels. Some writers thinking in a neo-Malthusian way have extended the argument to apply to physical resources other than food in the face of the rising demands for fuel and minerals in the twentieth century. In the nineteenth century Malthus' predictions were not confirmed, despite very rapid population growth in what are now termed the more developed countries. Economic development associated with the emergence of industrial capitalism made a rise in living standards possible without facing a Malthusian check, while colonial expansion made it possible for the industrial countries to draw on food and other physical resources from the less developed countries. At the same time, and for reasons which are discussed later, fertility levels dropped substantially.

Box 1.2   Famine in Ethiopia

Ethiopia had an estimated population of 42 million people in 1988. Of these, five million were at risk from famine. One in five children die before their fifth birthday and life expectancy for the population as a whole remains amongst the lowest in the world at 45 years. The population is spatially concentrated in the highlands (i.e. above 1500 m) which dominate the centre of the country and in which varied farming systems have been practised. (See Plate 1.) Surrounding the highlands are lower areas which, although they account for almost 50% of the area, support only about 18% of the population. Population densities are particularly high in the central part of the Northern Highlands (including parts of the provinces of Wollo and Tigre) where densities are over 150 persons per square kilometre and where the cultivable area falls below 0.2 hectares per person: see Figure 1.6. Rising population pressures have led to more frequent cultivation of the land, but in this fragile environment, and given limited capital and technological assistance, this has led to serious damage to the soil structure and to soil erosion and land degradation. Unfavourable ratios of people to cultivable land led to spontaneous resettlement away from the most densely settled areas of the highlands (Wood, 1985). This involved the movement of hundreds of thousands of people in the period since 1950, but these movements also caused ecological problems with resettlement often leading to poor land-use practices or attempts at cultivation on unsuitable slopes. During the 1970s the continued rapid growth of population, combined with the very low average income per head of less than £60 per annum, contributed to a variety of resource crises including inadequate access to sufficient firewood. Large numbers of trees were cut down in many environmentally fragile areas, despite the devastating consequences in terms of the topsoil being blown or washed away. In the 1970s the price of wood in the capital city, Addis Ababa, rose tenfold and cost the average household about 20% of its income.

Rainfall in Ethiopia, as in other countries in this semi-arid region, is of very great importance in sustaining food production since agriculture depends on precipitation rather than either on wells tapping groundwater resources or on irrigation distributing water from rivers. Almost everyone in the region eats non-irrigated drought-resistant crops such as millet and sorghum. Wollo and Tigre provinces in particular have had a long history of being affected by drought, but the succession of drought years which lasted over more than a decade from the early 1970s to the mid 1980s produced extreme crisis conditions in these areas,

most notably in 1984 and 1985. The death of tens of thousands of people from starvation as a result of the Ethiopian famine could as a result easily be interpreted as evidence that in any given area, the ecosystem can only support a limited number of people, and that when this limit is exceeded famine results, leading to a 'Malthusian' check on population growth.

While the drought conditions of the Ethiopian crisis were unavoidable, it is far from clear that widespread famine was inevitable. The current Ethiopian government came to power in 1975, having overthrown the previous Imperial Regime primarily because it failed to pay attention to the plight of the rural poor. For example, there was widespread discontent that the famine victims of the droughts of the early 1970s had been forced to sell their lands to buy food and that pastoralists had been evicted to make way for commercial agriculture. Having a socialist orientation, the revolutionary government was not well received by western governments. It also found itself faced with two civil wars in Eritrea and Tigre and chose to devote a high proportion of its very limited resources to military expenditure. As a result it continued to look to socialist countries for technical and military assistance. Attention to agricultural production was certainly inadequate, with food production per head of the population actually dropping by about 5% in the six years preceding the famine. At the same time the country's main commodity export to the world market, coffee, faced falling prices thus adding to the country's trade deficit and worsening its international debt. In the mid 1980s approximately one third of the country's annual budget was used on debt repayments. Faced with these extremely adverse circumstances it is perhaps less difficult to understand why the Ethiopian government was slow to seek international assistance, not doing so until 1982. There was only a minimal response to these initial requests from western nations. For example, in 1984 Ethiopia received only £7 million of relief assistance from the British government, a figure which compares unfavourably with the £37 million given to Kenya. Some have suggested that since the famine was most severe in the very areas in which the population was actively at war with the government, it was not in the Ethiopian government's interests to seek famine relief for these regions. The distribution of blame for the tragedy is, however, hard to allocate. It nevertheless remains a cruel irony that at the same time as famine was resulting in acute starvation for the people of Ethiopia, the US government was paying farmers billions of dollars to take vast tracts of arable land out of production.

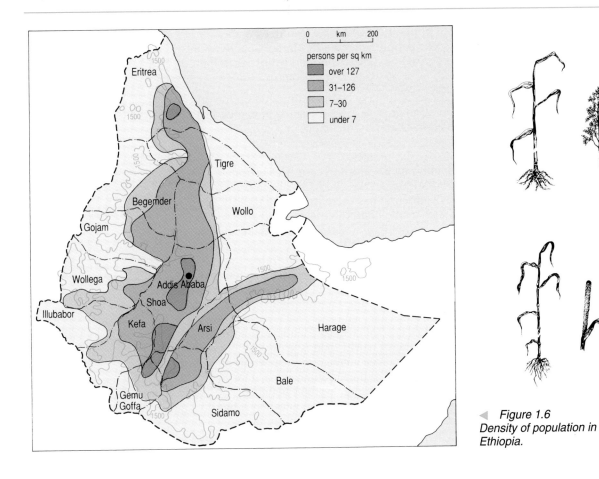

◀ Figure 1.6
Density of population in
Ethiopia.

▲ Migrants in Tigre. Drought leading to crop failures in 1984 and 1985 resulted in mass migrations.

Q   If Malthus was wrong about population and food resources in the western world, why have some people returned to Malthusian views in the face of rapid population growth in less developed countries in the twentieth century?

A   This trend in thinking would seem to have been encouraged on the one hand by the very high population growth rates recorded in some third world countries, and on the other hand by recurrent problems of famine in these countries.

The advent of recurrent famine in many developing countries is not, however, adequate proof of neo-Malthusian trends. As has been shown earlier in Table 1.4, world food production has increased more rapidly than population, and organisations such as the United Nations Food and Agricultural Organisation (FAO) predict that at a global level there should be no problem in expanding production further to meet the needs of the 6.3 billion people who are expected to be alive in the year 2000. Table 1.4 identified the fact that if there is a population resource problem in terms of food provision, then it is a regionally specific problem. Let us examine this distribution aspect in more detail by considering the different types of food in world trade.

No part of the world grows every type of crop in adequate quantities. Furthermore an ever smaller proportion of world food is grown by subsistence cultivators and as a result an ever increasing proportion of the world's population buys the food it needs, rather than growing it for themselves. The result is a growth in the exchange of food between producers and consumers and between one country and another. It has been estimated that 10% of all grain crosses international boundaries between the time of harvesting and consumption. Table 1.5 shows the scale of world trade in grains, meats and milk. A negative sign indicates that a region needs to make net imports of a particular food type. It shows for

Table 1.5 World trade in grains, meat and milk, by world region. 1980–82 (one-year average, million metric tons)

	Grains	Meat	Milk
North Africa/Middle East	−29.2	−1.18	−7.9
Sub-Saharan Africa	−5.8	−0.08	−1.3
South Asia	−1.6	0.05	−0.6
East Asia	−28.8	−0.82	−1.9
Asian Centrally Planned Economies	−18.6	0.23	−0.1
Latin America	−7.6	0.56	−2.2
North America	132.8	0.18	−1.1
European Community	4.3	0.12	11.3
Other western Europe	−11.0	−0.03	1.0
Eastern Europe	−11.4	0.49	0.3
USSR	−36.5	−0.80	−4.3
Oceania [1]	15.6	1.47	5.6
World total [2]	152.7	3.10	18.2

Notes: Negative amounts are imports. Positive amounts are exports.
[1]Australasia, Polynesia. Micronesia and Melanesia.
[2]Exports only.

Source: World Resources Institute (1986) *World Resources*, 1986, p. 51.

example that the Middle East and North Africa bought large quantities of grain, meat and milk, while at the other extreme North America made major exports of grain, but needed to import small quantities of milk. Only the EU countries and the Oceania region, according to this table, had no need to make net imports of any of these commodities.

Q   If the countries of sub-Saharan Africa are the ones in which food shortages are the most acute, why were food imports to these countries not greater?

A   The pattern of the world food trade is determined not by the absolute need for particular produce, but by the ability of consumers to pay the price demanded for the products they require. Many of the countries with the most chronic food shortages cannot afford to buy the food their populations need. The oil-producing countries of the Middle East were by and large able to purchase the foods which their populations needed and which could not be grown domestically, given the ecological and other constraints on their agricultural systems. The oil-producing countries, in particular, had no difficulty in making the necessary purchases of grain, meat and milk because of the exchange value of their natural fuel resources. The countries of sub-Saharan Africa were not, however, in this fortunate position. In the twentieth century the key question of population–food relationships was, therefore, whether the populations of countries with food deficits could afford to buy the food they needed. The continued occurrence of famine was not so much a population/resources problem as a poverty problem. This is not to say that population crises are any the less real or that population factors are not intertwined with patterns of poverty.

Summary

Malthus attempted to analyse the relation between population growth and resources in a highly selective fashion. The arguments which he presented were to a certain extent internally consistent and consequently his conclusions include certain elements of truth. However, the limiting assumptions of his work mean that his conclusions are not universally applicable. In particular he failed to pay sufficient attention to the spatially interdependent nature of economies and societies and to the rules governing interaction and exchange within and between societies.

3.3   Industrialisation and the demographic transition

By the time Malthus wrote his essays a process was under way which was both to prove him wrong (for a period at least) and to provide another 'grand theory' of population. The industrial revolution which transformed the British economy and society in the nineteenth century also had an unprecedented effect on the British population.

This change in the population, subsequently known as the **demographic transition**, is simplified in Figure 1.7(a). In stage 1 both fertility and mortality are high, so population growth is slow. One indirect effect of British industrialisation and economic growth was a reduction in

PUNCHLINE
by **@CHRISTIAN**

GUESS WHAT LITTLE PEASANT?! WE'VE DISCOVERED THAT BECAUSE OF THE "GREEN REVOLUTION".

WHEAT PRODUCTION HERE IN INDIA IN JUST 20 YEARS HAS SOARED FROM 11 MILLION TO 47 MILLION TONS A YEAR!!

AND NOW YOU NEED NOT GO HUNGRY BECAUSE OF PROBLEMS OF FOOD PRODUCTION!!

BUT SIR, MY FAMILY AND I ARE **STILL** SUFFERING FROM MALNUTRITION!!

ER YES... UNFORTUNATELY THATS A PROBLEM OF DISTRIBUTION...

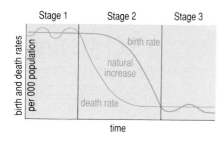

◁ *Figure 1.7(a)*
The demographic transition model.

mortality as a result of better nutrition, sanitation and medicine. The wide
gap between high fertility and low mortality in the early part of stage 2
implies a rapid growth in the population. The crucial and unprecedented
event occurred in the latter part of stage 2 with the reduction of fertility.
This produced slower population growth and ultimately led to the re-
establishment of a balance between crude birth rates and death rates in
stage 3. Initially these critical changes were attributed to industrialisation
producing better living standards and lower levels of infant mortality,
which it was believed had a direct influence on the reproductive behaviour
of married couples. More recently it has been suggested that the timing of
fertility decline in England and Wales was associated with a substantial
change in public opinion in favour of the limitation of family size, aided by
the advent of mass education and only later influenced by a growing
awareness of the reductions in infant mortality.

The earlier interpretation of fertility decline led to the expectation by
some that industrialisation would produce falling death rates and later
falling birth rates in other countries and in particular in the less developed
world, as so-called 'modernisation' spread to new geographic areas. In
practice, not only did knowledge and fruitful application of the medical
advances which had increased life expectancies in western Europe diffuse
very rapidly to other parts of the world, but they did so independently
from the spread of industrialisation. As a result the transition model as
initially formulated was demoted by some to no more than a descriptive
model with no explanatory power (since the demographic experience of
many countries as defined by the model did not relate to phases of so-called
'modernisation'), while others sought alternative explanatory frameworks.
For example, it was suggested that the demographic transition was
internally triggered, with fertility not falling until mortality had fallen, and
with couples gradually altering their perceptions of desired family size
once infant mortality levels had fallen to a sufficiently low level to ensure a
reasonable chance of their children surviving into adulthood. Empirical
evidence certainly shows that in no society has fertility declined prior to a
sustained decline in mortality. There are, however, many examples from
less developed and indeed from the Arab oil-rich states to show that neither
rising income levels nor increased life expectancy necessarily stimulate a
downturn in fertility. In some less developed countries advances in hygiene
and medical care have now reduced crude death rates to levels as low as or
lower than in the more developed countries, yet crude birth rates remain
much higher. Population experts have therefore moved in the direction of
seeking separate explanatory models of mortality and fertility trends, with
the term **fertility transition** increasingly being used. This recognises that a
radical change in fertility regimes has been experienced in many countries,
which in descriptive terms resembles the trends outlined in the
demographic transition model, but which in explanatory terms rests on a
very different basis. Woods (1982) has therefore proposed a variable model

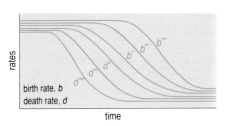

rates

birth rate, b

death rate, d

time

◀   *Figure 1.7(b)   A variable model of the demographic transition. In the variable model of the demographic transition, trends in death rates are not necessarily accompanied by any specific trends in birth rates. Thus, for example, a country on trajectory d''' could have b''' or b'' or b'. Thus not only are death and birth rates variable between countries, but so also are rates of population growth.*

of the demographic transition (shown in Figure 1.7b), with different determinants producing a reduction in death rates from those producing fertility decline.

3.4   Summary

This section has argued that certain demographic features such as infant mortality trends are closely related to levels of eonomic development (as distinct from economic growth) and to patterns of poverty. Attempts to explain all demographic processes as direct responses to economic stimuli would, however, be simplistic and misleading. Models such as the demographic transition are helpful descriptive generalisations but they often lack an adequate explanatory basis. Economic growth was undoubtedly vital in western Europe in the achievement of the so-called demographic transition, but only because economic change produced new social conditions favourable to lower fertility levels.

4   Fertility, mortality and migration

The previous section has shown that attempts to identify single explanations or 'grand theories' of population change have been unsuccessful. They do show that many factors are at work and suggest that more detailed analyses of fertility and mortality are needed. This section follows up this suggestion, providing more detailed information on variations of fertility and mortality and seeking more evidence on the factors which influence them. It also looks in more detail at the third component of population change – migration. As well as helping to clarify what influences fertility and mortality, migration is closely linked, through urbanisation, to economic development, which has been identified as a major, though complex, influence on population change.

4.1   Infant mortality

One demographic indicator which is often used to measure the ability of a particular environment (in physical, economic and political terms) to adequately provide for its population, is the proportion of babies surviving their first year of life. It remains a statistical fact that the chances of death are higher during the first year of life than in any other specific year. The reasons for this are in one sense obvious, in so far as infants are totally incapable of looking after themselves, and depend on the abilities of their parents, set within a specific social environment, to care for them. In other words the social environment (in the broadest sense of this term) into which a baby is born is of critical importance. Where this environment for one reason or other is unable to provide adequate support, the lives of the youngest members of society become very vulnerable.

Infant mortality rates are measured in terms of the number of deaths in a year of children under the age of one year per thousand live births. Clearly life chances are much greater in some places than others with some environments much more able to sustain the most vulnerable members of society. In North America and western Europe there are relatively few children being born, but they have an excellent chance of surviving (fewer than 10 infant deaths per 1000 births). Conversely, infant mortality is very high in Africa and Southern Asia with infant mortality rates in many countries exceeding 150 per 1000. In Africa few countries could offer their infants more than a 9 in 10 chance of surviving their first year. Why is this?

The survival of infants depends on a wide range of factors. A particularly strong relationship has been found to exist between the educational status of mothers and the life prospects of their children. This immediate relationship should not, however, be allowed to cloud the broader environmental circumstances affecting infant mortality rates. These

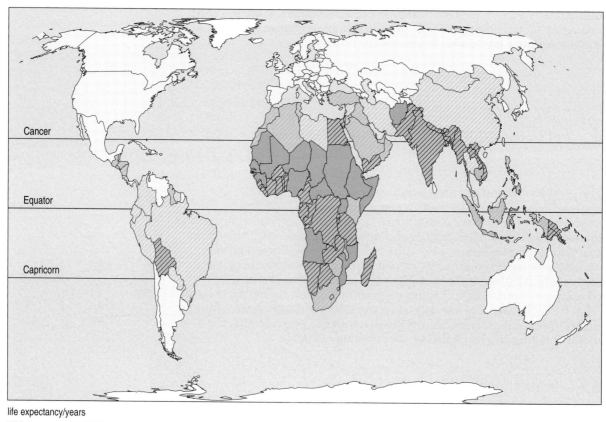

life expectancy/years

▨ less than 50	▨ 65–69	
▨ 50–59	☐ 70–89	
▨ 60–64	☐ no data	

▲ *Figure 1.8   Life expectancy at birth, 1990 (years).*

include provision of adequate medical care for mothers and children in an accessible form before and after childbirth, the nutritional status of the mother and child, adequate shelter and fuel for the immediate environment in which the child will live, the provision of clean water and in most cases the relative stability of household income necessary for the achievement of many of the other conditions. Most demographers agree that population increase in itself is not a cause of infant mortality, and indeed if there is a causal relationship that it runs in the opposite direction. Factors associated with the distribution of wealth and poverty play a much more important role in providing ultimate explanations for the uneven world patterns of infant mortality.

Spatial analysis of infant mortality patterns carried out at different scales shows surprisingly constant associations between high infant mortality and areas of low income. A recent cross-national study of the socio-economic factors affecting child and infant mortality, by the demographer John Hobcraft, stressed the importance of both mothers' and fathers' educational and occupational status. These factors are likely to influence household income. Hobcraft *et al.*'s (1984) work is of special interest since it shows that the spatial variations which exist between countries are also found *within* countries, with for example stark contrasts between the health of children belonging to wealthy urban families from those of the rural poor. For child mortality the former social group was twenty times less at risk than the children of the rural poor. Studies even within cities have shown stark contrasts in infant child mortality rates between social groups and consequently between the urban environments in which different social groups live.

As with agricultural technology with regard to food production, so also with medical knowledge, the twentieth century has been a period of remarkable advances. This has permitted **life expectancies** in the more developed countries to rise rapidly (see Figure 1.8). Japan, for example, has now attained a life expectancy of 78 years. It was anticipated that the much lower life expectancies of the less developed countries would catch up

▲ *A clinic held weekly for Fulani women and children near Dori, and collecting water for domestic use in Ofrey village, both in Burkina Faso. Provision of health care facilities and clean water are two significant factors in reducing infant mortality rates.*

rapidly with those of the more developed states, as medical knowledge and health-care facilities diffused to lower-income countries. Although there has been significant progress, life expectancies have not, however, risen as rapidly as anticipated. Initial progress occurred largely through the introduction of public health campaigns such as spraying against malarial infection and vaccination against diseases such as smallpox. Medical advances could not, however, remove poverty-related diseases. Diarrhoea-related illnesses and malnutrition have kept child and infant mortality rates high in a great many developing countries and have as a result also kept life expectancies lower than was originally predicted by demographers. The very slow rate of improvement in infant mortality rates has been a cause of particular concern, emphasising the inability of parents in the fragile social and economic environments of many developing countries to be able to earn an adequate living to provide for themselves and their children.

4.2 Fertility

As with mortality trends, so also with fertility, it was anticipated incorrectly that the western experience would transfer to less developed countries. The expected decline in fertility levels in developing countries has not occurred as soon or as quickly as population experts initially predicted. To understand some of the reasons why these expectations were formulated and why they have not been well founded, it is necessary to return to considering the demographic transition as it occurred in western Europe.

What explanations have been offered for declining fertility levels? The central pivot of early demographic transition theory was that all aspects of demographic regimes were fundamentally influenced by industrialisation and modernisation, yet this fits uncomfortably with the empirical data for western Europe. Demographers suggest that it was France rather than the then more industrialised nation of England which was first to experience fertility decline. In a similar vein, most regions of western Europe, despite their diverse economic circumstances, had experienced some reduction in fertility between 1890 and 1920. Figure 1.9 shows that in Germany a

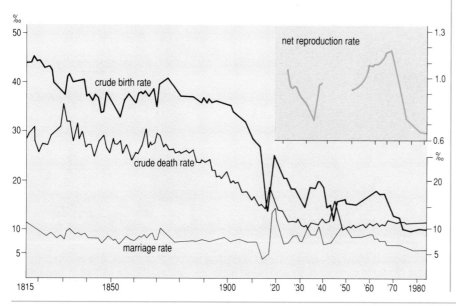

◀ *Figure 1.9 Natural population change in the Federal Republic of Germany, 1815–1983.*

sustained fall in fertility was evident from 1890. The diagram also shows that, despite the inevitable impacts of two major wars on both the crude birth and death rates of Germany, fertility in the long run has continued to fall until very recently, when it stabilised at a level slightly below mortality. In Germany, as in other west European countries, some of the reduction in fertility levels can be explained by the changing age of marriage, the increased involvement of women in the waged labour force, increased knowledge of and changing attitudes to contraception and new attitudes to family size, structure and formation. All of these factors are undoubtedly important, but the key issue is what determines these attitudes, with a continuing academic debate about the relative importance of cultural influences relative to the role of socio-economic structures and political contexts. Similar piecemeal explanations have been offered for the continuing high levels of fertility in many of the less developed countries.

Q   What socio-economic circumstances might encourage couples in the third world to have large families?

A   In the absence of adequate state welfare schemes to care for the elderly, one way to reduce the hardships of old age is perceived to be having many children who can provide for you. In the short term children may be seen to be potential wage-earners. This is particularly the case where parents do not have to support children throughout an extended education system and where agriculture depends on labour-intensive techniques. In urban situations in countries such as Brazil, India and Thailand, children are often significant wage-earners for the household. This situation occurs because of the difficulties that adults have in finding waged employment by comparison with the many jobs open to children in sweat-shops.

◄   Children often have to contribute to the family income by working – as here in the rice harvest in the barrio of Lower Rugak, Agusan del Sur, Mindanao in the Philippines.

An interesting extension of these ideas which can only be mentioned very briefly here is the work of the Australian demographer, J. Caldwell. In defending the view that it is economically rational for people in third world countries to have large families, he has proposed a theory of fertility change linked to the net direction of inter-generational wealth flows within the household. Where the economic context of a household results in a net transfer of wealth from children to parents because of the factors listed above and others, then it is entirely rational for parents to seek to have large families, while, as Caldwell (1982) points out, the increasing costs of having and supporting children in more developed countries has led to a reversal of the net wealth flows from older to younger generations. This in turn, he argues, has produced a progressive reduction in family sizes.

Caldwell's ideas remain to be fully tested, but they are interesting because unlike other work they focus the explanation neither on the individual decision-maker, nor on economistic arguments about the state, but rather on the fundamental demographic unit of the family. They lead naturally to an investigation of the relative significance of culture and economy in determining the roles of persons within the household and of the relations between production and reproduction within the household.

4.3   *Migration*

The relationship between household behaviour and the economic and cultural context is also explored by students of migration. Workers have often not been able to make an adequate living in agriculture and have transferred their labour to other sectors of the economy. To do this has usually also involved spatial relocation from a rural to an urban environment where there was the possibility of employment in industrial or service activities. Industrial developments in the more developed countries in the eighteenth and nineteenth centuries encouraged rural to urban migration during the historical phase where these countries were experiencing rapid population growth, although there was, of course, also massive migration of surplus population to settle in the 'New World'. Although in developing countries today massive population redistribution away from rural areas has been recorded, the character of urban growth has been rather different from that in western Europe and North America. In the discussion which follows, treatment of the labour migration process in contributing to urban growth is therefore separated into its effects in more developed and less developed countries. It should also be noted that although human migration from one environment to another happens for many different reasons (for example, marriage migration, residential relocation to find better accommodation) and at many different scales, what follows concerns only the category of movement known as **labour migration**.

The industrial revolution which occurred in Britain and in other countries in western Europe in the eighteenth and nineteenth centuries resulted in a redistribution of labour from agricultural to industrial activities. There was associated with this a geographical redistribution of population to the new centres of industrial activity, as labour was drawn into the industrial process. The rapid physical expansion of towns and cities was, according to A. Weber writing at the turn of the century, 'the most remarkable social phenomenon of the century'. The concentration of capital in production in urban areas required a parallel growth of the urban labour force. Industrial production favoured the increased division of labour with specialisation of economic functions, and this process accelerated as the scale of production rose from regional to national and then international levels. The famous Glasgow economist, Adam Smith, whose analysis of the nature of the division of labour was to earn him international renown, was also well aware of the consequences of industrial capitalism. Indeed he predicted that the workers needed to sustain the industrial process would also be 'brutalised' by urban living.

The pace of the industrialisation process was such that accommodation for the migrant workers had to be constructed very quickly. The exact form of the housing solutions provided for the rapidly growing industrial cities of western nations varied greatly from one city to another, creating geographical variety in the forms of 'problem' housing which these cities would have to deal with in the twentieth century. As the case study on Glasgow in Box 1.3 shows, the newly built housing stock for the city's Scottish migrant populations were mainly one-and two-room tenements, which were to make the city notorious in the early twentieth century as having one of the most overcrowded housing stocks in western Europe. Ironically in the eighteenth century the tenement housing stock was far from the worst available and Irish immigrants, because of their relative poverty and desire for accommodation close to the city's casual labour market, showed a 'strong predilection for the densely packed warrens of cheap housing in the central districts' (Gibb, 1988, p. 18).

Insanitary conditions in Glasgow and other industrial agglomerations clearly arose from a complex web of forces associated with the low incomes of the new industrial force and the inadequate physical environment in which they were expected to live. Figure 1.10 shows that Glasgow was not alone, and in some respects was much better than other comparable British cities, in terms of the high levels of infant mortality recorded as a result of the unsatisfactory physical environment which it was able to offer its citizens in the nineteenth century. It is ironic that the history of society's occupation of the natural environment has been one of progressive conquest towards a position where the physical forces of nature present relatively little threat to health and well-being, while at the same time there are many examples which show that nineteenth-and twentieth-century built environments have often proved alien, dangerous or deadly to significant portions of society.

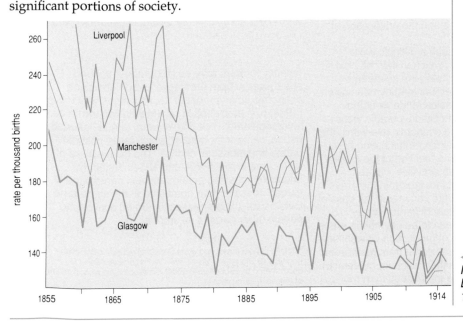

◀   *Figure 1.10*
Infant mortality in Glasgow, Liverpool and Manchester, 1855–1914.

Box 1.3   Migration and the growth of industrial urban environments: the case of Glasgow

The case of Glasgow, second city of the British empire in the nineteenth century, serves to illustrate these trends. Glasgow's early industrial and commercial growth was based on its trade in tobacco, linen and cotton textiles and sugar. By the second half of the nineteenth century, employment in shipbuilding and engineering began to rise in importance.*

In 1851, 56% of Glasgow's population were immigrants as defined by their place of birth. In addition to migrants from local labour sources, migrants also came from Ireland and by 1845 formed over a quarter of the city's population. Overall the city grew from having a mere 83 000 people in 1801 to a population of 761 000 a century later. Most of this growth, as in other industrial cities of western Europe, was due to the migration process.

The rate of growth was such that urban services were seldom able to keep pace with the demographic expansion which was taking place. Gibb (1988) notes that on the south side of Glasgow in 1831 there was only one doctor for the entire population. Similarly the urban environment was ill-equipped to house the immigrants. In the mid eighteenth century most working-class migrants lived either in unhealthy and overcrowded apartments in the 'backlands' around the city's central area or else in the city's infamous lodging houses. In the second half of the century purpose-built tenements were erected and to a certain extent delayed the city's housing crisis, since most of the new two-room tenement properties had to be designed to be affordable for the city's low-income population. Overcrowding and poverty provided an urban environment conducive to child neglect and was an obstacle to hygiene and the eradication of disease.

If Glasgow provided such a difficult, if not dangerous, environment in which to live, why did so many people move there during the nineteenth century? Part of the answer has already been given in terms of the rapid expansion of the industrial base of the city as a result of its favourable locus relative to Britain's growing economic role within the world economy. But population redistribution here as elsewhere was not only a function of the growth of labour demand in certain activities, but also a change in the requirements for labour in the areas from which the migrants were coming. In the case of Glasgow in the early nineteenth century three migrant streams can be seen to be sourced in areas where the capitalist mode of production was shifting its emphasis. Many migrants came from rural areas of Central Scotland where the enclosure movement was threatening to change their status from multiple

▲   A nineteenth-century Glasgow tenement yard.

tenancy farmers to hired labourers. A second stream came from the fragile and deteriorating economic environment of the Highlands which Gibb (1988, p.7) describes in neo-Malthusian terms as 'overpopulated'. The final stream, already alluded to above, came from Ireland, which faced political unrest and industrial decline. The main waves of Irish came in the 1840s and 1850s during the country's appalling potato famines, paradoxically 'forced' to leave their homeland at the same time as Britain was shipping wheat in armed convoys to certain other parts of its empire.

*Details of how these and other industrial activities attracted migrants to Glasgow from all over Scotland have been documented by Gibb (1988) and it is based on his work that most of this case study is built.

If the industrial cities of western Europe in the eighteenth and nineteenth centuries provided precarious environments for human life and reproduction, the large cities of developing countries in the late twentieth century, as a result of very rapid rates of natural population increase and high levels of in-migration, have become the insecure living environments for an ever growing proportion of the world's population. It has been estimated that if current urban growth rates are sustained, then by the year 2000 there will be 25 cities in the world with more than 10 million inhabitants and of these 20 will be in the third world. By contrast in 1970 there were only four urban agglomerations of this size and three of them were in the more developed countries. If the projections are correct, then the world's largest centre in the year 2000 will be Mexico City with a staggering population of over 30 million. Migration has been the principal mechanism fuelling urban growth in many of the largest cities. 75% of the population increase of Lagos in the 1950s and 68% of the urban growth of São Paulo in the 1960s were due to migration. In the oil states even higher contributions have been made to urban growth both by rural–urban migration and international labour migration. For example, in the 1970s over 80% of the populations of the Saudi cities of Dammam, Jeddah, Riyadh and Taif were migrants. Most migrants to third world cities have concentrated in just one or two cities in each country. Usually these cities have been the former locus of colonial administration or major ports for the export of raw materials to the more developed countries.

Q   Can you think of any reasons why migration to third world cities in the twentieth century should be seen as a different phenomenon from migration to western cities in the nineteenth century?

▲   *The scale of international labour migration is demonstrated by these queues of refugees, waiting for food and water, in the Sha'laan 1 camp between Iraq and Jordan. Over 80 000 migrant workers from Bangladesh, India, Pakistan, Sri Lanka, Thailand and the Philippines from more than 2 million foreign nationals trapped in Iraq and Kuwait – engineers, sales executives, technicians etc. – were stranded in two camps in September 1990 after Iraq's annexation of Kuwait.*

A   There are of course many differences of context, but one of the most important is that while rapid urbanisation in the west was related to a phase of concentration of capital in urban environments associated with the experience of industrialisation, in the third world many cities have little industrial base. Urban growth in the third world has not been closely associated with industrialisation, and, as a result, urban environments have developed in a very different way from those in the west. For many migrants the third world city provides no secure waged employment in the so-called 'formal' sector, and instead they have been forced to seek a living in what have been termed 'survival' or 'informal sector' occupations. Such jobs might, for example, involve small-scale production of clothes, street vending of food or providing low-income transport systems such as rickshaws or tricyles.

4.4   Summary

Population change is produced by a combination of fertility, mortality and migration. The variation of each of these components is influenced at household level by both cultural and economic factors. The highest rates of population growth occur in the cities, whether in nineteenth-century Europe or in the twentieth-century third world. The growth of these cities is boosted by migration, which seems to reflect both urban–rural and international differences in economic development.

5   Production, reproduction and population surplus

So far in this chapter attempts to explain population change have focused mainly on the possible causal relationships between population change and the physical and social environments. Economic arguments have been introduced in looking at the demographic transition but it now becomes pertinent to consider whether there might exist some wider link between systems of economic production and patterns of demographic reproduction.

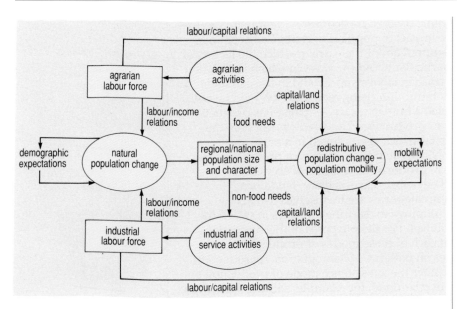

◀ *Figure 1.11*
Population needs and
population change.

Figure 1.11 seeks to examine some of the ways in which the dynamics
of population change and economic development are interrelated. The
overall impression which it should give you is that ultimately the
demographic and economic systems interact in a circular and
interdependent fashion. The reasons why this is so will only become
apparent as you consider the individual linkages in the diagram.

Consider first the central box of the diagram relating to population size
and character. From this box flow two arrows indicating that a population
has certain food and non-food needs. Both Malthus and Boserup identified
this linkage between population and economy, but disagreed about how a
change in population size would impact on economic structure. Figure 1.11
suggests that the impact should be assessed in terms of at least three factors
– labour, capital and land. These three factors are sometimes taken, along
with enterprise, and described as the **factors of production**. Although an
abstract economic concept, what is meant by this is that these are the basic
building-blocks of economic activity. For example, population growth will
contribute to the labour force through increasing the number of potential
workers, and it will introduce new pressures on land availability through
the need to produce more food and non-food goods as well as through
increased need for residential, transport and recreational space.

Economic organisation in any one place or time may be differentiated
from that at another by the way that the factors of production are
combined. Both Malthus and Boserup investigated population change in
relation to some of the factors of production while holding other factors of
production constant. For example, Malthus writing at the beginning of the
nineteenth century could not conceive of the way in which human
enterprise in the form of the industrial revolution would transform the
potential productivity of the economy. As a result rising population and
rising living standards no longer proved self-checking in the way in which
Malthus expected. Boserup assumed that labour immobility in relation to
finite land stocks would produce new forms of human enterprise, while in
practice, as has been shown, population redistribution has often permitted
labour to combine with capital at new locations and within an industrial

rather than agricultural economic context. It becomes evident therefore that in terms of understanding the impact of population change on economic systems, greater thought is required about the ways in which the factors of production combine and in particular about the ways in which population growth results both in an increase in labour supply and in the demands placed on economic production through rising food and non-food needs.

The particular way in which a human society organises its productive activities is termed the **mode of production**. This involves not only the amount of labour, land or capital which is employed in a particular productive activity, but also the way that these factors of production are organised. Five major modes of production, or forms of economic organisation, are often identified: subsistence, slavery, feudalism, capitalism and socialism. These categories are themselves usually subdivided in more precise groupings to describe the ways in which the factors of production are combined. For the purposes of this discussion, however, what matters is that each of these modes of production has potentially different influences on patterns of demographic reproduction. Return again to Figure 1.11. Under the subsistence mode of production of the kind studied by Boserup in relation to, for example, shifting agriculture, agricultural labour is controlled by the cultivator who is working to produce for his or her own family's food needs. Production is controlled within the family with net wealth flows being from children to parents in the way proposed by Caldwell. High fertility may prove highly logical since more children may equate with more labour in the future and hence the potential for either cultivating a greater land area or of cultivating the land more intensively. Therefore under this mode of production and in the context of one particular agricultural environment – shifting cultivation – it becomes possible to see how organisation of the labour force may have implications for demographic circumstances, which in turn will have an influence on future population size and character.

Under capitalism the labourer owns none of the other factors of production but is free to sell his or her labour. Organisation of the labour force reflects the tension between those wishing to sell their labour for an adequate wage in order to be able to meet their food and non-food needs, and those controlling the other factors of production whose concern is to accumulate capital from the production process: this is achieved through the differential between the value of the work received from labour and the wages paid to labour, that is the 'surplus value'. Furthermore because labour is seldom producing food, industrial goods or services directly for its own needs, there may be a significant spatial separation between place of production and consumption allowing economic and geographical specialisation of production. Finally, this mode of production implies an exchange of 'value' between producers and consumers, and control of this exchange will be critical in determining whether a population's food and non-food needs are adequately met. (Consider again the previous discussion of population growth and famine in Ethiopia.)

A number of writers on population issues have recently begun to explore population–resource relationships from the perspective of their understanding of the forces underpinning the organisation of capitalist society. From this perspective the growth of population can be interpreted as having implications for society as a whole, and may be a trend which will either be encouraged or discouraged by those who mediate power within society. This search for a political-economy explanation has inevitably led to a consideration of the writings of Marx, who was first to recognise the need to understand how capitalism worked.

Marx was well aware of the writings of Malthus, but dismissed his work in a rather disparaging fashion. Marx's interest in population growth related to the fact that in his view it led to the creation of 'surplus population' as opposed to the Malthusian concept of 'overpopulation'. Surplus population created a potential expansion of the labour force.

> Surplus population . . . also becomes a condition for the existence of the capitalist mode of production. It forms a disposable industrial reserve army, which belongs to capital just as absolutely as if the latter had bred it at its own cost. (Marx, 1976, p. 784)

The significance of **reserve armies of labour**, whether they be defined in terms of surplus population groups within a country or between different countries, is that they can be used to exert pressure on workers within the capitalist mode of production to keep wages low and to assure high levels of capital accumulation. From a Marxist perspective a fall in income per head does not arise simply because of population increase leading to a growth in labour supply relative to demand, but is sustained by economic and social structures operating to exploit tensions between owners of capital, employees and the unemployed. In the twentieth century the geographical control of most of the world's capital has been concentrated in the developed economies of North America, western Europe and Japan. The capitalist mode of production has, however, been 'exported', establishing production systems which operate on a world scale and which also involve the internationalisation of labour markets and finance. Each of these interrelated events can be argued to have had direct and indirect effects on the demographic circumstances of the less developed countries. For example, one phase in the emergence of international labour markets was the introduction of so-called 'guest workers' from less developed countries as a new 'reserve army' to be tapped by the industrial economies. In the 1980s there were estimated to be no less than 20 million international labour migrants mainly from the less developed countries working in either the more developed economies such as in the western European labour markets or in the oil-rich states of the Arab world.

Another example of the type of interrelationship which may exist between the changing scale of the capitalist mode of production and demographic patterns is the way in which the internationalisation of finance and unfavourable terms of trade with the more industrialised countries has led many developing countries into a situation of growing indebtedness. The combination of dropping oil prices and rising dollar interest rates in the early 1980s produced what has come to be termed the 'debt crisis'. A large number of developing countries (including Argentina, Brazil, Ivory Coast, Mexico, Morocco, Nigeria, Peru, the Philippines and Venezuela) had debts that in 1987 were, on average, valued at 60% of their Gross National Product. According to the World Bank, the third world's debt had risen still further by 1989 to record levels with the poorer states of the world being forced to use an increasing proportion of their limited incomes simply to service their debts. Since debt servicing can only be done in foreign currency, less developed nations can only meet their creditors' demands by cutting back on much needed imports (such as food), by diverting goods produced on the home market into exports, or by replacing production for the home market by production for export (as for example in the replacement of domestic food production by cash crop production). The need to pay for their increasing debts has therefore forced many less developed countries to reduce production geared to meeting the needs of their own populations (hence reducing living standards) and to reorganise the production of their economies (and labour forces) in such a way as to be able to earn more foreign currency (consider again the links in Figure 1.11). The International Monetary Fund has estimated that output in the world's debtor nations rose by only 1.1% in 1989 while the populations of these countries rose by 2.0%. One consequence has therefore been a lowering of living standards in these nations and consequently a reinforcement of inequalities in life chances. This may be one factor sustaining the continued spatial unevenness in demographic patterns such as in the global pattern of life expectancies.

International differentials in life expectancies do not, of course, arise because of deliberate malevolent actions, and there is ample evidence to show that people living in western capitalist nations do not desire people in the less developed countries to have shorter lives. Rather, differentials exist because of 'the **core–periphery structure** of the world economy, whereby the life chances of the population of the periphery are subordinated to those of the population of the core' (Johnston, 1989, p. 222).

Differentials in life expectancy and other demographic indicators do, of course, arise from a whole range of intertwined forces and not only because of the outworkings of any one economic process, such as the international system of indebtedness discussed above. It does, however, seem plausible to suggest that the way economic systems are structured by the capitalist mode of production has had a significant influence on demographic processes. If this statement is true it would also seem logical to expect that states which have sought to follow rather different economic courses such as radical socialist economies would also have produced distinctive demographic conditions as a result. Before examining this through considering the case of China (in Box 1.4), it is important to balance the arguments which have been presented by stating that most students of development issues, while accepting the thesis that economic structures have an influence on demographic processes, also recognise that demographic processes are not determined solely by the economic environment in which they occur. As shown earlier in Figure 1.11, economic processes influence demographic factors, but these in turn affect economic organisation. Thus 'population has an absolute dimension which in turn has implications for accumulation and the distribution of capital' (Corbridge, 1986, p. 103).

5.1   *Summary*

This chapter has explored some of the relations which exist between population processes and environmental circumstances. It has been pointed out that selective conceptions of population environment and population–resource relationships posited by Malthus and others have only limited validity. Population issues seem increasingly to relate to the social and economic contexts of demographic processes, rather than to circumstances determined by physical or ecological forces. Conversely it is not so much population numbers as the technological and sociological bases by which specific population groups seek to meet their needs which determines the impact they have on their immediate physical environment, as well as on environments in other parts of the globe. Understanding the social and economic environments appears to be increasingly critical in exploring and explaining the associations which exist between economic production and demographic reproduction in both rural and urban areas.

Box 1.4   China's population policy

The absolute dimension to population was well recognised by the Chinese communist leader, Mao Zedong. His interpretation of the role of demographic reproduction relative to the economic functions of production within a socialist state experienced an interesting reversal through time. Mao Zedong moved from an early stance favouring high population growth rates to a Chinese population policy which later forcefully endorsed a strict birth control programme. This perhaps reflected a shift from a view of people as being equated to labour with productive value, to the perception of people as also being consumers requiring food and shelter which a socialist state should seek to provide and distribute in an equitable fashion. To achieve this goal was clearly going to be easier if China's population growth could be slowed, and if, as a result of having fewer children, women's energies could be engaged in the labour force. In the 1970s birth control became a national priority in China, if the country's very rapid population growth was not to exacerbate (as distinct from causing) the country's problems in producing sufficient food, housing and jobs for the 1 billion Chinese. The government established its policy in the early 1970s using the slogan *wan-xi-shao* (later, longer, fewer). This meant later marriage, a longer interval between children and fewer children per couple. Initially a two-child family was advocated in urban areas and three in rural areas, but in 1979 China instituted a 'one couple, one child' policy: see Plate 2.

This policy has in fact encountered severe difficulties with more than 90% of rural families wanting two or more children and only about half the urban population being willing to conform to the one-child norm (Jowett, 1989). Inevitably data on births in a country with a policy of this kind are difficult to interpret. The data which do exist suggest that in 1987 only 51.7% of births were in fact first births and that the remaining 48.3% were second, third or subsequent births. Enforcement of the one child policy via a series of incentives for those complying with government policy and a range of penalties for those couples failing to do so has therefore so far failed to achieve the target of the one-child family. The policy has faced the greatest difficulties in the rural areas of China and amongst minority ethnic groups. Despite this it must be acknowledged that China has succeeded in greatly reducing its fertility levels. During the 1970s the birth rate was halved and by the early 1980s China's TFR stood at 2.4, only slightly above the replacement level. By contrast developing countries with a similar level of GNP per capita to China had an average TFR of over 6.0. China therefore represents the case of a socialist regime which has as a result of its policies produced demographic changes out of character with those found in market economies at the same level of economic development (Jowett, 1989). These changes have, however, only been achieved at the expense of the loss of considerable personal freedom. It is also important to note that even in a radical socialist regime willing to impose its policies in a very forceful fashion, there would appear to be limits to the extent to which demographic change can be engineered by those in control of a particular socio-economic environment.

References

BOSERUP, E. (1980) *Population and Technology*, Oxford, Basil Blackwell.

CALDWELL, J. (1982) *Theory of Fertility Decline*, London, Academic Press.

CORBRIDGE, S. (1986) *Capitalist World Development*, London, Macmillan.

GIBB, A. (1988) 'The demographic consequences of rapid industrial growth', Occasional Papers, No. 24, Department of Geography, University of Glasgow.

HOBCRAFT, J., McDONALD, J. and RUTSTEIN, S. (1984) 'Socio-economic factors in infant and child mortality', *Population Studies*, Vol. 38, pp. 193–224.

JOHNSTON, R. (1989) 'The individual and the world economy', in Johnston, R. and Taylor, P. (eds) *The World in Crisis*, Oxford, Blackwell.

JOWETT, A. (1989) 'China's one child programme', *Applied Population Research Unit Discussion Papers* 89/3, Department of Geography, University of Glasgow.

KURIAN, G.T. (1982) *Encyclopaedia of the Third World*, London, Mansell.

MARX, K. (1976) *Capital*, Vol. 1, Harmondsworth, Penguin Books (*Das Kapital*, written 1867–94).

WOOD, A. (1985) 'Population redistribution and agricultural settlement schemes in Ethiopia', pp. 84–111, in Clarke, J. *et al.* (eds) *Population and Development Projects in Africa*, Cambridge, Cambridge University Press.

WOODS, R. (1982) *Theoretical Population Geography*, London, Longman.

Further reading

FINDLAY, A. and FINDLAY, A. (1987) *Population and Development in the Third World*, London, Routledge.

JONES, H. (1990) *Population Geography* (2nd edn), Ch. 7, London, Paul Chapman.

LIVI-BACCI, M. (1992) *A Concise History of World Population*, Oxford, Blackwell.

Answer to Activity

Activity 1

Your graph should look like this:

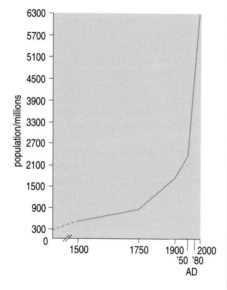

Chapter 2 World agriculture: productivity and sustainability

1 Introduction

The previous chapter has shown that there has been progressively faster growth of the human population through history, with the fastest rate occurring over the last few decades. This increase has been made possible by equally rapid growth in the supply of food, largely as a result of transformations in agriculture. However, these developments raise a number of questions for environmentalists, of which this chapter considers three (set out in the margin).

As so often happens, the questions are not as straightforward as they appear when one begins to attempt to answer them. It is certainly the case that environmental factors constrain what can be grown, but, as *Simmons** (in Sarre and Reddish (eds), 1996) has shown, cultural differences and technology also play important roles: there are great differences between the products of subsistence farmers and of commercial producers even where environmental conditions are similar. These differences also arise when considering productivity, because productivity can be defined in a number of ways: by relating output per worker, or per hectare, by cost of inputs and so on. Finally, in considering problems brought about by agriculture, it turns out that not only do some highly productive systems generate serious problems like soil erosion, water pollution and poisoning of wildlife, but they are also dependent on high inputs of energy and materials which may not be sustainable in the long term. There may be problems for the future being stored up, especially in relation to feeding a growing world population.

The structure of the chapter will be dictated by the three key questions. Section 2 will show how the current variety of agricultural systems has resulted from a combination of environmental and social factors. Section 3 will explore the different kinds of productivity and Section 4 will consider how the concept of sustainability brings together the impacts of agriculture and problems of its dependence on energy and materials inputs.

As you read this chapter, look out for answers to the following key questions:
● How have agricultural systems adjusted to the variety of ecosystems, soils and climatic zones which exist in different parts of the world?
● Which systems have been and are the most productive?
● What problems have these agricultural systems created?

*An author's name in italics indicates another book or a chapter in another book in this series.

2 The origins and diversity of agricultural systems

2.1 Introduction

From the array of flora and fauna particular to the world's biomes (that is, the natural major vegetation types) and the ecosystems which existed about 12 000 years ago, the first 'farmers' selected their crops and animals for domestication and began the first agriculture. The distribution of the world's biomes reflected the distribution of temperature, rainfall and solar radiation, resulting in the spread of vegetation types from equatorial forest to the tundra of the sub-Arctic and the high mountains (as described by *Silvertown* (1996a), Sections 2.2 and 3.2). This broad climatic framework is still the main influence on the pattern of agriculture, although it has been affected by vegetation change and human influence has changed the limits of growth of particular crops.

The section will look first at the origins of agriculture with the domestication of particular plants and animals, the global dispersal of selected plants and at the limitations imposed on this by these climatic features. Some of the ecological factors relevant to the development of agriculture are also outlined.

2.2 Agricultural beginnings

The first evidence of animal and plant domestication dates from about 10–12 000 BP.* Remarkably rapidly – in the context of human history – hunter–gatherers were displaced or became sedentary cultivators or pastoralists. By 2000 BP farming had penetrated to most parts of the globe and the hunter–gatherers were pushed into the areas least suitable for cultivators.

Although plants and animals were domesticated at a number of places on the Earth's surface, five areas are particularly important to the subsequent development of agriculture (see Figure 2.1). First is south-west Asia and the eastern Mediterranean where the first agricultural communities appeared in upland areas; cattle, sheep and goats were domesticated and wheat, barley and later lentils, peas, flax, olives, vines and figs were grown. These crops and animals were later adopted in the irrigated floodplains of the Tigris and Euphrates and the Nile. They were subsequently carried westwards along the shores of the Mediterranean, and wheat and barley were taken north of the Alps. Farming was first practised on the shores of the North Sea and the Baltic by about 6000 BP.

A second major area of plant domestication was in south-east Asia. Two types of agriculture appeared here: vegeculture, where parts of the growing plant were cut and planted, using the taro, the Asiatic yam, breadfruit, sago and bananas; second, and more important, was the domestication of rice, which spread into India, China and later Indonesia, where it has become the dominant type of farming in deltas and river valleys. Pigs, chickens, ducks and geese were domesticated in this area.

In China there was a third centre in the north where millet, soya beans and the mulberry were domesticated. Little is known of the early history of

This book uses the convention of time 'before the present' (BP), by adding 2000 years to BC dates.

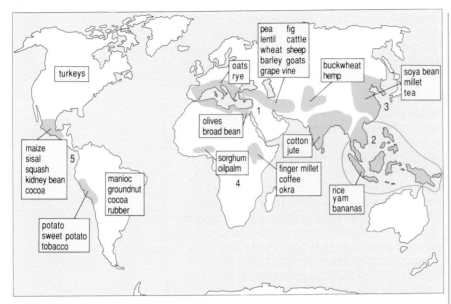

▲   Figure 2.1   Regions of plant domestication. The areas numbered 1–5 are the earliest and most important. The other areas were later in time or involved fewer and less important varieties.

agriculture in the fourth area – Africa – since evidence of early farming systems (seeds etc.) are preserved better in semi-arid conditions, but African yams and oilpalms were domesticated in West Africa by about three thousand years ago, and some crops were domesticated in Ethiopia.

The Americas were the fifth area of early domestication and a quite different set of plants and animals were domesticated there: maize, squash and beans in Central America, manioc (or cassava) in the lowlands of the Amazon basin, and potatoes in the Andes. Only the turkey was of much importance as a domesticated animal.

Despite all the developments since, humankind is still basically dependent on choices made by primitive peoples in particular climatic zones thousands of years ago. There are still only about 20 major crops in agricultural production in the United States and Europe, compared with an estimate of 200 000 species of wild plants, animals and micro-organisms in the US alone. Even now, six species of domestic animals predominate as food sources and all were domesticated at least 4000 years ago.

2.3   Underlying ecological principles

Even from this brief description of the origins of agriculture, it is clear that different centres of domestication involved very different combinations of plants and animals, even though all societies had the same goal of providing an adequate diet. In part, the need to select from local wild species explains the differences, but the variety in various parts of the world historically shows that many combinations of roots, cereals, animals and fruit can provide the essential dietary components – carbohydrates, proteins, fats, minerals and vitamins. In making these initial selections, early farmers were influenced by a number of ecological principles.

• *It was necessary to strike a balance between staple crops that provide mainly carbohydrates and those foods which provide greater inputs of proteins, fats, vitamins and so on. Such decisions were influenced by the productivity of particular crops and the efficiency of substitution of foods 'higher up' the food chain.*

Plants use solar energy to synthesise simple carbohydrates (sugar, starch, cellulose etc.) from carbon dioxide and water; this is the process known as **photosynthesis** (see Section 2.3 of *Silvertown*, 1996b). Green plants are the only primary producers of foodstuffs; humans cannot photosynthesise atmospheric carbon compounds for themselves, so they are dependent on plants for all their food, either directly or indirectly.

The average energy requirement is about 10.4 MJ per capita per day, most coming from carbohydrates and fats (see Box 2.1). Crops vary in the amount of energy they supply but starch is the principal component of some very productive staple crops, especially in the wet tropics where bananas and root crops such as cassava, sago and yams mean that carbohydrates supply four-fifths or more of the total energy supply. These starchy crops are, however, low in protein (lower than cereals) and therefore need to be supplemented by a protein-rich food such as beans, fish or meat.

There is a second stage in the **food chain**. Grass, grains and root crops can be eaten by livestock, as herbivores, and the products of the animal – meat, eggs or dairy products – eaten by humans, who are omnivores. However, the energy available in meat is a very small proportion of the energy stored in plants. Thus the energy in meat from cattle fed on corn in the United States is one twentieth of the energy in the corn; on ranches in the western United States cattle feed upon a very sparse scrub, and the meat produced per square hectare is only one sixtieth of the energy in this plant cover. Consequently, growing crops is always a far more efficient way of producing food calories than raising livestock: there is always a loss of energy in conversion from plant to animal.

The difference between ruminants and non-ruminants is of great importance. Ruminants – sheep, goats and cattle – can digest cellulose and hence feed upon grass and much of the natural vegetation. They can consequently utilise land which is not climatically suitable for growing food crops and so produce food for human beings from inferior land. Single-stomached animals such as pigs and poultry, in contrast, cannot digest the cellulose in grass and feed mainly upon grain or scraps. In poor countries

---

Box 2.1   Measuring energy

You may be familiar with seeing the amount of energy that can be derived by oxidizing food given in kilocalories, for instance on food packaging. However, the international standard 'SI' units are being increasingly used, so the joule (J) is being adopted as the unit of energy.

A joule is the amount of energy supplied by a watt acting for a second. The rate at which the body consumes energy varies according to what is being done: 70 W when resting in bed, 180 W when walking, 380 W for fast walking.

It is convenient for most purposes to use the total energy expended in 24 hours, that is the average wattage multiplied by 86 400 seconds.

Because the resulting number is large, it is usually expressed in megajoules (MJ = 10^6 J = 1 000 000 J): thus resting in bed expends 6.05 MJ and fast walking 32.83 MJ per day.

The same unit is used to measure the energy content of food. Some forms of manual labour require about 12.5 MJ (or 3 000 kilocalories) per day, while an agricultural worker may need 15 MJ. Adults in well-fed societies consume about 14 MJ a day, or 31% above their probable requirement, while the average person in the third world may have only 8.9 MJ, well below their needs.

In Chapter 4, Figure 4.9 you will come across gigajoules (GJ):

1 GJ = 10^9 J = 1 000 000 000 J or 1 000 MJ

they are scavengers and are very useful in getting rid of such wastes, utilising ground unsuitable for anything else, and producing meat and eggs as well. When such animals are stall-fed on food which could have been used for human consumption or which was grown on land which could have been used to grow such food, then it is a very inefficient form of production: it takes about 8 kg of grain to produce 1 kg of pork. Intensive production of beef is even less efficient, taking 20 kg of grain for 1 kg of beef. Over 90% of the grain consumed per head of the US population is in the form of meat, poultry and dairy products; two-thirds of their daily protein intake is of animal origin, compared with a world average of 25%.

- *The second decision which had to be made was how to maintain soil fertility.*

For growth, crops need from the soil nitrogen, potassium and phosphorus compounds and traces of another thirteen elements. The most important nutrient is nitrogen, and an insufficiency of nitrogen in the soil is the factor most commonly limiting crop growth.

In a natural ecosystem, nitrogen is obtained from the atmosphere, which is 79% nitrogen, but in this gaseous form is not usable; it has to be formed into nitrogenous compounds – or fixed – by the action of bacteria in the soil. Some of these are free-living, and **fix nitrogen** at a low rate; others exist in a symbiotic relationship with leguminous plants and fix at a much higher rate: clover can fix 400 kg of nitrogen per hectare per annum.

Where a crop is eaten, by people or animals, on the ground where it was grown and when all the waste including their excretions are spread again on the ground, there is more or less a closed cycle: most of the elements are returned for use by the next crop. But when the crop is eaten elsewhere, the soil is steadily depleted of nitrogen, phosphorus, potassium and so on. In order to preserve the nitrogen content of soil and hence its fertility, farmers must find a means of returning to the soil the nutrients which have been taken out by the crop. This can be done in a number of ways:

- cropland is abandoned after growing crops for 2–3 years and the field colonised by the local vegetation (**shifting cultivation** with **bush fallowing**): thus a cycle between vegetation, organic litter and soil is re-established and after a long period (10–40 years) soil fertility is restored;

- crops can be grown alternately with grass which increases the soil's organic matter and improves soil structure (**fallowing**);

- leguminous crops can be grown alternately with other crops (**crop rotation**): these include peas, beans, lentils, soya beans and clover; they all aid fixation of nitrogen in the soil;

- the dung of livestock kept by the farmer can be deposited on land used for growing crops (*organic fertiliser*); dung contains nitrogen, potassium and phosphates;

- the cultivation of 'paddy' rice is a special case: it requires the crop to be grown partially submerged in water and blue-green bacteria thrive in these conditions and have the ability to fix nitrogen (this is discussed in detail in Chapter 3).

Farmers have always used a variety of fertilisers such as human excreta, marl, horse dung and crop residues. Bones were ground to provide phosphorus. Although traditional means of providing nitrogen are still used, the use of manufactured nitrogen fertilisers has had a great impact on farming and has increased dramatically in the developed countries in the last fifty years (see Section 3.3).

- *The third decision was how to deal with weeds and pests.*

Within a natural ecosystem a large number of plant species occupy the land. A natural ecosystem is always moving to equilibrium and it is this diversity of species which directly contributes to the stability of the system. In farming, in contrast, the aim is to simplify this system and to produce only one crop from each field, and often hectare after hectare of the same crop. Farmers therefore set out to reduce the number of species of plants, animals and micro-organisms: any that are not economically useful are eliminated. After such an alteration, 'successional change' begins and the ecosystem slowly accumulates additional species; under natural conditions, gradually a new complex and stable ecosystem would evolve. The improved conditions that are made available to the crop are attractive to other plants – that is, weeds – which will also colonise the field and use water and plant nutrients that could be used by the crop plants. Farmers thus spend much of their time trying to remove weeds. This is done by ploughing and hoeing in traditional farming systems and, in modern farming, by spraying with herbicides.

An equal problem for the farmer is that of pests, which include small animals like rabbits and birds, but more importantly insects and bacteria. In traditional agriculture, growing crops in rotation and alternating crops and grass partially reduce the level of plant disease; in the tropics **intercropping**, where several crops are grown on the same plot, does occur. However, where one crop predominates, and is grown year after year, there is obviously the danger that if a disease takes hold, the consequences are going to be much more serious, as it will attack the entire crop. It thus becomes increasingly vital to protect the crop from such risks and chemical pesticides have to be used on a habitual basis. Pesticides kill most 'pests', including any natural predators; even if the original pest is killed, this may leave an ecological niche into which another may move. Pesticides have thus removed farming from having to attempt to fit in to some kind of ecological balance; they can be crude weapons in a very complex system and thus always risk unforeseen consequences.

◀ *Subsistence agriculture in Rwanda, showing a farmer and her child hoeing their land. Crops grown on these steep hillsides include cassava intercropped with beans, bananas and, in some areas, tea.*

2.4   Farming methods

The earliest farmers in every region used the simplest tools – digging sticks, hoes and the sickle – with shifting cultivation methods. The first fundamental technical advance was the ox-drawn plough, developed in the Middle East 4–5000 years ago; it later spread into Europe, India and China, but not into tropical Africa until the twentieth century, where it is still less important than the hoe, and not into the Americas until after 1492.

After the invention of the plough, the most important changes were the wider growth of legume crops to fix nitrogen in the soil and the use of cattle dung. Implements were slowly improved: in northern Europe the mouldboard plough allowed the cutting of grass sods, more effective destruction of weeds and effective ploughing of heavy clay soils.

Plant dispersal

Until the nineteenth century perhaps the most important way of improving agriculture was by the exchange of plants. Crops domesticated in particular regions were slowly taken to other regions; thus by 2000 BP wheat, domesticated in the Near East 10–11 000 years ago, was grown in Europe, North Africa, northern India and northern China. The exchange of plants was greatly accelerated by the European voyages of discovery that began in the fifteenth century when the Portuguese reached China and Columbus the Americas. Farming was transformed by the crops that European settlers took with them together with the plough that was unknown to the Amerindians. More dramatic was the introduction of horses, cattle, pigs and chickens. But plants indigenous to the Americas had equally remarkable effects upon other parts of the world. The potato and maize increased the food output of Europe in the eighteenth and nineteenth centuries; in the nineteenth and twentieth centuries maize and cassava became staple crops in tropical Africa.

The exchange of crops did not greatly affect the staple food crops of Asia, although maize, peanuts, sweet potatoes and cassava became important supplementary food crops, but cash crops were introduced by Europeans which profoundly altered local economies. In the late nineteenth century rubber proliferated in Malaya and Indonesia; earlier in the century tea from China had been adopted in India and Ceylon, and groundnuts from the Americas became a major cash crop in West Africa; coffee was taken from Africa to Brazil and to Indonesia. The process has continued into the twentieth century: improved rice varieties bred in the Philippines have supplanted the indigenous varieties in much of the rest of Asia, and wheat of Mexican origin is now widely grown in northern India.

Q   Would this trade in plant varieties increase or decrease the variety of plants being grown?

A   While it would potentially increase the variety of useful plants which could be grown in any particular area, the overall effect has been to reduce variety in several ways. Of the huge numbers of naturally occurring plant species, only about a thousand have been domesticated and most of the area under cultivation is devoted to quite a small number, as shown in Table 2.1. Finally, the breeding of improved varieties means that increasingly large proportions of the areas devoted to each major crop are sown with a few types of seed (discussed in Section 3.3).

Table·2.1   Areas under cultivation[1] of the ten most widely grown types of crop, 1991–2

	Million hectares	% of total arable area
Cereals	699.7	48.4
Oilseeds, nuts and kernels	127.0	8.8
Pulses	69.0	4.3
Roots and tubers	47.7	3.3
Cotton[2]	33.1	2.3
Vegetables and fruit	18.1	1.3
Sugar-cane and beet	13.1	0.9
Coffee	11.1	0.8
Tobacco	5.7	0.4
Cocoa	5.3	0.4
Total arable area[3]	1441.6	
Total percentage		71.3

Notes: [1] For many crops only volume of production is available, so these figures may understate area. [2] Data for 1971. [3] Total area for all arable crops, not just these ten.

Source: FAO (1993) *Production Yearbook 1992*, Vol. 46, Rome.

The industrialisation of agriculture

As well as the dissemination of crops, there were other profound changes in agriculture from the middle of the nineteenth century as traditional methods were supplemented and then replaced by techniques which depended upon scientific advances and the industrial production of inputs.

First was the substitution of power based upon commercial energy for human and animal muscle. Threshing by steam power had largely replaced the flail in Britain by the 1860s, and steam was being applied, although not very successfully, to ploughing. In the 1890s the first tractors were produced in the United States, although it was not until after the Second World War that they finally replaced horses in North America and western Europe. Machinery began to replace labour, especially in harvesting grain: the reaper, which cut grain, was invented in the United States in the 1830s and slowly adopted in western Europe; later the reaper-binder cut and bound the grain in stooks, and finally the combine harvester both cut and threshed. The latter was first used in the 1880s in California, but it is only since 1950 that it has replaced the reaper-binder. Milking machines, cotton pickers, sugarbeet harvesters and a variety of other machines have greatly reduced labour needs, mainly since 1950.

The invention of the internal combustion engine made the machinery less bulky and easier to use. Tractors with their multitude of attachments have proliferated. Their increasing size and power have allowed vast areas to come under cultivation.

The use of chemical fertilisers dates from the 1840s, but the critical advance was the discovery in 1910 of a way to fix nitrogen. Similarly herbicides and pesticides were used in the late nineteenth century, but only since 1950 have they transformed farming in North America and western Europe. Breeding of improved varieties of crops began after 1900; advances in plant genetics have allowed breeders to produce crops immune from specific diseases and extremely responsive to fertilisers.

◁   A thresher powered by a steam engine. The portable steam engine was steadily improved through the 1850s and 1860s and gave great flexibility as a source of power on the farm.

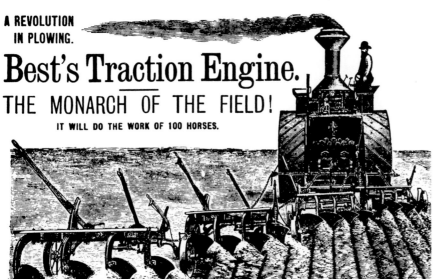

A REVOLUTION IN PLOWING.

Best's Traction Engine.

THE MONARCH OF THE FIELD!

IT WILL DO THE WORK OF 100 HORSES.

◁   An advertisement in Pacific Rural Press in 1890, when, it was claimed, twenty-two such engines were in operation.

◁   Lifting sugar beet at Sawston in Cambridgeshire: large-scale machinery has enabled more land to be brought into intensive cultivation while numbers of agricultural workers have declined.

◀ *A trailer full of grain at the 45 000 ton grain store in Wimblington, Cambridgeshire. Improvements in storage technology have been required to accommodate large surpluses.*

A critical contribution has been made by improvements in storage. Pre-industrial agriculture experienced massive losses of stored food through decomposition and attack by pests. Industrialisation allowed the development of structures more resistant to rodents and insects, drying, temperature and humidity control, freezing, pesticides and, most recently, irradiation. Better storage has allowed larger quantities of food from more varied sources to reach consumers in better condition, although there are still substantial post-harvest losses due to micro-organisms, insects and rodents, estimated to range from 9% in the United States to as much as 20% in some developing countries, especially those in the tropics. Recently, some storage technologies have themselves been questioned, but the net benefits have been great, with room for further improvement in the third world.

2.5   Climate and crop distribution

The movement of plants was in response to social, economic and scientific factors, but climatic factors remain a primary control of the particular location and the broad pattern of crop distribution. The two climatic factors of greatest importance in limiting the areas available for particular crops are rainfall and temperature.

Rainfall and crop growth

Rainfall provides the moisture in the soil that is essential for crop growth. Plants act as highly efficient pumping machines, taking water from the soil, incorporating some of it in their tissues and releasing the remainder into the atmosphere: this is **transpiration**. Every plant has a root system with an enormous total surface area to draw this water from the soil. Wheat takes only 350 to 500 kg of water to produce a kilogram of dry wheat plant, whereas the same amount of rice needs about 2200 kg; this is equivalent to about 1500 kg of water for one kilo of wheat grain, and 10 000 kg of water to produce a kilo of rice.

A sufficient amount of water is vital to plant growth: with too little water the plant wilts and dies. However, crop yields do not increase with every increase in water supply. For every crop there is an optimum amount, and with further increases yields begin to decline. The moisture requirements of individual crops vary greatly. Wheat and rye, for example, are grown mainly in areas where the annual rainfall is between 25.40 cm (10″) and 101.60 cm (70″). In contrast rubber needs over 178 cm and tea over 254 cm. Over half the Earth's land surface receives between 25.40 and 101.60 cm, so that wheat can be – and is – widely grown. But only 10% of the land has more than 178 cm, and only 5% over 254 cm, so that tea and rubber have a much more restricted distribution.

The amount of water available in the soil for the crop is not a function solely of rainfall since part of the rain is **evaporated**. The rate of evaporation increases with temperature, so that crops in the tropics need higher rainfall than those in the temperate zone (although even in Britain, from an annual rainfall of 90 cm, 38.10 cm evaporates before it can be used, and 46 cm flows back to the sea). Figure 2.2 was derived from the difference between mean annual rainfall and potential **evapotranspiration** (the loss from evaporation and transpiration). It shows where crop growth is impossible without

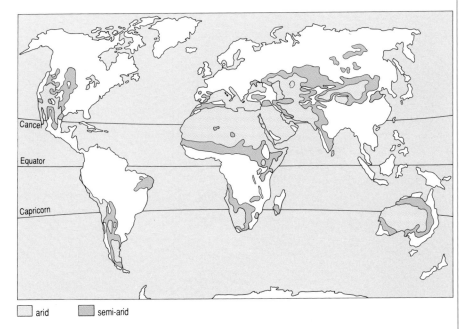

☐ arid   ▨ semi-arid

▲   Figure 2.2   The distribution of arid and semi-arid climates.

additional water – the arid zone – and where farming is possible without it but difficult – the semi-arid zone.

The semi-arid zone presents particular problems. Farmers there can only grow crops that are tolerant of low rainfall, such as millet, sorghum and wheat. But rainfall in both the arid and semi-arid zones is much more variable than in humid areas and consequently there is a higher probability that rainfall will fall below the minimum requirements for the crop and it will fail, or yields will be substantially reduced.

The total global supply of fresh water cannot be significantly added to, but it is possible to increase the efficiency with which it is distributed and thus to mitigate the effects of shortfalls or drought periods. This is done through **irrigation**, either from groundwater, as in parts of the south-western United States, or from rivers that rise in humid areas outside the desert and then cross it to the sea; the Nile, Colorado, Indus, Tigris and Euphrates are examples (see Box 2.2 on page 58). With clear skies and high temperatures these arid and semi-arid areas provide ideal conditions for crop growth.

Temperatures

Temperature is an important determinant of the distribution of crops, and they can be divided into two categories: those crops adapted to the temperature conditions of the tropics, and those adapted to the lower temperatures of the subtropics and temperate areas.

A number of crops grown only in the tropics are severely damaged if temperatures fall below 0°C and frosts occur; others are even more susceptible to cold and will die if temperatures fall below 10°C. For most tropical crops growth is most rapid when temperatures are 31° to 37°C.

Most crops grown in the subtropics and the temperate zone have both poleward and equatorial limits. Polewards, the **growing season** – the period between the last frost in winter and the first frost in autumn – diminishes (see Figure 2.3) so that the number of crops that can be grown

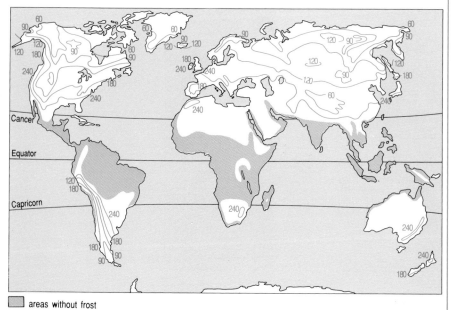

areas without frost

◄ Figure 2.3
Frost-free period in days.

declines (see Figure 2.4). North of the Arctic Circle only rye and oats have
any significance. Many temperate and subtropical crops also have limits of
cultivation towards the equator. Some of these crops require a cold period
to trigger growth, a process called **vernalisation**, will not thrive in higher
rainfall, and are susceptible to diseases found in the tropics. However, it
will be seen from Figure 2.4 that potatoes and most of the cereals are in fact
grown at or near the equator.

Q   How can temperate crops be grown near the equator?

A   The effect of altitude is to lower the average temperature. Because of
this, temperate crops can be grown at higher altitudes.

Arable land is absent from the cold, dry and mountainous areas of the
world; it is also uncommon in the humid tropics. Here the rate of net
photosynthesis is surprisingly low. This is because cloud cover reduces the
amount of solar radiation received; further, although gross photosynthesis
rates are high, respiration is also high, because of the high temperatures at
night, so that net photosynthesis is low, indeed lower than in middle
latitudes. Although this does not preclude cultivation, it is compounded by
difficulties with tropical soils, which are exposed to higher rainfall and
temperatures than are soils in the temperate zone, and have made it
difficult to use farming systems other than shifting cultivation, bush-
fallowing and the growth of perennial crops.

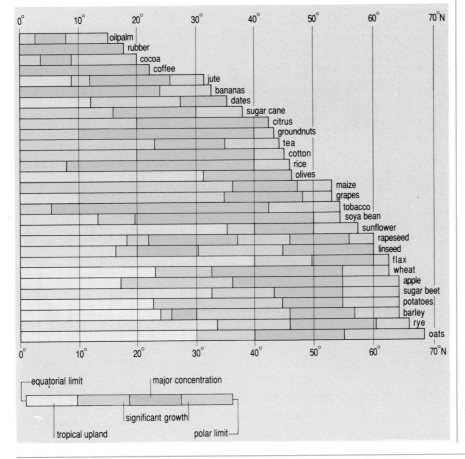

Figure 2.4
The latitudinal spread of the
major crops.

2.6   An overview of world agriculture

In spite of millennia of effort by subsistence farmers, several centuries of world trade in agricultural products and a century of commercial farming using machinery, chemicals and selective breeding, Figure 2.5 shows that the present area of cultivation is rather limited. This area could be extended, and perhaps doubled, though this would be extremely costly. However, constraints of climate, slope, soil and pests will continue to limit arable uses to a small minority of the land surface. Table 2.2 shows that much larger areas are useable as pasture and forest, as well as showing variations between continents. A third of the total land area is too cold and too dry to be useable for the growth of food, fibre or fuel.

Farms differ from each other in a great number of ways. There are variations in the crops that are grown, in the type of livestock kept, in the relative importance of crops and livestock in the farmer's income, in the level of technology, the amount of labour used per acre, and in the proportion of the output retained upon the farm. Differences in farm size and the type of land tenure are also important. It is thus exceedingly difficult to group farms into types of agriculture even for a small region, and a satisfactory classification of world agriculture has never been made. However, Derwent Whittlesey's classification of the major types and distribution map (Figure 2.6), made over fifty years ago, is still a useful guide.

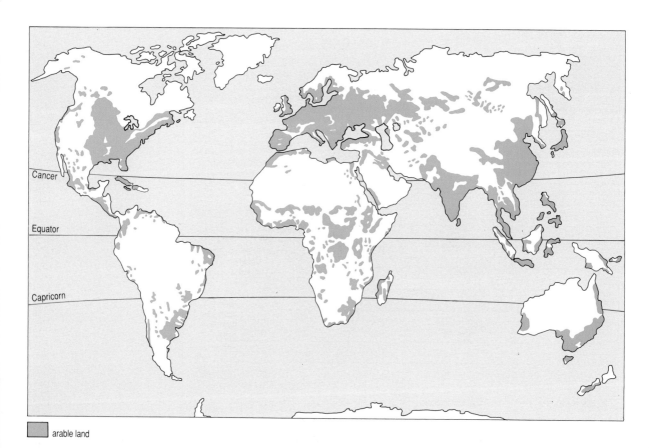

Cancer

Equator

Capricorn

▨ arable land

▲   Figure 2.5   The world distribution of arable land.

Q   Compare Figure 2.6 with Figures 2.2 and 2.3 and pick out at least two major climatic effects on agriculture.

Table 2.2   Patterns of world land use, 1991 (million hectares)

	All land uses		Arable		Pasture		Forest		Other	
	Land area	%	Area	%	Area	%	Area	%	Area	%
North America	1838	100	233	12.7	267	14.5	646	35.1	691	37.6
Europe	472	100	138	29.2	82	17.4	157	33.3	94	19.9
Oceania	845	100	48	5.7	430	50.9	157	18.5	209	24.7
USSR	2190	100	229	10.5	326	14.9	800	36.5	834	38.1
Africa	2963	100	181	6.1	900	30.4	681	23.0	1199	40.5
Latin America	2017	100	153	7.6	590	29.3	887	44.0	386	19.1
Asia	2678	100	457	17.1	759	28.3	530	21.7	931	34.8
World	13041	100	1441	11.0	3357	25.7	3861	29.6	4381	33.6

Source: FAO (1993) *Production Yearbook 1992*, Vol. 46, Rome.

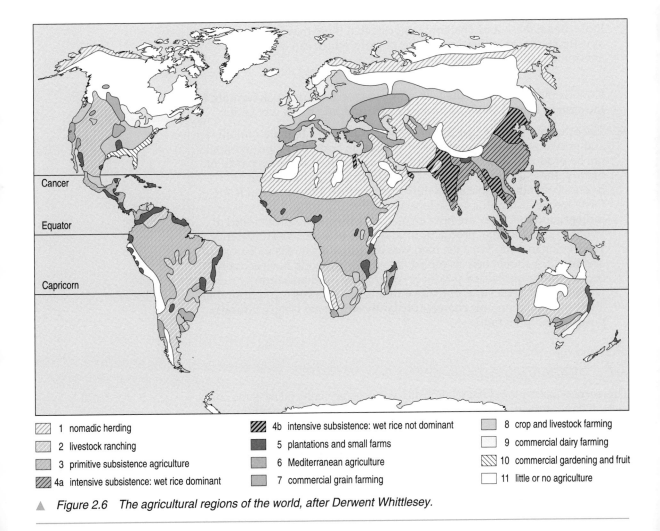

1 nomadic herding
2 livestock ranching
3 primitive subsistence agriculture
4a intensive subsistence: wet rice dominant
4b intensive subsistence: wet rice not dominant
5 plantations and small farms
6 Mediterranean agriculture
7 commercial grain farming
8 crop and livestock farming
9 commercial dairy farming
10 commercial gardening and fruit
11 little or no agriculture

▲   Figure 2.6   The agricultural regions of the world, after Derwent Whittlesey.

A   The two most obvious limits are those of cold, especially in the belt from North America, Greenland and Norway to the Soviet Union, and aridity, especially from the Sahara through Arabia, Iran, Afghanistan and into Mongolia.

Q   Why are some arid areas characterised by nomadic herding and others by ranching?

A   The fact that ranching occurs in the western United States and Australia as well as South Africa and Latin America suggests that this is an effect of economic development. Both use extensive livestock as a response to aridity, but ranching is commercial and nomadism not.

Q   Is the distribution of primitive subsistence agriculture dictated by environmental problems or differences in development?

A   The fact that the major areas are the equatorial forests suggests that environmental problems play a strong role: it has been shown above that, even apart from problems of flooding and dense vegetation, soils tend to be poor and net photosynthesis low. However, the area of primitive subsistence agriculture in the highlands of south-east Asia in close proximity to areas of intensive subsistence agriculture in China, India, Bangladesh and Thailand suggests that there may be developmental or cultural processes at work which can overcome environmental constraints.

Q   How would you explain the belt of mixed crop and livestock farming in the eastern United States, western Europe and into the Soviet Union?

A   The other maps show that this is a major area with adequate rainfall and growing season plus good arable land, so both crops and livestock *can* be produced. But the fact that they *are* produced depends partly on history: these were areas where a mixed farming regime (using livestock manure to sustain fertility and produce a range of foods for fairly local consumption) developed before and during industrialisation. They are now areas with a high enough standard of living to afford animal products, unlike areas of intensive subsistence production in Asia. Since the map was drawn, many areas in Europe and the USA shown here as mixed farming or dairying have actually become more specialised producers of grain as mechanisation, fertilisers and price support made animals less necessary and desirable. Technical and economic change constantly shift what is agriculturally possible and preferable.

2.7   Summary

This section has explored the relationship between agricultural systems and natural environments. It has considered the development of agriculture over time, from natural biomes, through domestication of plants and animals to the development of industrial societies and has stressed societies' increasing capacity to modify natural processes to produce more agricultural products, with concomitantly greater impacts on the environment, but still subject to ecological and climatic limits. Finally, this section looked at an attempt to classify the major types of agriculture around the world.

3   World agricultural production and productivity

3.1   Introduction

Although 45% of the world's workforce is engaged in agriculture, forestry and fisheries, the value of agricultural output is dwarfed by the value of other goods – from the manufacturing and mining sectors – and services. The developed countries produce half the agricultural output by value although they have only 23% of the world's population and 4% of the agricultural labour force.

Q   From what you learned in Section 2 above, how would you explain the relative performance of developed and less developed countries in agriculture?

A   Several factors contribute:

- the developed countries contain nearly half the good arable land
- their agricultural technology is more advanced, in the sense of using more energy, machinery and chemicals
- because their consumers are more affluent, they can produce higher value products as well as staples
- many developed countries subsidise agriculture to produce more than a free market would
- the figures for the developing world are an understatement because they do not include production for the farmers' own subsistence.

Table 2.3   Value of agricultural gross domestic product, 1988

	Agricultural GDP (US$ millions)	Agricultural GDP as percentage of world agricultural GDP
All developed	532 129	53.1
Western Europe	193 824	19.4
North America	114 380	11.4
Eastern Europe and former USSR	119 186	11.8
Other developed	90 921	9.1
Australasia	13 818	1.4
All developing	470 303	46.9
Asian CPE[1]	125 402	12.5
Latin America	85 100	8.4
Far East	160 432	16.1
Africa	55 227	5.5
Near East and North Africa	42 946	4.3
Other developing	1 196	0.1
World	1 002 432	100.0

Notes: [1]Asian CPE indicates Asian centrally planned economies: China, Kampuchea, Vietnam, Mongolia, North Korea.

Source: Grigg, D.B. (1992) 'World agriculture: production and productivity in the late 1980s', *Geography*, Vol. 77, pp. 97–108.

Thus developed countries, although having few of their workforce engaged in agriculture, are the leading agricultural producers. Western Europe is first, closely followed by the Soviet Union and North America. It is also worth noting that agricultural output, even in the developing countries, accounts for a small proportion of their total Gross Domestic Product. It is most important in south and east Asia, where it accounts for one-third of the value of all output; in Latin America and the Near East it is only one-tenth.

Nevertheless, agricultural production is of prime importance to us all and, to many, the main criterion of 'successful' farming is that of increasing productivity. This section will look at ways of measuring productivity and thus of evaluating the success of various agricultural systems.

3.2   Measures of productivity

Measures of **productivity** should be a useful means of comparing different agricultural systems and evaluating their appropriateness in the global agricultural system. In practice there are difficulties as productivity is not easy to define in principle and comparable data are not always obtainable. This section will explore the various ways of measuring productivity which can either be a measure of output per area or of the efficiency of outputs in relation to inputs.

In ecological terms, biological productivity is measured in terms of annual **biomass** per hectare: 3.58×10^3 MJ of light energy (approximate amount fixed per hectare per crop season) produces about 3500 kg/ha of dry biomass: this ranges from 200 kg/ha for crops such as beans to 11 000 kg/ha for maize and sugar-cane; average agricultural (crop) ecosystems produce an annual biomass per hectare that is slightly greater than the average of natural ecosystems (but should do considering inputs from added moisture, fertiliser etc.). Crops in the US contribute slightly more than 20% of the total plant biomass produced annually (Pimentel and Pimentel, 1979).

Looking more specifically at agricultural outputs, an obvious starting-point is the weight of useful crops produced (i.e. grain rather than total crop plant including straw for example), that is the *yield by weight per hectare*. This is only appropriate when comparing similar crops, such as staple cereals, since it would make little sense to compare output weights of systems producing cereals, meat and lettuce. In these cases, and when comparing output of varied commercial systems, it makes more sense to compare the *value of output* rather than its weight. This too is not without problems, since value depends on effective demand and is influenced by the spending power of the customers as well as the effectiveness of producers, but is often preferable to yield. It may be more appropriate to measure **value added** by agriculture and so to allow for the fact that high outputs may be largely accounted for by manufactured inputs, like fuel, fertiliser and pesticides. Unfortunately, value-added figures are available only for a limited number of countries. The measure of value used instead is **Agricultural Gross Domestic Product** (AGDP), which can be obtained from the United Nations' *National Accounts*. It is that part of national income received by labour, landlords and farmers, and, therefore, not strictly the value of output.

A similar point is made when contrasting intensive and extensive systems. In *intensive* farming systems, such as horticulture, the cost of inputs – labour, machinery, pesticides, seed and so forth – is high per hectare, but so is net income (the difference between sales and costs) per hectare. In contrast the cost of labour – and total costs – per hectare in extensive livestock farming systems is much lower, but so too is net income per hectare. Thus to have a satisfactory total income, farmers in these extensive agricultural systems have to have much larger farms. Such distinctions are to be found throughout the world: the sheep farms of Australia are extensive, the rice farms of Java intensive; the ranches of the western USA extensive, the corn farms of the Mid West relatively intensive.

In discussing intensiveness, inputs as well as outputs have been brought into the calculation. In economic terms the measure of productivity which best relates inputs to outputs is **total factor productivity**, since this relates the value of outputs to the value of land, labour and capital used in production. However, there are rarely statistics available on capital investment and these are certainly not available for international comparison. In practice, most measures of productivity relate output to only one input, either dividing total output by the land area used to give *land productivity* or dividing by the labour force to give *labour productivity*.

In the absence of any single ideal measure, the discussion below will consider and compare several measures of productivity, starting with yield per hectare, then relating AGDP per hectare to AGDP per capita. The source of data for international comparisons of agricultural productivity is the United Nations Food and Agriculture Organisation (FAO).

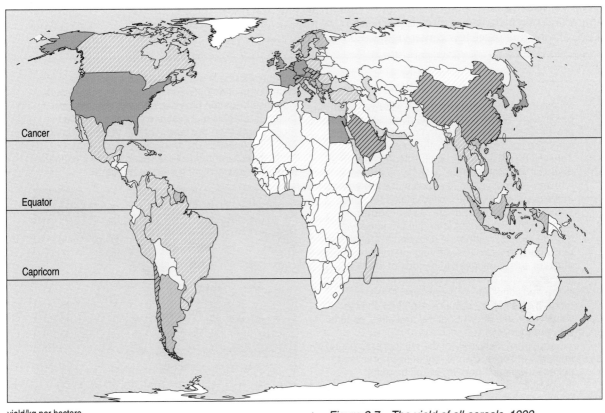

yield/kg per hectare

▓ 5000 and over	▨ 3000–3999	☐ 1000–1999	☐ no data
▨ 4000–4999	▨ 2000–2999	☐ 999 and less	

▲ *Figure 2.7   The yield of all cereals, 1992 (kilogram/hectare of arable land). Since figures are by country, output in small areas (such as the Nile Valley) may be generalised to much larger ones (in this case Egypt); similarly, Alaska is included in the average for the United States.*

Cancer

Equator

Capricorn

3.3   Land productivity

Crop yields are the most obvious measures of **land productivity**. Each crop ought to be considered, but as cereals occupy a high proportion of the total area under cultivation in all but a few countries, they are a good guide to the general level of yields.

Q   According to Figure 2.7, where are yields highest and lowest?

A   Cereal yields are highest in north-western Europe, east Asia (Japan, China, North and South Korea), the United States and Egypt; they are low, not only throughout most of the developing world, particularly in Africa, but also in several developed regions, such as the former Soviet Union, Canada and Australia.

Q   How would you explain these variations?

A   An explanation of this requires some consideration of the causes of yield variations; the major causes are climate, the supply of plant nutrients, the crop varieties used and the control of weeds and disease.

Climate

Crop yields are a function of a number of climatic factors, the most important of which, as has already been seen, are rainfall and temperature. Lack of moisture gives low cereal yields and much of Canadian and

---

Box 2.2   Irrigation

Irrigation is a term which covers a wide variety of practices. In some parts of Asia rivers at high water are allowed to flood fields, and crops are planted when the floods recede; in other places, such as Tamil Nadu or Sri Lanka, each village has a small reservoir with earth dams, called a 'tank', that stores monsoon rainfall, to be released in the dry season. In the south-west of the United States, in Australia and the oases of the Sahara, water is obtained by pumps from fossil water stored in the underlying rock. In the most elaborate irrigation systems, large dams contain water in reservoirs on the upper parts of rivers; this is released in periods of low water, and distributed by canals. Irrigation is not confined to the traditional arid and semi-arid areas; in modern times spray systems are used on high-value crops in eastern England and other parts of western Europe.

Approximately 15% of the world's arable land is irrigated. In some regions, irrigation is practised because without it farming would be quite impossible, as in the deserts of Egypt and Saudi Arabia; elsewhere irrigation is used to supplement rainfall and to overcome the high variability of rainfall in semi-arid regions, for example, where rainfall is concentrated into a short period of the year. It is most common in Asia; yet surprisingly a minority – 36% of non-Communist Asia – of the continent's rice is not irrigated although there is a clear connection between irrigation and rice yields. Because irrigation provides security against crop failure, it attracts high-value crops and receives more inputs than dry land. Thus in the developing countries only 20% of arable land is irrigated, but receives 60% of the fertiliser used in these countries and produces 40% of the value of output.

▲   The modern version of the 'Persian wheel' is a basic means of lifting irrigation water from wells in the Indus Plains of Pakistan and north-western India.

Australian cereal output is in semi-arid regions as is that of the Soviet Union, Argentina and the Near East. This limitation can be overcome by irrigation: see Box 2.2.

There is little that farmers can do about modifying temperatures, although in exceptional cases high-value crops – vegetables, fruit and flowers – are produced in glass-houses. Elsewhere, farmers have to adapt to the prevailing temperature regime.

The supply of plant nutrients

Until the mid nineteenth century, fallowing, manuring and rotation with legumes were the main means of maintaining crop yields. However, with the industrial fixation of nitrogen, with oxygen to produce nitrates, or with hydrogen to produce ammonia, farmers soon turned to commercially produced chemical fertilisers: in 1900, 91% of nitrogen applied to farm crops was organic; by 1913 this was only 40%. Since 1945 their use has boomed: the early 1970s saw the cost of fertilisers fall in real terms and their ease of handling compared with farmyard manure has led to their even greater use. In the 1970s chemical fertilisers provided 75% of the nitrogen, 70% of the potassium and 59% of the phosphorus for plant growth in the UK; the consumption has risen eight-fold between the 1930s and the 1980s.

Equally dramatic rises in the consumption of chemical fertilisers have occurred in nearly all developed countries in the post-war period. Nor has the increase in fertiliser consumption been confined to the developed countries; indeed the rate of increase has been greater in the developing countries, largely because of the very low usage in 1949–51: see Table 2.4.

Table 2.4 Fertiliser usage per hectare of arable. 1949–81 (kg of nutrients/hectare)

	1949–51	1980–1	Increase factor
Latin America	3.1	46	14.8
Far East	1.6	38	23.8
Near East	2.4	34	14.2
Africa	0.4	10	25.0
Asian CPE[1]	–	146	–
Developing	1.4	49	35.0
Developed	22.3	116	5.2
World	12.4	80	6.5

Note: [1]Asian Centrally Planned Economies: China, Kampuchea, Vietnam, Mongolia, North Korea.

Source: FAO (1971) *The State of Food and Agriculture, 1970*, Rome; FAO (1983) *The State of Food and Agriculture, 1982*, Rome.

There are still marked variations in their usage of chemical fertiliser: see Figure 2.8. The highest usage is in western Europe, east Asia and some irrigated areas in the Middle East, the lowest in Africa, Latin America and south Asia; this reflects the fact that fertilisers are a product of the heavy chemical industry, concentrated in the industrial countries where the real cost of fertiliser is low and, furthermore, agriculture is highly subsidised. The developed countries produce 75% of world fertiliser output, many developing countries relying upon imports from the developed world. But

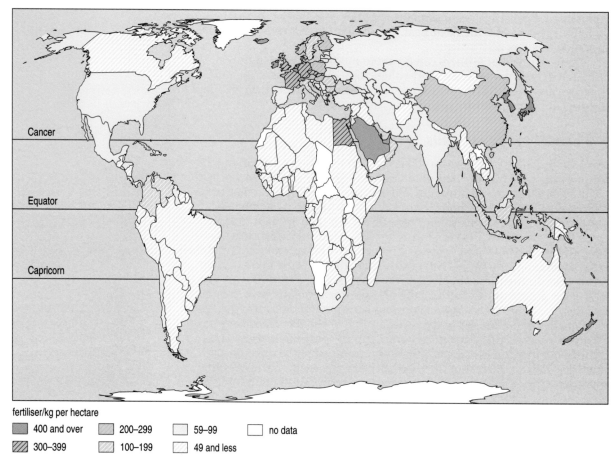

fertiliser/kg per hectare

▓ 400 and over   ▨ 200–299   ☐ 59–99   ☐ no data

▨ 300–399   ▨ 100–199   ☐ 49 and less

▲ *Figure 2.8   Synthetic fertiliser use, 1989 (kilograms per hectare of arable land).*

there is not simply a distinction between developed and developing countries: fertiliser usage per hectare is low in the Soviet Union, North America, Australia and Argentina. Much of the crop acreage in these countries is in semi-arid zones, where the absence of moisture impairs the uptake of fertiliser by plants. Further, much of the farming in these regions is carried out on a large scale, with the aim of maximising output per capita rather than per hectare. Although the use of fertiliser is not the only, indeed not the most important, cause of yield variations, there is a relationship between use per hectare and crop yields: this is shown in Figure 2.9. MAFF has concluded that nitrogen was indicative of technological progress between 1975 and 1984 and argues that nitrogen and fungicides exploited the potential of new wheat varieties which were introduced in the mid-1970s.

New plant varieties

In traditional farming farmers selected their seed for the next crop from their harvest and, until the nineteenth century, one quarter or more of each harvest was retained for this purpose. They were able to select the better seed and slowly improve yield and resistance to disease in this manner. In addition, seed from other regions could be imported. Hence Russian wheats which developed in a short growing season were taken to Canada

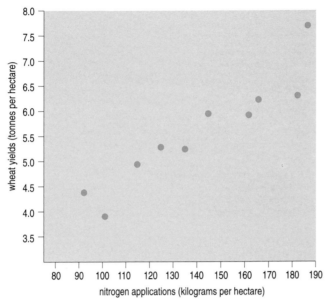

▲   *Figure 2.9   Changing nitrogen applications (kilograms per hectare) in relation to yields of winter wheat (tonnes per hectare) for England and Wales, 1975–84.*

Source of data: ADAS, Cambridge.

in the nineteenth century and made possible the cultivation of the Prairies. But it was not until the principles of genetics were applied to crop breeding in the early twentieth century that it was possible to breed varieties with specific characteristics: tolerance of drought, more rapid maturing, the ability to absorb fertiliser and immunity to specific disease have all contributed to higher yields.

The first major impovement came with the breeding of hybrid maize in the United States in 1922, and by the 1950s it had largely replaced the earlier varieties. In Europe there were comparable advances, particularly in wheat, and since 1945 a succession of new varieties have contributed to higher yields. Indeed approximately half the yield increase in the United States and western Europe since 1945 is attributed to new crop varieties. Less advance was made elsewhere except in Japan and its colonies where new rice varieties were bred and widely adopted in the inter-war period. Advances made in Mexico and the Philippines in the breeding of semi-dwarf wheat and rice have led to major changes in the varieties used in Asia. Although initially only adopted by those with large farms, the new High Yielding Varieties have spread to small farms, and now substantial proportions of the farm area are sown with HYV wheat and rice, and hybrid maize (see Table 2.5), particularly in China where the new varieties were locally bred rather than introduced from Mexico and the Philippines. Africa has least of its area in the new varieties, partly because rice and wheat are not widely grown.

The substantial increases in yield are only possible, however, under optimum conditions, and require liberal fertiliser use, irrigation and – because the new varieties lack the immunity to disease which older varieties had acquired – the use of pesticides. They thus tie the farmer into new farming methods which require much greater outlays of capital, not only for the seed, but for the fertilisers and pesticides which must be used as well in order to reap the benefit. This favours the large land-owners, or

Table 2.5 Percentage of area in wheat and rice sown with HYV
in 1977, and in hybrid maize. 1983–5

	Wheat	Rice	Maize
Asia	72.4	30.4	6.8
Near East	17.0	3.6	32.0[1]
Africa	22.5	2.7	16.2
China	25.0	80.0	72.0
Latin America	41.0	13.0	50.2

Note: [1] Including North Africa.

Source Dalrymple, D. (1978) *Development and Spread of High Yielding Varieties of Wheat and Rice in the Less Developed Nations,* Washington, DC: Dalrymple, D. (1988) *Development and Spread of improved Maize Varieties and Hybrids in Developing Countries.* Washington, DC.

necessitates high levels of borrowing for the smaller farmer. This is a great problem in Africa where, although women produce most of the food, they have great difficulty in being granted loans, because they do not own title to the land; they are also usually excluded from development programmes and from membership of co-operatives.

The control of pests and weeds

As we have seen, the conditions that encourage the growth of crop plants also encourage other plants – weeds – as unwelcome competition for the supply of nutrients and moisture. The control of weeds has always been a major aim of farmers and once occupied much of the year. In Europe ploughing before sowing and ploughing in the fallow year were undertaken, whilst horse and hand hoeing were practised during growth when crops such as potatoes, sugar-beet and fodder roots were grown. In most farming systems the control of weeds was only possible with very large amounts of hand labour. However, during the late nineteenth century, chemical sprays that killed weeds – herbicides – were tried in France. But it is only since the 1950s that herbicides have been used in large quantities in North America and western Europe. Their use has greatly reduced the amount of labour in crop production. This aspect is not so crucial to third world countries, where labour costs are low and supply is plentiful.

Traditional farmers had great difficulties controlling crop diseases that were spread by insects and fungi. Rotations may have reduced the incidence of plant disease, as did intercropping in the tropics. Early chemical sprays used inorganic materials to control the diseases of potatoes and the vine in the 1860s, and the spraying of fruit was common in the 1920s, when aircraft were first used. But as with herbicides, the rapid growth of usage of insecticides and fungicides has come only since the 1950s, and is greatest in North America and Europe, which consume half the world's output. In 1982 there were 4000 registered pesticides in the UK, representing 1000 different types, and worth nearly £400 million to the chemical industry. Chemical sprays have had adverse effects on wildlife, and recent work suggests some foods may contain unhealthy amounts of some pesticides. In the developing countries their use is largely confined to export crops. The lack of control over pesticide use and the continued use of chemicals banned in first world countries give cause for concern over health risks in the third world (see, for example, Figure 6.12 below).

Summary

The differences in yields between different parts of the world – the contrast is largely between Europe, the United States and east Asia and the rest of the world – is thus explained by a variety of factors. The non-tropical location of these regions is one such factor, but more important are the intensity with which inputs are used. In east Asia – China, Korea, Taiwan and Japan – labour is still intensively used in preparing seed beds, weeding and transplanting rice; in Europe, and particularly western Europe, high yields are due more to the use of chemical fertilisers, pesticides and improved crop varieties. In western Europe cereal yields are now over three times those of the 1930s, before the widespread use of chemical treatments or improved varieties. In contrast fertilisers and new varieties are little used in Latin America, Africa and much of Asia.

Yields will be a major contributor to output value, but existing measures of the value of agricultural produce cover a much wider range of produce than just cereals and include livestock as well as crops. The explanation of variations in yields which was given suggests that in many cases a high value of outputs will rely on a high level of purchased inputs. A better measure of the efficiency of farms is therefore value added – the difference between the value of outputs and the value of inputs. Estimates of value added have been published by FAO and the World Bank. Unfortunately, these are not widely available and Agricultural Gross Domestic Product (AGDP), as calculated by the World Bank, has been used instead; these figures are a rough guide to the value of agricultural output but not, of course, to the value added. Nevertheless, they do present a striking picture, as shown in Figure 2.10. overleaf.

Q   What countries stand out as high or low in AGDP per hectare?

A   Not surprisingly, east Asia and western Europe are high in value per hectare but are matched by Egypt, Somalia and Papua New Guinea. Irrigation and plantation crops provide the explanation for the surprises. At the low end of the scale, Ethiopia and Sudan might be expected but Australia, Argentina and North America are all major food exporters. The explanation of their low value per hectare is partly minimisation of labour costs and partly the low productivity of semi-arid pastoral production, which covers large areas of these countries.

Activity 1

Compare Figures 2.7 and 2.10 and pick out examples of countries performing differently in terms of yield and AGDP per hectare. Why do the differences occur?

The most dramatic contrast has already been mentioned: Papua New Guinea is in the lowest cereal yield category but the highest AGDP per hectare. This is the extreme of a more general phenomenon – the generally better performance of countries near the equator in terms of the value of output than in terms of cereal yields. This is a reminder of the implicit bias of the cereal yield measure since cereals, with the partial exception of rice, are better suited to the mid latitudes than the tropics and in the case of Africa and Latin America are less preferred than root crops like yams and manioc.

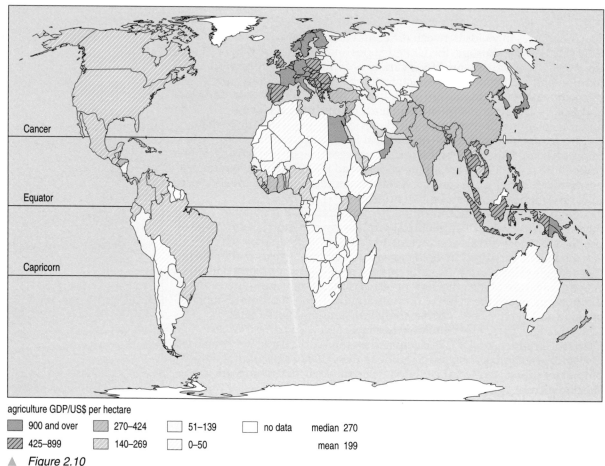

agriculture GDP/US$ per hectare

▓ 900 and over	▨ 270–424	☐ 51–139	☐ no data	median 270
▨ 425–899	▨ 140–269	☐ 0–50		mean 199

▲ *Figure 2.10*
Agricultural Gross Domestic Product per hectare of arable and permanent pasture, 1988 (US dollars).

Source: World Bank (1991) *World Development Report 1990*, Washington; FAO (1990) Product Yearbook 1989, Vol. 43, Rome.

A striking feature of Figure 2.10 is that China is in a lower AGDP per hectare category than the UK and is much superior to the USA. This certainly departs from most people's images of the productivity of these countries, suggesting that those images are based on a different measure.

3.4   *Labour productivity*

Although land productivity is an important measure of agricultural productivity, it is not the only one. Indeed labour productivity is more usually used as an index of overall productivity in the developed world. Some indication of international differences can be obtained by dividing the AGDP by the agricultural workforce of each country: this is shown in Figure 2.11. This pattern of productivity is different from that of crop yields and AGDP per hectare. The highest figures are for western Europe, the United States, Canada, Australia and Japan. Conversely output per head is remarkably low throughout all of Africa and Asia (except Japan); it is higher in Latin America, but still much below Europe and the United

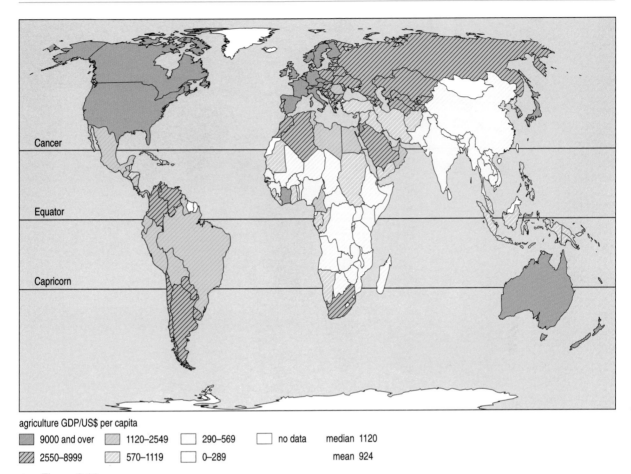

agriculture GDP/US$ per capita

■ 9000 and over	▨ 1120–2549	☐ 290–569	☐ no data
▧ 2550–8999	▨ 570–1119	☐ 0–289	

median 1120

mean 924

▲ *Figure 2.11*
Agricultural Gross Domestic Product per capita of the agricultural labour force, 1988 (US dollars).

Source: World Bank (1991) *World Development Report 1990*, Washington; FAO (1990) Product Yearbook 1989, Vol. 43, Rome.

States. This reflects priincipally variations in labour per hectare: areas
which use a great deal of labour and little machinery have low labour
productivities; those with low labour usage and much machinery have high
outputs per capita. To understand world differences in labour densities, it
is necessary to examine briefly the history of the agricultural labour force.

Until the eighteenth century the agricultural labour force made up 75%
or more of the population of nearly all countries except Britain and the
Netherlands. However, with the spread of industrialisation and the growth
in the number of workers in manufacturing, mining and services, the
proportion of agricultural workers began to decline, and has declined
continuously to the present day. As industrialisation spread from western
Europe to other parts of the world so a similar decline occurred there; even
in Afro-Asia and Latin America there has been a substantial fall in this
proportion since 1950, so that the agricultural workforce exceeded 75% in
very few countries in 1992 (see Figure 2.12 overleaf).

Whilst the decline of the *proportion* of the workforce engaged in
agriculture has been universal, changes in the *absolute numbers* engaged in

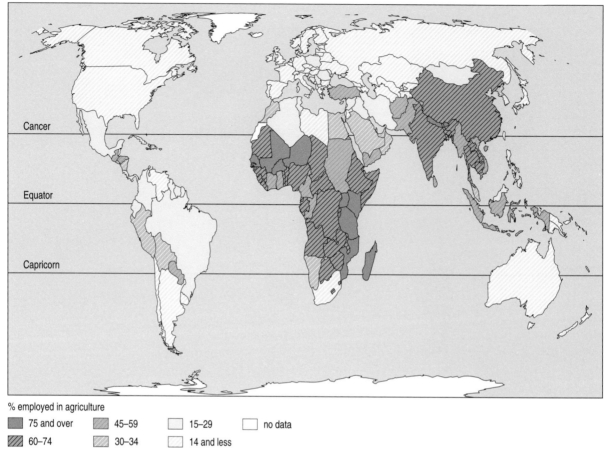

% employed in agriculture

| ▓ | 75 and over | ▨ | 45–59 | ☐ | 15–29 | ☐ | no data |
| ▨ | 60–74 | ▨ | 30–34 | ☐ | 14 and less | | |

▲ *Figure 2.12   Percentage of the economically active population engaged in agriculture, 1992.*

agriculture have followed a different course, because industrialisation was accompanied by an increase in population, including the rural population, mainly due to a fall in mortality, but in eighteenth- and nineteenth-century Europe to a rise in fertility as well. As industrialisation got under way in nineteenth-century Europe, the difference between wages in agriculture and in industry attracted people from the farms to the towns, a process that has continued to the present day. But initially the rate of migration out of agriculture was exceeded by the rate of natural increase in the agricultural workforce, and so employment in farming slowly increased until the late nineteenth century, except in the British Isles where decline began in the 1850s. Elsewhere the decline of the agricultural workforce began later, either because the onset of industrialisation was later, as in eastern Europe, or because new areas of land were still being colonised, as in Australia, Canada, Argentina and the United States. But by the 1950s the agricultural workforce of the developed countries was everywhere in decline; between 1970 and 1985 alone it fell by over one-third.

In contrast agricultural labour in most parts of the developing countries has been in continuous increase down to the present; in Latin America, Africa and Asia the cities have attracted many migrants from the

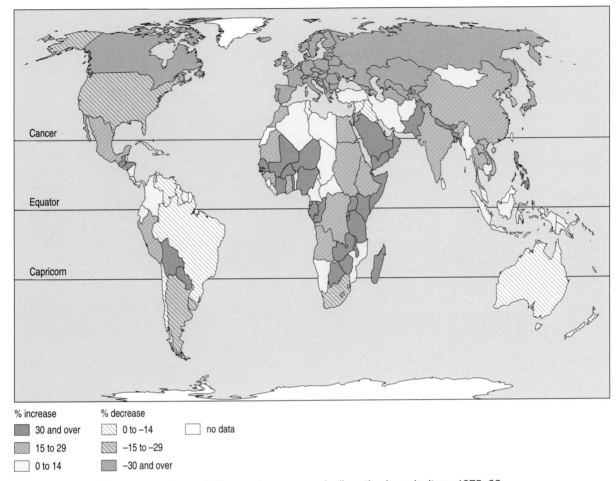

% increase   % decrease

■ 30 and over   ▨ 0 to −14   □ no data

▧ 15 to 29   ▨ −15 to −29

□ 0 to 14   ■ −30 and over

▲ *Figure 2.13   Percentage change in the numbers economically active in agriculture, 1975–92.*

countryside because of the prospect of better services and higher wages, in spite of the prevalence of urban unemployment. In many parts of Asia rapid rural population growth has led to the subdivision of farms, and this has accelerated out-migration. But the growth of rural populations by natural increase has also been very high, and exceeded the rate of out-migration, so that the agricultural labour force has continued to increase in spite of out-migration; indeed the agricultural labour force of the developing countries as a whole increased by over 25% between 1975 and 1992. Only in a few developing countries has a decline in the agricultural population begun – in parts of Latin America and the Middle East: see Figure 2.13.

In the developed countries agricultural populations have declined because of the opportunities for better incomes in the cities; also because in some cases farmers have replaced labour by machines. The reaper, the threshing machine and other labour-saving implements began to be widely used in North America and Britain in the late nineteenth century; perhaps as important was the tractor, introduced in the early twentieth century, and the combine harvester which was first used in Britain in the 1930s. But it was not until after the Second World War that the tractor finally replaced the horse, and the mechanisation of the United States and western Europe

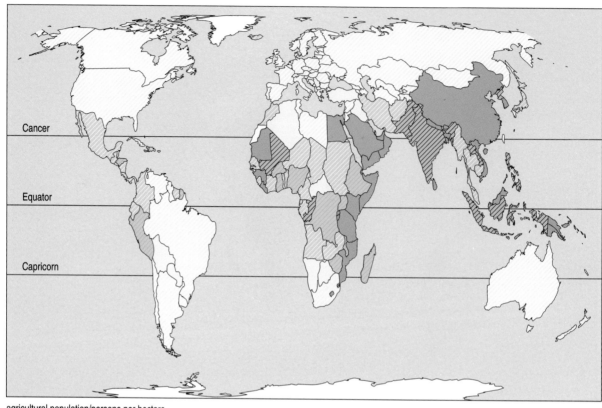

agricultural population/persons per hectare

�+	4.0 and over	▨	2.0–2.9	☐	0.6–0.9	☐	no data
▨	3.0–3.9	▨	1.0–1.9	☐	0.5 and less		

▲ *Figure 2.14   Agricultural population (per hectare of arable land), 1989.*

has since proceeded at an unparalleled pace. Mechanisation has proceeded less rapidly in Afro-Asia and Latin America, for there is less incentive when agricultural populations are increasing, and thus wages are low. There are very striking differences in the labour used in agriculture. These are shown in Figure 2.14: agricultural population per hectare is lowest in the developed countries – Europe, the USSR, Australasia, North America and much of Latin America, where mechanisation has made more progress than elsewhere in the developing world. Densities are highest in Africa and Asia but particularly high in east Asia and in the dry lands of the Middle East and eastern Africa.

Activity 2

Use the data on productivity to separate countries into four categories:
*   high productivity per hectare and per capita
*   high per capita and low per hectare
*   low per hectare and per capita
*   high per hectare and low per capita.

▲ *Campesinos spraying fields near Lima, Peru. Even in less developed countries, large-scale monoculture is well established and requires chemical pest control.*

3.5   Summary

By combining the two measures of productivity – value of output per hectare and per capita – it is possible to recognise several classes of countries. First are those where both labour and land productivity are high. These are in western Europe, with Denmark in the top category for both measures. A second category includes those countries where yields are low, but output per head high – these include Canada, Argentina and Australasia. In these countries agricultural population densities are low, and farming is highly mechanised; but crop yields are relatively low, reflecting the limited use of fertilisers. A third category includes tropical Africa and South Asia, where the productivity of land is moderate to low and labour productivity is very low; although agricultural population densities are high, few modern inputs – fertilisers, new crop varieties or pesticides – are used; nor has mechanisation made much progress. Finally, a small group of countries have comparatively high yields but lower labour productivity. In east Asia and parts of south-east Asia, yields are high because of the intensive use of labour, whilst in the last twenty years an increasing proportion of land is irrigated, fertilisers used and HYV varieties are replacing the traditional rice varieties. These kinds of system, led by Japan, may well be the most relevant to the future need to feed growing numbers of low-income people. (This is discussed further in the next chapter.)

4 Sustainability

4.1 Introduction

The previous section has shown the high levels of productivity which can be achieved in modern farming systems. Growing populations and rising living standards constantly demand that agricultural output also increases. However, concerns have grown about the effects of intensive farming methods and whether they are sustainable in the long term. The World Commission on Environment and Development (or Brundtland Report) (1987) defined sustainable development as . . . development which meets the needs of the present without compromising the ability of future generations to meet their own needs'. This brings into consideration rates of resource consumption and long-term environmental impacts.

This section begins by looking at the widespread problems that threaten the maintenance of ecological balance which is vital to sustaining output. The dependence of modern agriculture upon large energy inputs will then be explored. Thirdly, the growing public opposition to chemical farming methods and their long-term effects on both land and people will be considered. Lastly, when considering sustainability it is necessary to look at the interdependence of the global agricultural system and the prospects for the future.

4.2 Farming and soil

As this chapter has shown so far, farming changes the natural ecology of an area: it requires the removal of natural vegetation, and the cultivation of a limited number of crop plants and/or keeping a limited number of animals, rather than the great variety of species which occur naturally. Land clearance and crop cultivation remove plant nutrients and, when bare, the soil is exposed to higher wind speeds, greater temperatures and higher intensity of rainfall. The overall effect of this can be to substantially alter the local environment, often in ways detrimental to farming productivity. Furthermore, many farming techniques, such as irrigation and the use of chemical fertilisers, have unexpected adverse consequences.

The term **land degradation** has been used to describe physical and chemical changes which reduce the long-term productivity of the soil. But this is difficult to measure; removal of top soil by wind in the American Great Plains in the 1930s was obvious, as is gullying in parts of Nigeria now, but slow changes in the chemical composition of soils is less easy to monitor. However, some ten years ago, the Food and Agriculture Organisation of the United Nations estimated that one quarter of the world's arable land was subject to degradation, through salinisation, soil erosion and desertification. They subsequently raised this estimate to one third.

Acidity, alkalinity and salinity

One of the chemical characteristics of a soil is its acidity or alkalinity. These are at opposite ends of a continuum which is measured on the pH scale. This runs from pH 0, which is highly acid, through neutral at pH 7 to pH 14 which is highly alkaline. Soils with a pH rating of 5–7 are suited to the growth of most crops, but outside this range only a few crops thrive and the yields of others is reduced. Ultimately conditions become so extreme that crops will not grow. Although both acid and alkaline soils occur in natural conditions, their pH may be changed by some agricultural practices.

Soil acidity occurs widely in both agricultural and non-agricultural soils and is a particular problem in cool, high rainfall areas. For example, most soils of upland Britain are acid. Soil acidity has several adverse effects. It reduces the capacity of bacteria in the soil to fix nitrogen and it limits the ability of crops to take up plant nutrients from the soil. In Europe it is claimed that deforestation increased soil acidity in the Middle Ages, but soil acidity can be reduced by applying lime, a technique long known to farmers. More recently there has been some concern that acid rain may be affecting crop yields. Industrial smoke emitted over the last two centuries has increased the amount of sulphur dioxide and nitric oxides in the atmosphere; they are converted in the atmosphere into a mixture of sulphuric and nitric acid that returns to Earth in rain. Their effect upon trees has been much discussed, but little is yet known of the effect upon crops. Acid rain may increase soil acidity which slows the growth of most crops, but it also supplies sulphur which promotes growth. Although no generalisations can be made about the net effect, acid rain is thought to have reduced crop yields by 10% in parts of eastern Canada.

Soil alkalinity is encouraged by the build-up of calcium, magnesium, potassium and sodium in the soil. This occurs mainly in arid conditions where there is insufficient leaching to remove these metals in solution. Alkalinity may be worsened by irrigation with water containing dissolved metals, especially if the irrigated land is inadequately drained so that the metals concentrate in the upper layer. The worst cases of alkalinity, with pH as high as 10, are those involving sodium carbonate. No crops will grow under such conditions. In principle, alkalinity can be reduced by flushing with water and/or treatment with gypsum to convert sodium carbonate to sodium sulphate, which is less caustic and more easily leached. Such treatment may often be too expensive to be adopted in practice. However, even if alkaline soils can be reduced to neutrality, they may remain unproductive because of the concentration of salts in the soil water – a condition known as **salinisation**.

Saline soils are a major problem in many parts of the world and occur under three conditions. First, in some semi-arid areas there is saline groundwater which moves to the surface by capillary action, and the salts are deposited by evaporation on the surface as a salt crust. In Australia the removal of deep-rooted forests and their replacement by pasture or crops has led to the salinisation of 5 million hectares, mainly in Western Australia. Second are potential arable soils near coasts in the humid tropics of south and south-east Asia. Some 20 million hectares currently supporting mangrove swamps could be drained to provide excellent cropland if it were not for their high salinity (and other ecological arguments for their conservation). Third, and more important, is the occurrence of saline soils in irrigated areas. (See Plate 3a.)

In these arid regions water is obtained from rivers or reservoirs and carried to fields by canals; before irrigation the water table – the upper surface of water in the ground – was at a considerable depth, but after several decades of irrigation, water that has leaked from unlined canals and floodwater have led to a slow but continuous rise of the water table. This can be substantial: modern irrigation in the Punjab has lifted the water table 7–9 metres above the level of 1895, and in much of India and Pakistan the water table is now within a few metres of the surface. These underground waters may become increasingly salty with the salts taken down with the water; where the groundwater reaches the root zone, its brackishness will stunt or prohibit crop growth, or even if not salty will kill crops by cutting off the oxygen supply.

The salinisation of irrigated areas is not new; it was widespread in the Tigris and Euphrates valley some 2500 years ago. Nor is it confined to developing countries: 25–35% of the irrigated land in the south-west of the United States is affected by salinisation. But it is in developing countries that these problems are most acute. In the late nineteenth century modern irrigation systems were constructed in many parts of south-west and southern Asia: large concrete dams formed reservoirs from which water was distributed to cropland by canals. These regions are now suffering from a high degree of salinisation and waterlogging. Indeed some authorities believe that half the world's irrigated land is salinised, at best reducing crop yields, at worst causing land abandonment. In India 13 million of the 43 million irrigated hectares are salinised. In Pakistan, with a far greater dependence on irrigation, some 70% of irrigated land is salinised, in Iraq half, in Egypt 30%, in Iran 15%. In Egypt construction of the Aswan Dam has increased the availability of water, allowed more double cropping and reduced the risk of crop failure. But the excessive use of water and the spread of salinisation has in places reduced crop yields by 20%.

In irrigated areas the problems are of excessive water, and the cures are, firstly, more careful management of water, second the lining of canals and, third, the underdrainage of land. All these are expensive, but less expensive than bringing new land into cultivation.

Soil erosion

Soil erosion is regarded as a major threat to land productivity, and there have been many pessimistic statements about its extent and impact upon crop production. But it is difficult to measure and even more difficult to estimate its effect upon crop yields.

Soil is constantly being eroded by the action of running water and wind. Soil erosion is a natural process, and it only becomes a problem for the farmer when erosion substantially exceeds the formation of new soil. Soil forms in Britain and the United States at about 0.3–1.3 tonnes per hectare per year, or at 0.02 mm–0.1 mm per year.

A net soil loss has a number of adverse consequences. The removal of soil reduces the organic matter, plant nutrients and water retention in the soil and restricts the development of roots. Consequently crop yields may fall unless fertilisers are used. When soil erosion becomes serious, its effects are obvious. Wind erosion removes much of the upper soil, and with water erosion deep gullies appear, which make it physically impossible to work the land. (See Plate 3b.) Although there is historical evidence of soil erosion, it was not until the 1930s that there were efforts to measure soil erosion and to develop soil conservation techniques. The United States Department of Agriculture pioneered research into soil erosion and remains active today. In northern Europe soil erosion has had less apparent effect although it is now accepted that some modern farming techniques are having adverse effects upon the soil.

The United States Department of Agriculture has estimated T-values for the major soil types. A T-value is the level of soil erosion that will permit a high level of crop productivity to be sustained economically and indefinitely. T-values do not exceed 11.2 tonnes per hectare per year on any soil type yet 27% of the cropland in the United States has rates above the T-value, and 10% has rates above 22.4 tonnes per hectare.

The relative importance of the action of wind and running water depends upon local climatic conditions and soil type. In England and Wales *wind erosion* does not occur until wind speeds exceed 9 metres per second. Although such wind speeds occur at least once a year in nearly every part of England and Wales, the south-west and west of the country have the highest frequency of such speeds; but they also have the highest rainfall. As this binds soil particles together, wind erosion is unusual. It is much more common in the eastern lowlands on dry friable soils; peats in the Fens and sands in the Vale of York have most recorded instances of wind erosion, but as yet few farmers see it as a major hazard. It has been estimated that 14% of the arable land of England and Wales is susceptible to wind erosion.

To some extent modern farming practices have increased the risks of wind erosion, notably removal of hedgerows to enlarge fields and thus allow the easier use of machinery. The decline of grassland in the east of England has also increased the risk of both wind and water erosion, for the continuous cover throughout the year protects the land, whilst the practice of bare fallowing increases the risk of wind erosion since bare dry soil is highly susceptible to erosion. This is a particular problem where farming has penetrated into semi-arid areas. It was, of course, in the Great Plains in the 1930s that the most celebrated of all wind erosion events took place and this – the Dust Bowl – persuaded the United States government of the need to finance and promote soil conservation.

In much of the world's arable regions erosion caused by running water is a more serious problem than wind. A number of factors determine the extent of *water erosion*. First is the intensity and duration of rainfall. The size of raindrops and intensity of storms is greater in the tropics than in temperate areas, where rainfall comes mainly as drizzle. There is a marked difference in the erosivity of rainfall between the tropics and temperate regions, and even between northern and Mediterranean France, the former being only 25% of the latter.

◀ Ploughing parallel to the steepest slope on even a very gentle gradient can lead to run-off concentration and initiation of gully erosion.

Second is vegetation: the greater the cover, the less the soil is exposed to the impact of rain or wind. Thus experiments in England have shown that erosion rates are least under woodland, and then increase progressively under grass, cereals, and are highest on bare soils. The latter occur not only on fallow land but in short periods between crops. In England autumn-sown cereals are at risk partly because there is little plant cover in winter but also because this is the period of highest rainfall. Research in the hilly watersheds of the Mississippi, where the natural forest cover had been stripped away, shows the effect of human land uses on erosion rates: from the forested area, only a few hundredths of a ton per hectare per annum was lost; from grassy pasture 4 tons; from areas cultivated for corn an average of 54 tons (ranging from 8 to 106); and on abandoned farmlands where gullies had formed, 450 tons were being lost in a year.

A third and very important factor is the angle of slope: the steeper the slope the greater erosive capacity of running water. Not surprisingly some of the more dramatic erosion occurs on steep slopes in uplands where deforestation has taken place. Population pressure has driven farmers to clear forest and plough higher-angle slopes in many parts of the world. Serious erosion has been recorded in uplands in Nepal, Bolivia, Kenya, Ethiopia and Central Java. In Colombia slopes of 45° have been cultivated and soil loss has reached 370 tonnes per hectare per year. In Britain the high prices paid to cereal farmers, and the development of tractors that can operate on steep slopes – up to 21° – have led farmers to cultivate grassland on 20° slopes on the South Downs. The effect of slope on soil erosion is well illustrated by an example from Nigeria: when cassava was grown on land with a 1° slope, the rate of soil erosion was 3 tonnes per hectare per year, on 5° slopes 87 tonnes, but on slopes of 15°, 227 tonnes per hectares per year were lost.

Fourth, soil texture plays an important role in determining the rate of soil erosion. Generally the coarser the particles in a soil the more easily

▲ *A highly organised fuel collection service in Niger. The fuel needs of the increasing urban populations in third world countries can lead to large-scale deforestation.*

eroded it will be, and the greater the clay content, the lower the erosion. In England erosion is greatest on sand, moderate on silts and lowest on clays.

Finally, research in both the United States and Britain suggests that some modern farming practices increase the risk of erosion. Before the use of herbicides and pesticides, farmers in Britain and the eastern United States commonly rotated crops and grass and combined crop and livestock production. This increased the amount of organic matter in the soil as well as providing more continuous vegetation cover. There is evidence that organic manures improve soil structure, and so reduce the erodability of soils. Levels of organic matter are reduced under cereal monoculture using artificial fertiliser. The use of heavy machinery compacts soils and reduces infiltration, thus increasing run-off. Continuous crops of cereals leave the land bare for much of the winter, whilst stubble-burning reduces ground cover.

There have been many attempts to persuade farmers to adopt conservation methods. In the United States farmers now practise *contour ploughing* on slopes, ploughing along the slope rather than down it, but in Britain few farmers have adopted this practice, whilst as long as chemical farming is profitable there is unlikely to be a return to rotations or mixed farming. One new technique that does reduce soil erosion is *no-tillage or zero-cultivation*. Instead of ploughing the land and harrowing several times before sowing, farmers only lightly till and drill seed directly into the soil. This is rapidly increasing in the United States, but has made little progress in Britain. Other conservation techniques include planting shelter belts and intercropping, where crops with different growth rates are planted on the same land. These methods are discussed in Chapter 4.

The precise effect of erosion upon the world's arable land is unclear; no one doubts that it is a serious problem, but there are few reliable estimates based on careful research. Some estimates of its extent seem excessive, but a few illustrations will show that there is cause for concern. Thus 61% of India's cultivated area is said to be undergoing some form of degradation including salinisation; 80% of Bolivia's crop and grass area has had a fall in productivity from soil erosion; 80% of Madagascar's crop land is affected by severe erosion; half of Australia's agricultural land is said to be in need of treatment for land degradation.

Traditional farming practices would seem to offer better protection against soil erosion than modern farming methods, but in the United States the cost of accepting conservation methods is extremely high and it has been argued that adoption of the full range of recommended techniques would bankrupt most farmers.

The obverse of the soil erosion problem is that of silt, and this is another problem which besets irrigation schemes. Areas where irrigation systems are vital for agriculture are often those where deforestation has proceeded apace. All the soil from upstream which is eroded by water finds its way into the water courses and is deposited as sediment in the riverbeds and onto the floodplains, bringing renewed fertility but also causing flooding. The silt also finds its way into the canals and other waterworks of the irrigation systems. This necessitates continuing funds and labour for maintenance, and even so drastically shortens the useful life of reservoirs supplying the systems, particularly where the reservoir is small in relation to the flow of the river feeding it. It also reduces their effectiveness for supplying hydroelectric power.

Silt can be seen as the major form of human-caused water pollution in the world and in fact exacts a heavier cost than any other water pollutant, possibly even all others combined.

◀ *Agriculture in semi-arid areas periodically affected by drought may have not only the short-term consequence of crop failure but also longer-term effects of land degradation. This is an example from Senegal.*

Desertification

In the 1930s a number of writers noted that vegetation in parts of West Africa was deteriorating: poor grass and thorn scrub in the Sahel vegetation zone was in places becoming like desert, whilst some drier savannas – areas of grass and trees – were being encroached upon by the scrub of the Sahel. This was publicised as 'the advance of the Sahara'. In the 1970s famine in the Sahel drew further attention to this problem, and the term **desertification** was coined to describe deterioration of vegetation on the edge of the desert. But the definition was very soon expanded: in 1977 a United Nations report defined desertification as soil stripping, gully erosion, salinisation and alkalisation. Thus desertification would seem to be the end-point of land degradation in arid and semi-arid regions.

Initially the expansion of a poor vegetation in the Sahel was attributed solely to climatic change. There is no doubt that there has been a decline in the mean annual rainfall of the Sahel and adjacent savanna areas, and an increase in variability of rainfall, which occurs almost entirely in a short period in the summer. But other factors have contributed. The northern and drier parts of the Sahel zone are occupied by nomadic pastoralists, the south by sedentary farmers. In the two decades before 1960 the combination of rising human populations and a run of above-average rainfall prompted the expansion of cultivation into parts of the Sahel. This reduced the area available to pastoralists, and their livestock no longer had sufficient land for grazing. Cattle, sheep, goats and camels selectively grazed the more palatable plant species; less palatable species spread, further reducing the capacity of the land to sustain livestock, and sometimes to soil erosion. At the same time the frequency of droughts was increasing after 1960, so compounding the problem.

▲ *Desertification in Burkina Faso, where climatic change has been exacerbated by overgrazing by livestock.*

Overgrazing in arid regions is not confined to the Sahel. It has occurred in Patagonia, where sheep numbers rose rapidly in the late nineteenth and early twentieth centuries; in semi-arid areas of Iran and Syria sheep numbers are estimated to be three to four times the carrying capacity of the meagre vegetation. In Africa overgrazing has often been the result of land appropriation, earlier in this century of tribal pastures for European agricultural settlement, notably in Southern Rhodesia (now Zimbabwe) and Kenya, where Masai grazing lands have been reduced by their inclusion in a wildlife park since independence. (See Plate 4.)

Problems of desertification are widespread, and principally result from farmers pushing into lands where rainfall is too low for crop cultivation without irrigation. Environmental refugees, the result of deterioration of agricultural land caused by unsustainable methods, now form the largest class of refugees in the world, estimated at over 10 million.

4.3   Energy and modern agriculture

One of the striking characteristics of modern agriculture is the importance of inputs purchased from the industrial sector. Traditional farmers used seed from their own harvest, fertilisers from animals and power from animal and human muscle – so that, for example, in Sweden in the 1860s, purchased inputs were only 5% of the gross value of agricultural production. In most developed economies now, where farming has been 'industrialised', this proportion is over half, and so farmers are highly dependent upon the price of inputs as well as the price of their products.

Many of these purchased inputs depend upon the use of energy. Obviously tractors and other machines need fuel, electricity is essential for drying grain, heating and ventilating broiler houses and piggeries and powering milking machines. But energy is also needed to produce chemical fertilisers and pesticides. There is thus a great difference in the amount of energy used in modern and traditional farming as can be seen in Table 2.6. Although crop yields in modern agriculture considerably exceed those in traditional farming, the use of commercial energy is far greater: modern production of rice requires 375 times as much commercial energy per hectare as traditional, and uses 80 times as much energy per kilogram of rice produced. These issues of use of energy and relative productivity are considered in more detail in Chapter 3 on the production of wetland rice, and Chapter 4 which looks at temperate agriculture.

Such use of commercial energy is comparatively recent. In the United States the amount of commercial energy needed to produce a given weight of food output rose fivefold between 1900 and 1970. Thus when oil prices rose dramatically in the early 1970s there was concern about the economic viability of modern industrial farming, and there were even suggestions that a return to the use of horses might be necessary. In the event, the real price of energy and fertilisers has since fallen. In the immediate future there seems no likely shortage of fuel, or mineral deposits for fertilisers – natural gas, potassium salts and phosphate rock – for developed countries. However, the economic viability of these types of farming might be in question if the system of protected prices in the EU ended; and if the use of chemical farming spread to the developing countries, prices might rise dramatically. In the longer term, the sustainability of high energy agriculture will depend on achievement of a sustainable energy policy, an issue considered in *Blunden and Reddish* (eds) (1996).

Table 2.6 *Commercial energy used in the production of rice and maize by modern and traditional methods (energy per hectare/10^6 joules)*

	Rice		Maize	
	Modern (US)	Traditional (Philippines)	Modern (US)	Traditional (Mexico)
Machinery	4 200	173	4 200	173
Fuel	8 988	–	8 240	–
Nitrogen fertiliser	10 752	–	10 000	–
Phosphate fertiliser	–	–	586	–
Potassium fertiliser	605	–	605	–
Seed	3 360	–	621	–
Irrigation	27 336	–	351	–
Insecticides	560	–	110	–
Herbicides	560	–	110	–
Drying	4 600	–	1 239	–
Electricity	3 200	–	3 248	–
Transport	724	–	724	–
Total	64 885	173	30 034	173
Yield kg/ha	5 800	1 250	5 083	950
Energy input per kg of output (joules × 10^6/kg)	11.9	0.14	5.91	0.18

Source: FAO (1977) *The State of Food and Agriculture, 1976*, Rome, p. 93.

4.4   Impacts of agriculture, public opinion and subsidy

The profound changes brought about by the industrialisation of agriculture, especially in the USA and Europe, have not only caused soil degradation and dependency on fossil fuels but also had a series of impacts outside agriculture which have in turn opened agriculture to more intense public scrutiny. These impacts stem from the greatly increased use of machinery, chemicals and bought-in animal feed. They range from changes to the rural landscape, through pollution of air and water to public questioning of subsidy regimes and international effects.

The impact on landscape will be explored in more detail in Chapters 4 and 5, but here it is enough to mention drainage of wetlands, ploughing of former pasture as mixed farms are converted to cereal monoculture, removal of small woods and hedgerows to create larger fields in which larger machines can operate and ploughing and reseeding moorland to create more productive pasture. The result is the elimination of wildlife refuges and the creation of visually monotonous landscapes with much reduced amenity value.

Pollution effects are increasingly spreading beyond farms. Concern at the effects of pesticides and herbicides on wild plants and animals was multiplied by Rachel Carson's book *Silent Spring* in 1962 and has continued, although many of the persistent toxic chemicals have been withdrawn from sale in developed countries. However, the increase in scale and intensity of farming has made even beneficial materials like manure into a problem as intensive animal feed-lots produce such large quantities of slurry that they can overwhelm the capacity of even substantial streams to dilute, disperse and decompose. Currently, there is growing concern about nitrates entering streams and underground water supplies in increasing quantities as levels of synthetic nitrates increase and as more animals are intensively reared. Whereas specific pesticides can be replaced by safer alternatives, nitrogen in some form is an indispensable part of intensive crop production, so nitrate pollution raises questions about the balance between intensive production and water purity. Since 1990 the UK Ministry of Agriculture, Fisheries and Food has taken steps to bring levels of nitrates in water supplies within the EC Directive of 50 milligrammes per litre. Through its policies of Nitrate Sensitive Areas and Nitrate Vulnerable Zones, incentives are available to farmers to reduce their use of this chemical.

In Europe and North America the 1980s were a time of increasing public and political concern about overproduction of certain agricultural commodities, including dairy produce and cereals. The spectacle of surplus produce being dumped, stored or sold at cut rates to the former USSR caused widespread questioning of the subsidies paid to farmers, a questioning exacerbated by cases where farmers sometimes seemed able to obtain subsidies both for producing more and for producing less of the same product. During this period, many have argued for less intensive forms of agriculture, including organic farming, which would reduce surpluses and harmful impacts on the environment. However, official policy changes, including the (then) EC's scheme for 'set-aside', have taken land out of production. The GATT agreement reached in 1993 should lead to a continuing reduction of agricultural protection in the EU and, indeed, in other parts of the world. (See Chapter 6 for further discussion of this agreement.)

In recent years the effects of intensive modern agriculture have been queried far beyond the farm gate. The urban population has become increasingly concerned at the effects on food quality. The residues of herbicides, pesticides and antibiotics are traceable right through the human food chain and are compounded by the effects of additives used in storage and food processing. Of course, such residues and additives are subject to government controls, but many members of the public query whether such controls are adequate. In the UK rising rates of salmonella poisoning led to revelations that salmonella was endemic in many large commercial flocks of chickens and had for the first time begun to affect eggs. The emergence of a new cattle disease, BSE, was attributed to cattle feed containing the offal of sheep infected with scrapie. Whether these are merely temporary setbacks in the process of intensification or indications of an agricultural system that has become over-industrialised is currently being debated.

Finally, there are those who raise ethical questions about world agriculture, focusing on two sets of issues. First, the ethics of intensive animal-rearing. The morality of veal production and battery chickens has long been disputed, and some animal rights activists have taken direct action. So far, such arguments have not made much headway against the cheapness and convenience of the produce. However, coupled with evidence of infection or of products used to resist infection, this is a balance which could shift in future towards a demand for higher quality.

The second set of ethical objections is at a world scale and may have partly contradictory implications. This is an objection to a world system that overproduces in rich countries and produces chronic malnutrition and

◀ *There are growing ethical objections to intensive animal-rearing and inhumane conditions of slaughter.*

frequent famines in poor countries. The simplistic suggestion that surpluses should be used to feed the starving still commands widespread support – immediate and practical in the case of appeals for disaster relief for Ethiopia – in spite of expert advice that such aid can harm food producers in recipient areas. More thoughtful critics point to the trade linkages which drive many third world countries to promote crops for export to the first world even while their own populations are malnourished (an issue taken up in Chapter 6).

Much of the best land which could be used for domestic food production is used instead to grow cash crops for the developed countries of the first world. The five most common are sugar, tobacco, coffee, cocoa and tea. The fluctuations – and often decline – in prices of such commodities make them unreliable bases for national planning and budgeting. Typically, countries are dependent on just one or two crops, which may represent 75% or more of the value of their exports. Further areas brought under cultivation, or intensification of cultivation, may in fact mean a lower return as prices fall. Labourers, working on these estates, are displaced from the land, unable to grow even subsistence crops for their families. The inequalities of land ownership are at their extreme in Latin America, where 80% of the land is owned by 8% of land-owners; smallholders who comprise 66% of all owners have just 4% of the land. However, it is a worldwide problem: 4% of land-owners in the Far East own 31% of agricultural land. Increasingly in Asia, indebtedness of small farmers leads to landlessness.

Prime land in the third world is also used for rearing livestock, chiefly for export. And the levels of livestock production in the developed countries are only achieved through imports from the third world: one-third of the world's grain, two-thirds of oilseeds, half the fishmeal and one-third of milk products are used to feed livestock, the majority of which are consumed by the developed world. This represents a large outflow of protein – in the form of groundnuts from Nigeria, soya beans from Brazil, fishmeal from Peru and so on – from the malnourished countries to feed the well-fed populations of the first world (Harrison, 1987).

Such issues of distribution will have to be addressed, as the current inequitable situation both within and between countries does not provide a good basis for sustainable development. As populations grow and the effective demand for 'westernised' foods – such as wheat in Africa – increases, the need for increased productivity will put further demands on often fragile ecological situations, and may well generate the same kinds of harmful impacts as those which are becoming increasingly obvious in the rich countries of the world.

Activity 3

Monitor the media for discussion and evidence of problems of sustainability of first and third world agriculture. You may well find:

*   evidence of the impacts of high industrial inputs on wildlife, soils or water
*   policy changes towards set-aside or less intensive use in first world countries
*   problems arising from commercialisation in areas of intensive subsistence agriculture
*   negative impacts of clearance of tropical forest for agricultural use.

PUNCHLINE by **@CHRISTIAN**

FATHER WHY IS IT THAT OUR PEOPLE AREN'T GETTING ENOUGH FOOD TO EAT?

I'M AFRAID THERE'S SOMETHING IN OUR FIELDS TODAY MY SON,...

THAT HAS COMPLETELY WIPED OUT OUR TRADITIONAL HARVEST.

FATHER, WHAT IS IT?! DROUGHT?! LOCUSTS?! RATS?!

CASH CROPS...

5   Summary and conclusion

Section 2 of the chapter showed that agricultural systems have had to adjust to environmental constraints, notably temperature, available rainfall, slopes and soils, so that only about a tenth of the Earth's land surface is cultivated and a quarter grazed. However, enough has been said about the ten-thousand-year history of agriculture to show that farmers have been extremely active in selecting preferred varieties and modifying ecosystems and water systems to affect what is produced. They have been extremely successful in increasing productivity rates, but in North America and Europe since 1945 there has been a dramatic increase in use of inputs like machinery and chemicals in order to increase outputs; these technologies are increasingly being adopted in less developed countries. The proliferation of non-sustainable practices seems as absurd a direction for policy as some of those which have been applied to encourage and then discourage production in developed countries.

The patterns clearly show a general resemblance, but not a very encouraging one. Broadly, the developed countries have agricultural surpluses and low expected population growth while the least developed countries have food shortages and high expected population growth, with Africa the scene of the highest population growth rates and the worst food shortages. Asia and Latin America have a more mixed picture, some countries having industrialised and/or improved agricultural productivity sufficiently to outrun population growth.

The current growth rates suggest that the future is likely to see a worsening of a trend which is already well established. Overall, the growth in cereal production has progressively outstripped world population growth. In spite of this, in 1990 there were more people in the world who did not get enough to eat than there were in 1980. Estimates put the number whose inadequate diet prevented them from leading productive working lives at in excess of 730 million, or 15% of the total world population. The present problem is that world agriculture produces sufficient food to feed everyone but that the system of distribution does not provide for those most in need. Without major changes in policy, famines and chronic malnourishment will increase.

Given the crucial role of agriculture in feeding the human population, it is vital to promote systems which are sustainable indefinitely. Such systems would have to have fewer harmful impacts on environments than some of those currently in use, though the problem of loss of natural habitats can be avoided only if growth in agricultural productivity can continue to keep pace with population growth in future.

The technology exists and provides the potential to feed a doubled world population, but a dramatic growth in production threatens further environmental damage and provides no guarantee to better distribution.

There have been some important international initiatives which are seeking responses which are efficient, equitable and environment-friendly. In 1980 the World Conservation Strategy produced an agenda for sustainable development linking agricultural development with the need to maintain ecological processes: see Box 2.3. The Brundtland Report, *Our Common Future* (1987), called for a series of policy initiatives concerned with 'sustaining and expanding the resource base of the Earth'. The Rio UNCED Conference (1992) pursued these initiatives in 'Agenda 21'. In *Global Environmental Issues* (Blackmore and Reddish, eds, 1996), they are analysed in more detail.

Box 2.3   World Conservation Strategy

Living Resource Conservation for Sustainable Development – a Report from the International Union for Conservation of Nature and Natural Resources (IUCN), Gland, Switzerland (1980) – was a diagnostic exercise on a global scale, not constrained by geographical boundaries, on the threat to ecosystems from the processes of agricultural and industrial development, and the basic need to survive.

It proposed *three* principal objectives of resource conservation:

1   The maintenance of ecological processes and life support systems.

2   The preservation of genetic diversity.

3   The sustainable utilisation of species and ecosystems.

Of these objectives, the first is the most critical, the other two being closely linked to it. To achieve stability and sustainability in terms of ecological processes, *three* specific requirements were laid down:

•   the utilisation of good croplands for crops, rather than the raising of cattle;

•   the ecologically sound management of crops;

•   the protection of watershed forests.

A quotation from the Strategy underlines the significance of these requirements:

Only 10% of the world's population live in mountainous areas but another 40% live in the adjacent plains; so the lives and livelihoods of half the world directly depend on the way in which watershed ecosystems are managed.

The Report predicted the irreversible damage of continued deforestation and of the increasing use of fertilisers and pesticides. The Report did not explain how these proposals were to be implemented.

From the global overview of this chapter we shall now take a more detailed look at the basis of productivity and environmental impacts of some of the major agricultural systems. Because rapid population growth makes further growth of output essential, we have chosen to analyse in depth two of the most productive forms of agriculture: the capital-intensive farming typical of the developed countries of temperate zones and the labour-intensive methods of subtropical zones, which reach their peak in the production of wetland rice.

References

BLACKMORE, R. and REDDISH, A. (eds) (1996) *Global Environmental Issues*, London, Hodder and Stoughton/The Open University (second edition) (Book Four of this series).

HARRISON, P. (1987) *Inside the Third World*, Harmondsworth, Penguin Books.

PIMENTEL, D. and PIMENTEL, M. (1979) *Food, Energy and Society*, London, Edward Arnold.

SARRE, P. and REDDISH, A. (eds) (1996) *Environment and Society*, London, Hodder and Stoughton/The Open University (second edition) (Book One of this series).

SILVERTOWN, J. (1996a) 'Inhabitants of the biosphere', Ch. 6 in Sarre, P. and Reddish, A. (eds).

SILVERTOWN, J. (1995b) 'Earth as an environment for life', Ch. 5 in Silvertown, J. and Sarre, P. and Reddish, A. (eds).

SIMMONS, I. (1990) 'The impact of human societies on their environment', Ch. 2 in Sarre, P. and Reddish, A. (eds).

WORLD COMMISSION ON ENVIRONMENT AND DEVELOPMENT (1987) *Our Common Future*, Oxford, Oxford University Press.

Further reading

BERNSTEIN, H., CROW, B., MACKINTOSH, M. and MARTIN, C. (eds) (1990) *The Food Question*, London, Earthscan.

GRIGG, D. B. (1974) *The Agricultural Systems of the World: an evolutionary approach*, Cambridge, Cambridge University Press.

Chapter 3 Farming a wetland ecosystem: rice cultivation in Asia

1 Introduction

As you read this chapter, look out for the answers to the following key questions:
● What are the key biological and chemical factors in the ecosystem supporting wetland rice cultivation?
● How have human efforts to increase rice production modified both rice ecosystems and the form of human society?
● What are the current problems of, and future possibilites for, paddy rice cultivation?

We saw in the last chapter some of the many ways in which people have tried to obtain useful products through varying degrees of 'management' of terrestrial ecosystems. In this chapter we will consider in more detail one of these managed ecosystems: that of wetland or 'paddy' rice cultivation.

This farming system is one of the most important in the history of human settlement, and has assumed a growing importance in global food supply in the second half of the twentieth century. This is because rice constitutes nearly half of the overall cereal output of developing countries, where three-quarters of the world's population live, compared to only 3% of the total cereal output of developed countries. The significance of this in world terms is twofold: firstly, rice is overwhelmingly a crop grown by poor, small-scale farmers; secondly, it is an important food source to many of the world's poorest consumers, who are most vulnerable to food shortages.

By the late 1980s the world average yield of rice per unit of land cultivated was nearly 50% higher than that of wheat. During the three decades from 1957 to 1987 rice output more than doubled, enabling many Asian countries to achieve self-sufficiency in rice in spite of high population growth rates. United Nations statistics show that in 1985 the total cereal output in developing countries overtook that in developed countries for the first time since records were begun, forty years earlier.

This rapid increase in rice production was one of the results of a number of changes in rice cultivation, often known as the 'green revolution'. In this chapter we shall analyse the nature of the green revolution in Asian rice cultivation in the twentieth century, and its impacts on environment and on society.

2   The ecology of wetland rice

2.1   Wetland rice

Rice cultivation in Asia has a history of at least 7000 years. The major characteristic which distinguishes the historic development of Asian rice production from all other agricultural systems is that the crop was grown principally on flooded or waterlogged land. By contrast, rice cultivation by farmers in Africa and South America was, until the latter half of the twentieth century, principally on 'upland' which was freely drained. (This is not to say that flooded or 'wet' rice cultivation was never practised by farmers elsewhere: systems for growing rice in seasonally flooded valleys were an important part of traditional farming in both West and East Africa. Furthermore, upland rice is important in some parts of Asia.)

Our focus in this chapter will be confined to Asian wetland rice because some 90% of the world's rice is produced in Asia, and 80% of rice production in Asia is associated with the exploitation of soils which have been flooded in a variety of ways: through irrigation, through impounding rainwater, or through the seasonal flooding of rivers. This practice is so different from the use of well-drained soils for all other cereals (and, indeed, the vast majority of other crops) that it is important to understand what advantages there might be in growing rice on flooded soils.

2.2   The wetland ecosystem

We can make a useful start by remembering the two principal processes which define biological activity within an ecosystem. These are firstly the **flow of energy**, or **food chain**, and secondly the **cycle of mineral elements (or nutrients)**. Figures 3.1 and 3.2 depict both these processes in a generalised ecosystem, showing how energy and minerals pass between

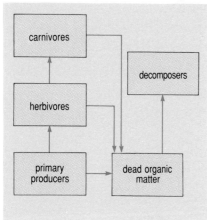

▲   Figure 3.1
Compartments of a 'model' ecosystem, showing the pathways of energy flow.

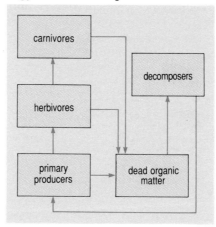

▲   Figure 3.2
The pathway of mineral elements in the model ecosystem.

different types of organisms: from primary producers (plants) to herbivores and then on to carnivores and ultimately on to decomposers. (Readers of Book One in this series (Sarre and Reddish (eds), 1996) will recognise these diagrams from Chapter 7 of that volume.)

To determine the significance of using wetlands (flooded or waterlogged land) for plant production, we need to concentrate our analysis on the mineral cycle, and in particular on the soil conditions which in all terrestrial ecosystems control the working of the critical link in the cycle where mineral elements are absorbed by the plants.

In order to identify 'wetland' soil conditions, let us consider the ecosystem in which wild rice originated – the swamps of the Himalayan foothills, between north-east India and south-west China.

Q   Using what you already know about the effects of rainfall and temperature on the mineral cycle in other ecosystems, what broad characteristics might be expected in the soils of this tropical swampland?

A   Firstly, tropical soils exposed to high temperatures and rainfall are most typically regarded as having had many of their minerals washed out of them. However, the lower-lying parts of the landscape receive water and minerals washed down from the surrounding hills and they therefore often contain higher concentrations of minerals and, where rain falls only in certain seasons of the year, remain moist for longer than free-draining upland soils. Secondly, unlike waterlogged areas in temperate climates, the high temperatures allow the activity of 'decomposers', the organisms which break down dead organic matter. However, and thirdly, the waterlogging of the soil for long periods means that air is excluded from the soil, which prevents the activity of **aerobic organisms** which need oxygen for respiration. Decomposition must therefore be carried out by **anaerobic organisms** which can live in the absence of oxygen.

For practically all cultivated plants, the lack of oxygen in waterlogged soil prevents respiration in the roots, which are quickly killed as a result. In contrast, the rice plant is able to grow successfully in flooded soil because it has an efficient system of air channels between the root and the shoot. This system (which is four times more efficient than that of barley and ten times more efficient than that of maize, both crops adapted to 'upland' conditions) allows air absorbed from the atmosphere by the shoot to be supplied to the roots growing in oxygen-deficient soil.

These general descriptions of soil in swamps suggest that, to early cultivators with few resources, rice offered the possibility of growing food in a niche in the landscape more fertile than the uplands to which they were restricted by other crop plants. Moreover, in the tropics and subtropics, where the length of the growing season is commonly limited by lack of rainfall during a dry season of several months' duration, the accumulated water in these low-lying areas is often sufficient to significantly prolong the period each year when crops can be grown. A longer growing period could increase the number of crops available for harvest each year.

Research carried out over the past twenty-five years on the chemistry of flooded soils has indicated a number of mineral-cycling characteristics in such soils which make them particularly resilient to continuous intensive crop-growing, and probably more so than many freely drained soils. Because these characteristics have some bearing on the development of rice farming considered in later sections, they will be briefly described.

Minerals and primary production

Before we explore specific characteristics of flooded soils, let us outline some of the general features of the absorption of mineral elements by plants growing in soil. This is fairly straightforward because, although plants differ widely in terms of the *quantities* of minerals that they absorb, all plants seem to need to absorb the same set of minerals, known as *mineral nutrients*. The principal ones are:

- nitrogen
- phosphorus
- sulphur

- potassium
- calcium
- magnesium

In addition very small quantities of the following nutrients must also be absorbed:

- zinc
- copper
- molybdenum
- boron

- manganese
- iron
- chlorine

Q   Where do these mineral nutrients come from?

A   The mineral cycle in Figure 3.2 shows that a part of the mineral nutrients in the soil are derived from the decomposition of organic matter from dead plants, animals and micro-organisms. However, all nutrients except nitrogen are ultimately derived from the rocks from which the soil is formed. These may be rocks underlying the soil, or mineral material such as clay or silt deposited after being transported from elsewhere by wind or water. Nitrogen in the soil is derived not from rock material but from the air, which is 79% nitrogen gas, through the agency of 'nitrogen-fixing' bacteria and algae which can incorporate atmospheric nitrogen into organic material.

The nutrient nitrogen is particularly important in the cultivation of cereal crops, because it is contained in large quantities in the harvested grain. A harvest of 2 tonnes of rice, for example, contains approximately 32 kg of nitrogen, as against only 6 kg of phosphorus and 16 kg of potassium. Because of this heavy demand for nitrogen, the roots of cereals must absorb large quantities of this nutrient from the soil. Moreover, because so much nitrogen is contained in the grain, little is returned to the soil by the decomposition of the dead plant leaves, stalks and roots. Cereal crops therefore move large amounts of nitrogen out of the soil and into other parts of the food chain. In order to maintain this flow, nitrogen must be replaced in the soil, and many systems of cereal farming have evolved to incorporate plants which support nitrogen-fixing bacteria on their roots (legumes) through rotations (for example, in temperate climates clover in pastures one year followed by wheat or barley the next) or through intercropping (that is, growing more than one crop in the same field, such as beans with maize in Latin America). As we shall see, similar techniques involving *Azolla* ferns and blue-green bacteria have been developed for flooded rice fields in Asia.

As Chapter 2 discussed, where farmers sought higher production of grain from the same area of land, they adopted the practice of adding extra nutrients to the soil in the form of organic or mineral fertilisers and, among these, fertilisers containing nitrogen have been the most important. However, when very large amounts of grain are produced repeatedly from the same field, the quantities of many types of nutrients may become insufficient for normal plant growth, so that in addition to the principal nutrients like nitrogen and phosphorus, fertilisers must also contain nutrients such as zinc or copper, which the crop plants require only in minute quantities.

We can see, therefore, that in order to maintain a high flow of energy up the food chain through large harvests from the primary production (crop) part of the ecosystem, farmers have to develop ways of increasing the rate at which mineral nutrients are cycled through the soil.

Q   We have mentioned that mineral nutrients in soil are derived from both organic and inorganic sources, but in what form are they actually *absorbed* by the plant roots?

A   All mineral nutrients are absorbed by plant roots in the form of simple inorganic chemicals which are soluble in water.

The final stage of the decomposition of organic matter in the soil is therefore its **mineralisation** to release mineral nutrients in this simple inorganic form. This is important because the solubility, and hence availability to plants, of these minerals is determined by the acidity and oxygen content of the soil. Flooding changes both of these soil characteristics and so profoundly alters the availability of many of the nutrients needed by plants.

Let us now consider some differences between flooded and freely drained soils.

Some comparisons of flooded and freely drained soils

A first effect of flooding on soil is to change its physical *structure*. The different organic and inorganic components of soil are combined in various physical ways to form *aggregates*: the solid material of the soil. The spaces or *pores* between the aggregates allow the movement of water and air in the soil.

When a soil is flooded, the system of air spaces and pores within it will tend to collapse due to:

•   the compression of the air trapped in the pores by the water, leading to small air 'explosions' which break down larger aggregates, or clods, into smaller ones.

•   the swelling of certain types of minerals, particularly clays

•   the dissolution of some of the substances that stick soil aggregates together.

As a result of this collapse of soil pores, water moves through the flooded soil much more slowly: that is, the soil will lose less water through drainage. When the soil surface is covered by water, the oxygen supply in the soil is quickly depleted (in less than a day) in all but a thin layer of up to 1 cm thick, at the soil surface, to which oxygen diffuses from the air through the water. Thus, a flooded soil typically consists of two layers, illustrated in Figure 3.3: a thin upper layer supplied with oxygen, and the underlying bulk of the soil in which oxygen is absent. In this latter layer, the aerobic microbes which function in the presence of oxygen become dormant, or die, and microbial activity is dominated by those anaerobic organisms which do not require oxygen for respiration. Instead of oxygen, these anaerobic organisms transform a wide range of other chemical compounds in the process of respiration, and it is these transformations which very largely account for the differences in chemistry between flooded and freely drained soils.

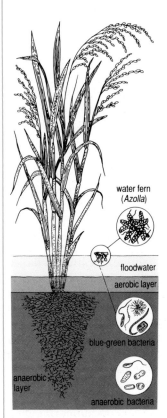
▲ Figure 3.3
Rice plant showing flooded growing conditions and aerobic and anaerobic layers of soil. Nitrogen is fixed by micro-organisms including those shown: blue-green bacteria, free-living or in a symbiotic relation with the water fern Azolla and a variety of other bacteria.

The consequences of these transformations are outlined as follows:

• *Increased nutrient availability for crops* When a soil is flooded, some important mineral nutrients, such as phosphorus, become more soluble, and hence more available for absorption by plants.

• *Soils become neutral* In contrast to the situation in cool and cold climates (where waterlogging inhibits decomposition and allows build-up of thick peat layers which acidify the water), in tropical or subtropical areas all soils, whether alkaline or acidic before flooding takes place, become neutral within a month of flooding. While flooded, therefore, soils do not suffer problems of excess acidity or alkalinity which can cause serious reduction in growth of crops growing in freely drained, upland soils.

• *Reduced decomposition of organic matter, but enhanced nitrogen supply* The decomposition, or mineralisation, of organic matter by anaerobic micro-organisms in flooded soils is slower than that by the aerobic micro-organisms in freely drained soils. Thus, organic matter tends to accumulate more in flooded than in freely drained soils. This might be considered a disadvantage of flooded soils, slowing down the mineral cycle and restricting the movement of minerals into primary (crop) production. However, although *quantitatively* slower, the anaerobic mineralisation has some *qualitative* advantages in relation to the all-important mineralisation of nitrogen, particularly when this nutrient is present in only low concentrations in organic matter.

Q We have already encountered one example of organic matter with a low concentration of nitrogen. What was it?

A Cereal straw.

In freely drained soils organic matter such as straw, in which the content of nitrogen is lower than about 1.5%, can only be mineralised by aerobic micro-organisms if they scavenge extra nitrogen from elsewhere in the soil, thus mopping up the nitrogen which would otherwise be available for absorption by plant roots. For this reason, addition of large amounts of straw to a field just before a cereal crop is planted may create a temporary 'famine' of nitrogen which will reduce the growth of the crop. In flooded soils, however, the activity of anaerobic organisms is less sensitive to the nitrogen content of the organic matter, so that nutrients contained in material such as straw are more readily available to a crop of rice than, say, a crop of wheat.

• *Advantages of continuously as against intermittently flooded soil* The collapse of pores and channels in a continuously flooded soil means that less water, and hence nutrients dissolved in the water, will drain through the soil into the groundwater. Flooded conditions effectively conserve water and nutrients in the rooting zone of the crop. By contrast, under conditions of intermittent flooding, large amounts of water will drain through the cracks which open as the soil dries out between periods of flooding. Furthermore, the drying out of the soil allows oxygen into the anaerobic layer, allowing the aerobic decomposition process to begin. An alternation between aerobic and anaerobic conditions in the soil has negative effects on the mineral supply to crops by allowing large amounts of nitrogen in the soil to return to the atmosphere through a process known as **denitrification**. This works in the following way: each time the soil dries out and aerobic micro-organisms can become active, mineral nitrogen will be converted from ammonium compounds to nitrates. Both forms can be absorbed by plants but the nitrate form has two disadvantages: it is highly

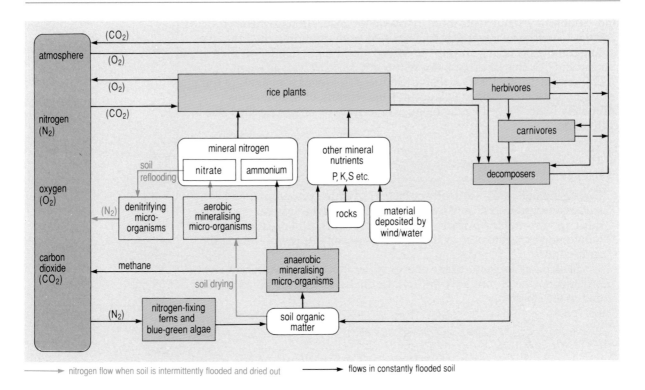

nitrogen flow when soil is intermittently flooded and dried out   flows in constantly flooded soil

mobile, and therefore easily lost in drainage water; also, each time the soil is reflooded the anaerobic organisms convert nitrates to the gases nitrous oxide and nitrogen, which bubble to the surface and are lost to the atmosphere. Inadequate control of water in rice cultivation, leading to intermittent flooding and drying out of the field, results in much lower efficiency in the use of both water and soil nutrients. See Figure 3.4.

Not all aspects of flooded soils are advantageous, however.

• *Disadvantages of continuously flooded soils*   In certain soils with high levels of organic matter and low levels of iron, flooding may cause the formation of hydrogen sulphide gas, which is toxic to rice plants. A major product of the anaerobic decomposition of organic matter in flooded soil is methane (commonly known as 'marsh gas'), which bubbles to the surface and into the atmosphere. Although of no immediate significance to rice farmers, the increasing content of methane in the atmosphere on a global scale has been claimed to be a contributory factor in the increased retention of radiation reflected from the Earth's surface (the so-called *greenhouse effect*). However, the relative contribution of flooded ricefields to increased methane in the atmosphere is as yet unknown. Natural swamps and the digestive system of cattle are other major sources.

Summary

From this brief listing of the chemical consequences of flooding soils we can summarise that, compared to freely drained soils, flooded or waterlogged soils in warm climates have a number of mineral-cycling advantages for intensive agricultural production. In particular, they conserve water and nutrients through reduced drainage, conserve organic matter, avoid acidity,

▲   *Figure 3.4*
Nutrient flows in a flooded rice field and in a field which is intermittently flooded and dried out.

and render certain nutrients more available for absorption by the crop. There is therefore an advantage to be gained from extending the principle from naturally occurring swamps to soils which are not flooded in their natural state. Such a strategy is not without risks, however, as when flooding is intermittent rather than continuous, losses of water and of the important nutrient nitrogen will occur. Close control of water is thus intimately linked with the maintenance of fertility of soils in flooded rice cultivation.

Activity 1

Our discussion of the wetland ecosystem in which rice originated has so far concentrated on only two compartments of the system: the decomposers and the primary producers. What type of organisms would you expect to occupy the other compartments (herbivores and carnivores)?

Sketch out a scheme of the energy flows between the different organisms. Check your answers against the list and figure given at the end of the chapter.

2.3   Variability, domestication and adaptation of the rice plant

It was stated above that rice is believed to have originated in the foothills of the eastern Himalayas: in north-east India, Indo-China, and south-west China. This geographical origin is suggested by the large concentration of wild rice species in that area. Wild rices include many perennial species, but, in spite of the advantages of permanent flooding for nutrient retention, it was the annual species which were of more interest to human cultivators as they could be more successfully grown in areas which were not flooded for the entire year. The archaeological evidence indicating the earliest date of rice cultivation was found in the Yangzi delta (eastern China), where excavation of a village situated at the edge of a marsh has yielded remains which were carbon-dated to 7000 BP. Other evidence of rice cultivation has been discovered at sites in north-east Thailand which have been similarly dated to 6500 BP.

Archaeological studies have indicated that the progressive spread of rice cultivation to the rest of southern and eastern Asia took place over a 6000-year period. Thus, the earliest indications of rice cultivation in Vietnam date from 5000 BP, and those in India from 3500 BP. Rice-growing arrived in Japan much later, in about 2400 BP, probably via Korea. In Malaya and Indonesia, rice did not become a major crop until after 1000 BP. See Figure 3.5.

In many of the areas into which rice cultivation spread, rice increasingly became a staple food, substituting millets and tubers which had previously been the principal food crops. While rice cultivation seems likely to have been originally carried out exclusively in swamps, the spread of the crop to new areas meant that it needed to be adapted to a wide range of environments differing in soil type, temperature, day-length and rainfall. It is clear that farmers were active in this adaptive process, by selecting varieties that were best suited to each local set of conditions. As a result, the range of rice varieties is very broad, varying from those at one extreme

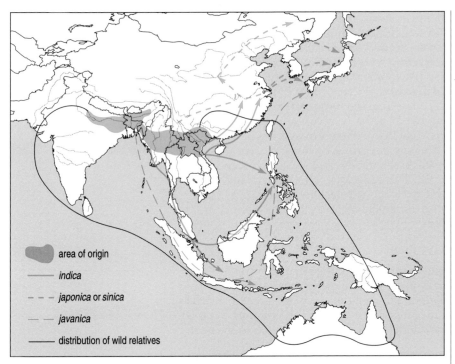

▲   *Figure 3.5   Spread of Asian rice* Oryza sativa *from its area of origin as a wild annual grass:* indica, japonica *(or* sinica*) and* javanica *are the three major geographic races of* Oryza sativa. *The* indicas *were originally cultivated in the humid parts of the Asian tropics and subtropics;* japonicas *developed in subtropical temperate regions;* javanicas *developed in equatorial Indonesia.*

suitable for upland (unflooded) conditions, where they are grown like other cereals, to, at the other extreme, varieties of 'floating' rice which can be grown in water up to 5 m deep.

Not only is the range of rice-growing conditions extremely broad but the discrimination of different characteristics has proved to be very subtle. When the International Rice Research Institute (IRRI) set out in the 1960s to make a collection of the existing rice varieties in the world, it recovered, over a twenty-year period, some 63 000 local varieties of the rice species cultivated in Asia (*Oryza sativa*). This was in addition to over a thousand species of wild rice, and some 2500 cultivated varieties of the African rice species (*Oryza glaberrima*). To give some perspective to this genetic diversity induced by rice farmers, it may be noted that the number of known varieties of wheat (*Triticum aestivum*) is estimated to total only 20 000. The enormous number of local varieties of *Oryza sativa* were practically all developed by Asian farmers before the beginning of the twentieth century. In order to better understand how this came about and the changes which subsequently took place in the twentieth century, we shall briefly consider the mechanics of selecting new varieties of rice.

Selecting new varieties of rice

As was discussed in *Silvertown* (1996, Section 6), in the process of 'domesticating' a wild plant species human cultivators select which seed will form the basis of the next generation of plants on their fields. They may, for example, choose to sow seeds of plants that were taller, or which matured faster, or which had bigger seeds. In so doing they reduce the genetic diversity of the plants on their fields in relation to the diversity of the wild population. This process, carried on over thousands of years by millions of different farmers selecting seed according to different criteria can fairly easily be seen to give rise to many different distinct populations, or varieties, of the same original species.

This selection process was integral to the harvest: the most common method of rice harvesting employed until recently in Asian rice farming was to cut each individual stem just below the grain-bearing head, or 'panicle', with a small knife. This system provided farmers with an opportunity to inspect each panicle separately, and set aside for seed any which were of particular interest. This technique, known by plant breeders as **mass selection**, was by itself sufficient to select and maintain a large number of distinct rice varieties due to the fact that the rice plant is predominantly **self-pollinated**.

Q What is meant by self-pollination?

A By self-pollination is meant that flowers of the rice plant are fertilised by their own pollen. That is, grain generally contains genetic material derived only from the plant on which it develops. As a result, a variety of rice resulting from many generations of selection will tend to breed true, and maintain its characteristics from one generation to the next.

However, within this predominant pattern, two factors operate to present some degree of genetic variability in each generation, which allows farmers to select plants which can form the basis of a new variety. Firstly, the genetic material of a population of rice plants will not be completely homogeneous (identical), so that plants will vary in many quantitative ways, such as height, length of growth period before flowering and so on. Secondly, where a number of different rice varieties are grown in the same area, or where wild rice species grow near cultivated rice, a very small amount of **cross-pollination** occurs, and this may also give rise to new types of plant from which farmers are able to select seed to reproduce as a new variety.

Until the mid twentieth century, all domesticated rice varieties in Asia were partitioned between three major subspecies of *Oryza sativa*. The two most important of these were the *indica* subspecies, grown in tropical Asia, and the *japonica* subspecies grown in the cooler and more northern rice-growing areas. A third subspecies, *javanica*, predominated in Indonesia: see Figure 3.5. The principal distinctions between the *indica* and *japonica* rices were:

● grain type: *indica* varieties were long-grain, while *japonica* were short-grain;

● flowering characteristics: *indica* varieties were sensitive to day-length, and so a given variety started flowering only at a particular date, irrespective of when it was planted, while *japonica* varieties are not sensitive to day-length but will tend to flower after a fixed period of growth, which would vary between varieties.

In the twentieth century the process of selecting new varieties of rice was radically changed by the development of new *techniques* and by the adoption of different *criteria* for selection. Simultaneously, the control of the selection process passed from farmers to research scientists.

At the beginning of the twentieth century, Japanese scientists developed a technique of artificially pollinating rice plants of one variety with pollen from another. This enhanced cross-pollination enabled them to generate a large number of new genetic combinations from which they could select new varieties, and also gave them the chance of combining the desired qualities of two or more parent varieties in a single new 'improved' variety. This work resulted in the production in the 1930s of *japonica* varieties that, when grown with large quantities of fertiliser, produced yields two to three times higher than those possible with traditional 'farmers' varieties'.

Cross-pollination between *japonica* and *indica* rices produces very few fertile plants and was practically unknown before the 1950s. It was at that time that breeding programmes were set up to promote cross-pollination between the two subspecies and to closely evaluate plants grown from the small amounts of fertile seed which such crosses produced. This process of crossing and selection resulted in the identification by IRRI of the so-called **high-yielding varieties (HYVs)** that were the basis of the introduction of the 'green revolution' to tropical Asia in the 1960s (to be explored in more detail in Section 4).

2.4   Summary

In this section we have explored some of the underlying biological mechanisms in the Asian rice cultivation system. We have seen that rice cultivation enabled early cultivators to exploit nutrient-rich swamp soils, and that agricultural advantages could be gained from reproducing flooded soil conditions elsewhere. We have also seen that the spread of rice cultivation depended upon farmers' ability to form new population, or varieties, of rice better adapted to new sets of growing conditions.

In Sections 3 and 4 we shall trace the wider socio-economic context of Asian rice cultivation in two stages: Section 3 deals with the history of rice-growing in Asia before the twentieth century; Section 4 analyses the origins and impact of the changes in rice-growing in the twentieth century.

3   Development of rice cultivation up to the twentieth century

In our discussion of the history of rice farming we will identify two basic trends: the *extension* of cultivated areas, and the *intensification* of production per unit area of land cultivated.

3.1   Organised farming: the development of water control

We saw earlier how in warm climates a flooded soil environment may present certain advantages for mineral cycling for primary biomass production, and that a central feature of wetland rice production systems was their unique ability to exploit flooded soil environments. Although rice cultivation appears to have been first carried out in naturally waterlogged marshy areas, the spread of rice farming required the creation of flooded fields in areas where they did not occur naturally. The ability to impose these flooded conditions depended upon water control, and the spread and development of rice-growing very largely reflected the development of technologies of water control.

At its simplest, water control involved impounding rainwater on fields by constructing low dykes, or 'bunds', around the field edges. On more sloping land retention of water was achieved by cutting terraces to provide a series of horizontal fields step-like up the slope. The steeper the slope, the smaller the field size necessary to stop water flowing downhill. Some of the earliest terraces believed to be associated with rice cultivation have been found at Gio Linh, in Vietnam, and these are thought to have been constructed in 4000 BP. (See also Plate 5.)

The spread of rice cultivation to areas with lower rainfall required the development of ways to supplement the supply of water to rice fields. Three basic techniques were important in Asian rice-growing. These were:

(a)   The construction of *tanks* to store water from diverted streams. This system was important in upper parts of river valleys, and formed the basis for very extensive irrigation systems that supported important medieval states in Sri Lanka, Thailand and Cambodia. The latter was once the Khmer empire, which flourished for over 300 years, and whose irrigated area is thought to have covered some 167 000 ha. The collapse of the empire in the fourteenth century is believed to have been due at least partly to the silting up of the tanks and canal system.

(b)   The excavation of *ponds* to collect and store rainwater. The construction of ponds in elevated locations for water storage to irrigate lower-lying fields is believed to date from the first or second centuries in southern and central China, and from the fifth century in Japan. The size of ponds was very variable, some serving a single farm while others might serve an entire village.

(c) Diversion of stream or river water using *contour canals*. In this case a barrier across a stream would partially divert the flow into a canal system whose gradient was designed to distribute the water by gravity to fields at a flow rate slow enough to avoid scouring and erosion of the canal but fast enough to avoid silting up. Systems using these contour canals were developed in medieval times in Indonesia, in the highlands of Laos and Vietnam, and in Japan. In China a very large system of this kind, at Guanxian, irrigating some half a million hectares with water diverted from the Min river, was built as early as 2250 BP.

Although these water control systems varied enormously in accordance with the local characteristics of landscape and social organisation, a number of common features may be identified. We shall note three. In all cases they rely upon *gravity* to distribute water, so that all were constructed in the sloping highland landscapes of upper watersheds, and not in flat river floodplains. A second common characteristic is that all construction was carried out using only *simple manual tools*, but using very *sophisticated design*. This is particularly apparent in contour canal systems, where canal gradients were set with great precision and water was sometimes carried through tunnels or across valleys in aqueducts made of wood or stone. The success of this technology is shown by the fact that in many cases it remains in use to the present day. Thus tank irrigation continues in Sri Lanka; ponds continued to provide as much as 20% of Japan's irrigation until the twentieth century and were subsequently incorporated as part of later irrigation networks; the Guanxian contour canal system remains in use more than 2000 years after its construction, unchanged

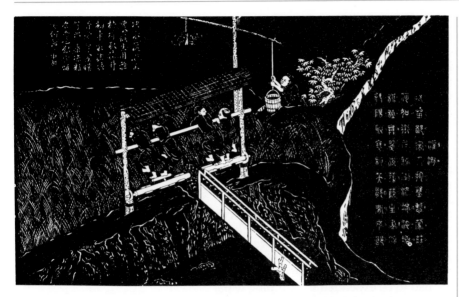

▲   *An example of the use of human labour to lift water: an illustration from a medieval Chinese treatise on agriculture showing two different lifting devices – a square-pallet chain pump, and a beam and bucket.*

except that it now irrigates an area double that for which it was originally designed. A third feature of these irrigation systems is that, although they sometimes reached a very large size, the large systems were essentially *multiples of small-scale systems*. That is, the technology employed by an individual farmer or village to build and manage a tank, pond or contour canal was essentially the same as that employed in large systems, which consisted of interlocking networks of these small units.

This characteristic of 'scaling up' irrigation through multiples of small units was reflected in the organisation of irrigation management in two distinct levels: the level of the unit, and the level of the multiple.

At the 'unit' level, the water source (tank, pond or contour canal) was directly run and maintained by those farmers who used it. The farmers' organisations which evolved for this purpose often (and of necessity) achieved a high degree of regulation of water use by individual members, but also constituted an important focus for lobbying on behalf of communities to which they belonged. These village-level irrigation organisations, of which the *subak* system in Bali, Indonesia, is one of the best known, still function in many parts of Asia where irrigated rice production has been long established.

The state was responsible for the organisation of water control at the second, 'multiple' level, indeed the management of irrigation was a prime function of the state in medieval Asia. Since rice-growing was the basis of a reliable food supply, and a taxable surplus, the continuation and growth of the power of a state depended on its ability to sustain or extend the control of water for rice. Consequently, state-appointed officials were responsible for resolving conflicts between different communities or groups, for establishing water use regulations, and for the investment of resources in improvement or extension of irrigation.

Q   What do the characteristics of early irrigation systems imply about the characteristics of the states which organised them?

A   The vital need for water, the need for all users to limit water use and to carry out essential maintenance, especially the need for massive expenditure of labour to construct large systems with simple tools, all point towards the need for a state with a high level of authoritarian power. These were often monarchies and the priesthood commonly played a role in legitimising state activities. *Simmons* (1996) also points to the vulnerability of such states to any breakdown in the irrigation system.

Further developments occurred as the need arose to bring under rice cultivation land on which water control required much larger investment than the systems described above. The most important of such investments were needed to establish water control for agriculture on flat river floodplains and deltas, and coastal swamplands. In practice, the increased investment needed in these low-lying areas amounted to greater labour input: to dig drainage ditches in swamps, to build dykes to prevent uncontrolled river flooding and to keep out the sea. Since there was little slope in the landscape, it was much more difficult to use gravity to distribute water, and so energy had to be expended in pumping water. Although medieval states were active in organising rice cultivation in these situations, for example the colonisation of the Yangzi delta through the digging of drainage works in fourteenth-century China, and the flood control measures built in the same period in the Tonkin delta in Vietnam, the emphasis in such projects was on organising conscripted labour. The eighteenth and nineteenth centuries saw a rapid expansion of rice cultivation on river floodplains and deltas in south-east Asia in which capital investment and hired labour played a much more important role.

Q   What do you think were the historical developments that promoted this rapid expansion of rice cultivation in the eighteenth and nineteenth centuries?

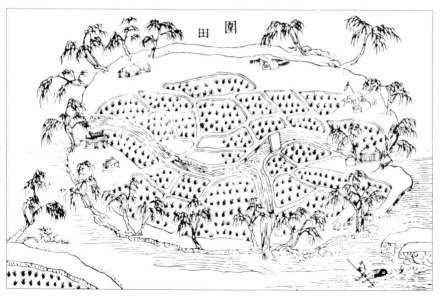

▲   *Water management for controlled flooding of rice fields in lowland areas.*
(Illustrated in the Chinese agricultural treatise Shoushi tongkao of 1742.)

A   They were the expansion of European colonialism and international trade. One of the first concerns of European colonists was to expand supplies of commercial products like spices, tea and rubber. This was achieved by establishing plantations managed by Europeans but using local labour. This labour had to be fed, so rice began to be traded over increasing distances and there was a need to expand rice production to produce a greater surplus. The colonial administrations themselves contributed to the increase in rice area by opening up previously sparsely populated deltas like those of the Irrawaddy (Burma), Mekong (Vietnam) and the Chau Phraya (Thailand).

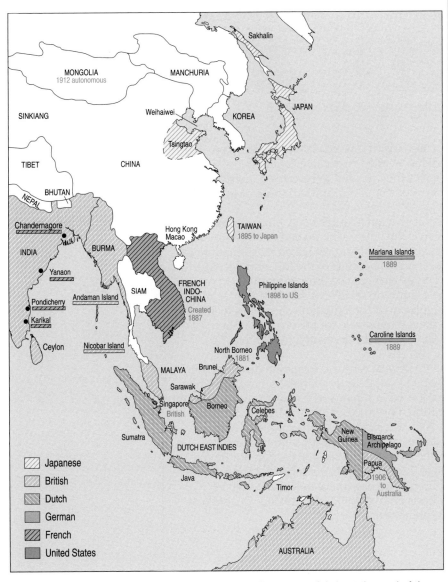

▲ Figure 3.6   Colonial government in rice-growing areas of Asia at the end of the nineteenth century.

▲   *A manual method of irrigating the flooded rice fields in Vietnam.*

Unlike the highland irrigation systems discussed above, the drainage and flood-control schemes of the deltas needed expensive capital items such as sluice gates and pumps. Moreover they needed to be carried out on a large scale to be effective: they did not consist of multiples of more or less self-contained small units.

Investment was therefore not only on a larger scale, but could not be spread as easily over time because the entire scheme had to be completed before production could be reliably started and return on investment begin. It is perhaps for these reasons, and also because of the low population density of the new lands, that investment by wealthy individuals in return for land ownership rights was a common feature of this type of irrigation. Although the new 'landlords' subsequently rented out their land to small-scale farmers, they often remained responsible for ensuring that their tenants provided the very considerable labour needed to maintain the irrigation system.

Since the latter half of the nineteenth century, irrigation development has continued to be largely shaped by state intervention, either through the consolidation of existing irrigation, as in Japan and China, or by the attempt to bring new land under irrigation. For much of the first half of the twentieth century the state in Asia was more often than not that of a European or Japanese colonial administration and this left its mark on irrigation development. Thus the irrigation expansion carried out by the Japanese colonial government in Taiwan and Korea inherited some of the organisational strengths of irrigation management in Japan, while in European-ruled colonies emphasis on hydraulic engineering was often at the expense of organisational coherence.

Summary

In this section we have seen how the spread of rice cultivation was dependent upon the development of water control techniques to allow flooded fields to be established in different types of landscape. The control of water had strong organisational requirements which for thousands of years profoundly shaped the political and organisational character of the Asian rice-growing societies. The expansion of European-dominated international trade resulted in a rapid expansion in rice cultivation in which capital investment played a larger role than before, and this has had an impact on the organisational aspects of water control for rice growing.

3.2   Intensifying production: from medieval China to nineteenth-century Japan

While the development and spread of water control technologies determined the extension of the area under rice cultivation, a number of other farming techniques were developed which allowed a sustained increase in the grain output from each unit of land through an increase in the amount of labour invested in growing the crop. These techniques were developed to greatest effect in China during the later Southern Song, Yuan and Ming dynasties (thirteenth to seventeenth centuries), and in Japan during the Tokugawa period (seventeenth to nineteenth centuries). The principal techniques were:

- the selection of earlier-maturing varieties
- transplanting
- wet-tillage
- the use of organic fertiliser.

The selection of earlier-maturing varieties was important in increasing the number of crops that could be grown on a single piece of land in a year. Where water control is effectively established, the main factor determining the length of the growing period becomes temperature. Rice requires a temperature above 15°C for pollination and grain formation to take place, so that it is more restricted by low winter temperatures than other cereals such as wheat and barley. Thus, in the more northern part of its range – in central China, Korea and Japan – a quick-maturing rice crop may allow another crop to be grown in the same land in the cooler part of the year. In warmer latitudes – for example, in Vietnam and south China – two rice crops could be grown in a year. The search for early-maturing rice varieties has been complicated because earlier maturity has often been associated with lower productivity, so that the development of high-yielding early varieties has had a significant historical impact. One of the earliest such instances was that of the early-maturing 'champa' rices which spread from Vietnam to China during medieval times, passing from farmer to farmer. Just under a thousand years ago the Chinese state actively promoted the use of these varieties to enable the growing of two crops a year in the Yangzi delta, where Chinese farmers were able in the next two hundred years to select higher-yielding strains of these early-maturing rices. The role of the Chinese state in this case is interesting for it may be seen as a forerunner of the role taken by the Japanese state in identifying and disseminating more productive varieties in the nineteenth century.

Transplanting rice means that instead of sowing the seed directly into the field, the rice is grown in a small nursery for between one and two months before the seedlings are planted into the flooded soil of the main field. The advantages of the system are that it uses less seed, that it occupies the main field for less time and hence makes growing two crops a year in the same field easier, and, perhaps most important, it reduces weed growth

▲ *Wet tillage, or puddling, reduces the loss of water from rice fields by greatly reducing the downward drainage of water through the soil. (Yunnan Province, China, 1986.)*

in the field. Not only does the waterlogged condition of the soil suppress many weeds, but seedlings can be planted in lines, which assists weeding of the growing crop. (See Plate 6.)

Wet-tillage (puddling), that is ploughing the submerged field, was carried out to promote the collapse of soil structure in the flooded field. In its simplest form, puddling was carried out by cultivating the field while wet, before the crop was planted. Cultivation could be done either manually with a hoe, or using draught animals to pull a shallow plough.

Q   Why should this be advantageous?

A   It reduces to a minimum the rate of drainage of water in the field. This not only improves the conservation of water and nutrients, but also lessens the risk of the field drying out. As we saw in Section 2, intermittent waterlogging results in reduced levels of nutrients, particularly nitrogen.

Organic fertiliser use is perhaps one of the most characteristic features of intensive rice-growing in China, where for centuries practically any organic waste has been traded commercially as fertiliser. Associated with the wide range of materials used, which included human and animal manure, food-processing waste (beancurd waste, fishmeal etc.), ashes and crop residues, were composting techniques to render the material suitable for application to fields before planting. A further method of enriching the soil to allow it to sustain continuous crop production is the technique of growing aquatic plants such as blue-green bacteria and the fern-like *Azolla pinnata* in the flooded rice fields. As was noted in Chapter 2, these plants are able to fix nitrogen from the atmosphere, and when they die, after about a month of growth, they provide organic matter from which nitrogen is released in a form which the rice crop may absorb. As we saw in Section 2, shortage of nitrogen in the soil is often the major constraint to grain yield, and this technique, developed particularly in China and Vietnam, is an important way of sustaining soil fertility.

Unlike agricultural development in Europe and America, which pursued the substitution of labour by machinery, many of these technical developments of rice-growing *increased* the amount of labour required to grow the rice crop. Growing rice in nurseries and transplanting the seedlings requires more work (20–30 person-days per hectare) than simply broadcasting, or scattering, the seed onto the soil (1–2 person-days per hectare). The application of organic fertiliser requires work in composting, transporting (several tons per hectare of land) and incorporating the fertiliser into the soil. Furthermore, in order to be effective, all these techniques require a good control of the level of water in the field, and this close control of water is in itself a labour-intensive activity, particularly where manually operated water-pumps are used, as in China and Japan until the early twentieth century.

By the application of large amounts of labour – often of the order of 280 work-days per hectare for a single rice crop – Chinese and Japanese farmers were able to produce between 1.5 and 2.5 tonnes of rice per hectare per season (often two seasons in a year) with considerable reliability at a time when yields from an equivalent area of wheat would have been only 0.5 to 1.0 tonnes. In tropical Asia rice yields were more restricted by lack of irrigation. Less control over water meant that availability of water for the rice crop was erratic, and the effectiveness of labour-intensive fertiliser use and transplanting techniques were diminished.

It is evident that for many of the tasks described above the skill content is high and the resulting productivity made the investment of labour worth while. However, it is perhaps more important that the productivity of rice-growing allowed a *diversification* of farm activities. A well-known system developed in central China involved growing mulberry trees on the dykes round flooded fields to provide food for silkworm culture: the waste from silk production was added to ponds and flooded fields not only to fertilise the rice crop but also to feed fish such as carp and tilapia which provided a valuable source of food and income. Other forms of diversification involved growing a second crop, such as wheat or sugar cane, after draining the rice fields, or growing cash crops, of which cotton and soya bean were the most important, on neighbouring 'upland' fields. This diversified farm output, coupled with primary processing (for instance, silk and cotton), allowed a high year-round absorption of labour in rural areas and a relatively high value of output.

The sensitive management of water and soil required to achieve high output in this type of wet rice cultivation meant that there were few opportunities for economies of scale. Thus, although in both China and Japan ownership of land was highly concentrated in the hands of relatively few landlords, they had little incentive to manage their lands as large-scale production units, and most rice land was divided into small plots which were cultivated by tenants. The relatively high degree of technical management under the control of the tenants placed them in a strong bargaining position in relation to their landlords, and there are indications that over quite long periods of Chinese and Japanese history they were able to improve the proportion of their output which they retained.

However, much depended upon the nature and politics of the state, as the divergent paths of development of China and Japan in the nineteenth century illustrate. As the Chinese economy stagnated in the aftermath of the opium wars, landlords resorted to physical extortion in order to extract rents from increasingly impoverished tenants. By contrast, in Japan the Meiji period which began in the mid nineteenth century saw the state intervene in a number of ways to increase agricultural output as part of its industrialisation drive. In order to sustain both the supply of cheap rice for the growing industrial workforce and a flow of taxation (commonly reckoned to have been about 30%) from agriculture to finance industrial development, the state intervened to limit the level of rents paid by tenants to landlords and pursued a vigorous programme of agricultural research and extension. Although landlords retained a role in some aspects of agriculture, notably that of developing irrigation infrastructure, the state-led research and extension focused inevitably upon those responsible for cultivating the fields, two-thirds of whom were tenants at the turn of the century. To the extent that landlords were not directly involved in cultivating their fields, therefore, they became marginalised in Japanese agriculture and more concerned with industrial investment.

3.3   *Summary*

In this section we have sketched out some of the major factors shaping Asian rice cultivation over the 6000 years to the end of the nineteenth century. Some of these factors are 'technological':

•   the extension of the wetland conditions suitable for rice through the development of water control techniques

•   the improvement of mineral cycling in rice fields through the use of composting and manuring

•   the selection of earlier-maturing varieties of rice allowing a larger harvest of primary production through the production of more than one crop in a year

•   the development of fish-farming and silkworm production allowing the 'harvest' to be extended to the herbivores and carnivores of the wetland ecosystem, as well as the primary production.

We saw, however, that these technological developments were driven by socio-economic forces and, in turn, produced socio-economic impacts at the level of individual farmers, at the level of communities, and at the level of the state.

Activity 2

Review Section 3 and note down:

(a) different ways in which social and economic developments can be regarded as promoting changes in rice cultivation methods;

(b) ways in which the technology of rice production shaped social and economic relationships.

Then compare your answers with those given at the end of the chapter.

4   The twentieth century: the green revolution in rice production

In this section we trace the transformation in rice cultivation which took place in the twentieth century, from its origins in Japan and China to its wider impact in tropical Asia in the second half of the century. A final subsection analyses the consequences of this transformation in environmental and socio-economic terms.

4.1   Japan and China

Despite an early interest in American and British farm equipment, the Japanese drive to increase rice productivity in the late nineteenth and early twentieth centuries resulted in little mechanisation. The foreign-made equipment was rejected as being inappropriate for the scale and conditions of wetland rice cultivation, and the only significant mechanisation to take place was the introduction of electric pumps for irrigation, which cut the labour required for irrigation by about two-thirds (from 70 to 22 days per hectare).

With the passing of interest in western farming methods, the major Japanese research effort went into a search for more productive rice varieties, capable of increased yield with large applications of fertiliser; this was becoming possible as cheaper steamship transport opened the way for large-scale importation of fertiliser material, such as soyabean cake from Manchuria. In the nineteenth century the search for more productive varieties had been led by 'veteran farmers' who were responsible for bringing together and testing the large number of rice varieties already developed locally by farmers in different parts of Japan. The role of the state was to provide facilities for these activities and an organisation for distributing seed of the better varieties. By the end of the First World War, however, the pace of improvement in yields was slowing, and the increase in output was falling behind that of consumption. Japanese rice supplies were becoming increasingly dependent upon imports obtained from Japanese-ruled Taiwan and Korea. This dependence provided the impetus for major reorganisation and investment in irrigation by the Japanese administration in those colonies in the inter-war years.

In the late 1920s breeding work was started by the Japanese ministry of agriculture and forestry to develop new rice varieties by artificially cross-pollinating between the existing 'farmers' varieties', and this produced the early-maturing dwarf *Norin* varieties. Because of the short, stiff straw of these varieties, the heavy grain panicles resulting from large applications of fertiliser did not cause them to fall over, or 'lodge', and hence farmers could use greater amounts of fertiliser to increase grain production than with taller, traditional varieties. The *Norin* rice varieties were quickly adopted by farmers – in 1935 they were planted on some 160 000 ha in Japan – and they were subsequently introduced to Taiwan and Korea. However, the impoverishment of agriculture during the Depression, and then disruption during the Second World War, meant that the benefits of the new varieties were not fully felt until some twenty years later.

The end of the Second World War saw the elimination of the landlord class as a result of land reforms carried out by United States military administrations in Japan, Taiwan and Korea, and by the communist government in mainland China. In all these countries this produced a rice-farming sector consisting of families farming very small landholdings of less than one hectare. In Japan, Korea and Taiwan this pattern of smallholding remained unchanged, while in China the subsequent thirty years saw a progressive combining of individual farming through mutual aid teams, then agricultural co-operatives, and finally communes.

Despite this difference, the technical development of wet rice production followed much the same pattern in all these countries: improved water control and use of high-yielding dwarf varieties grown with large applications of fertiliser were used to achieve very high land productivity. But in addition, from the 1960s onwards, the development of small motor-driven agricultural machinery – originally manufactured in Japan but later also in Korea and China – allowed a progressive *reduction* in the labour input by mechanising operations of heaviest labour demand, like tillage and transplanting. The labour released by mechanisation was absorbed by the growth of industry, much of which was sited in rural areas for this purpose. In China a significant proportion of the rural labour force (5%, or 17 million people) was also employed full-time on irrigation construction projects undertaken and paid for by the communes. The overall increase in land and labour productivity in Japanese rice-growing may be seen from Table 3.1; some figures for Taiwan are also given for comparative purposes.

Q   According to the figures given in Table 3.1, was the productivity increase in Japanese rice farming between 1900 and 1970 greater with respect to land or to labour?

A   Change in productivity per hectare: 5750 − 2870 = 2880, i.e. an increase of just over 100%.

Change in productivity per day: 39.4 − 10.7 = 28.7, i.e. an increase of nearly 300%.

Table 3.1 *Development of land and labour productivity in Japan and Korea in the twentieth century*

		Labour input (person-days/ha)		Productivity (kg rice)	
		in rice only	farming total	per ha	per day
Japan	1880	278	353	2360	8.5
Japan	1900	267	397	2870	10.7
Japan	1920	235	384	3940	16.7
Taiwan	1926	96	–	2110	22.0
Japan	1940	206	359	3740	18.1
Japan	1960	214	345	4990	23.3
Japan	1970	146	235	5750	39.4
Taiwan	1972	125	–	5700	45.6

Source: Bray, F. (1986) *The Rice Economies: technology and development in Asian societies,* Oxford, Blackwell.

Q How has the time spent in rice farming in Japan changed as a
      percentage of total time spent in farming over the same period?

A In 1900 labour input in rice was:

$$\frac{267}{397} \times 100\% \text{ of the total, or } 67\%.$$

In 1970 it was:

$$\frac{146}{235} \times 100\%, \text{ i.e. } 62\%.$$

Q The number of work-days required to cultivate one hectare of rice in
      Japan was lower in 1970 than in 1920, while the reverse seems to have
      been the case in Taiwan. Can you think of reasons for this?

A Japan was more industrialised at both times, so labour costs would
      tend to be higher and prices of manufactured products lower, both
      favouring mechanisation. In Taiwan irrigation increased from the 1920s
      onwards, allowing greater use of labour-intensive methods in rice-
      farming. Taiwan also experienced rapid population growth after the
      Communists took power on the mainland in 1949 and Nationalists fled
      to Taiwan. This would have increased both demand for rice and the
      supply of labour.

At this point we will compare Asian rice farming with mechanised cereal
farming developed in North America. Comparisons of inputs and outputs
of agricultural systems are complex and, where a diversified farm output is
concerned, a complete picture can scarcely be obtained by considering the
output of only one crop. Thus the data presented in Table 3.2 serve only as
a guide to the different magnitudes of input, output and productivity
measures for different methods of growing cereals. The figures for Japanese
farming are for 1960 and the others have been compiled from various
studies carried out in the 1960s and 1970s; thus these data are not strictly
comparable.

Table 3.2 sets out the inputs of four different cereal-growing systems,
two American and two Asian. For each system a series of productivity
measures (outputs) has been calculated:

•   cereal output per unit (hectare) of land area

•   cereal output per unit (hour) of human labour

•   cereal output per unit of 'commercial energy' (purchased inputs such as
fertiliser and machinery).

Finally, an overall measure of energy efficiency is obtained by converting
the cereal harvest, and all the inputs required to produce it, to energy-
equivalents (MJ: 1 megajoule = 1 million joules of energy). In this way all
the different inputs may be added together to give a single figure for 'total
energy input'. The energy value of the harvest is divided by this 'total
energy input' to provide a figure for the number of joules of output for each
joule of input.

First we shall compare system 1 (US wheat production) with system 3
(Japanese rice production). Note the dramatic divergence in agricultural
labour use between the United States, where mechanisation reduced labour
use to a mere seven hours for each hectare of wheat, and Japan, where the
same area of rice employed no less than 1729 hours of human labour. As a
result, although the output of rice *per hectare* of land in Japan was more than

double that of wheat in the US, the productivity of agricultural *labour* was over a hundred times higher in the US.

Q   Which of the two systems used more energy to produce a given weight of grain?

A   The amount of energy used to produce grain in the two systems was about the same because the Japanese put in about $2\frac{1}{2}$ times as much energy to produce $2\frac{1}{2}$ times as much grain per hectare. American mechanisation was more than compensated by the much higher energy inputs in Japanese wetland rice in the form of labour, fertiliser and irrigation.

The figures in Table 3.2 allow us to compare Japanese rice cultivation with two other systems of rice production: system 2 represents a 'western' rice-growing system from the southern United States which employs large-scale land-levelling equipment, combine harvesters, and aircraft to sow, fertilise and spray the crop; system 4 represents tropical Asian wetland rice cultivation with a low level of water control and a low level of purchased inputs. Activity 3 will allow you to make comparisons between these systems.

Table 3.2 Comparison of inputs and outputs and their energy equivalents in different cereal growing systems

System	1 Wheat (USA)		2 Rice (USA)		3 Rice (Japan)		4 Rice (Philippines)	
	quantity per ha	energy MJ/ha	quantity per ha	energy MJ/ha	quantity per ha	energy MJ/ha	quantity per ha	energy MJ/ha
Inputs								
labour	7 hrs	13	17 hrs	33.2	1729 hrs	3 387	576 hrs	1276
animal draught							272 hrs	3998
machinery, irrigation (incl. fuel)		5 084		32 145		8 645		174
fertiliser	141 kg	3 578	382 kg	18 178	305 kg	10 303	5.6 kg	357
seed	106kg	2 938	157 kg	4 787	112 kg	3 415	108 kg	1678
pesticides	0.5 kg	210	13.4 kg	5 503	11 kg	4 397	0.6 kg	183
total energy input		11 823		60 646		30 138		7667
Outputs								
per unit area kg grain/ha	2060	28 552	6160	93 915	4848	73 912	1654	25 217
per unit of labour kg grain/hr	294		362		2.8		2.9	
per unit of commercial energy[1]								
kg grain/MJ	0.17		0.10		0.18		0.69	
Energy ratio								
MJ out/MJ in	2.41		1.55		2.45		3.3	

Note: [1]Excluding labour and animal power.

Source: Table compiled from data in Pimentel, D. and Pimentel, M. (1979) *Food, Energy and Society*, London, Edward Arnold.

Activity 3

Use the figures given in Table 3.2 to compare the US (system 2), Japanese (system 3) and Philippine (system 4) rice cultivation systems. Rank the three systems (from highest to lowest) in terms of:

- labour input
- total energy input
- land productivity
- labour productivity
- overall energy ratio.

Which system do you consider to be the most efficient? Compare your answers with those given at the end of the chapter.

Let us now make two observations on these comparisons. Firstly, we can see that, with the technologies developed in twentieth-century agriculture, increased output of grain from a given area of land required an increase in *energy* use. The sustainability of such systems depended, therefore, upon the cost and continuing availability of energy for this input. Secondly, in rice-growing areas of Asia, human labour was for centuries the form of energy with which increased output from a fixed amount of land was achieved. In the twentieth century, other forms of energy have increasingly supplemented, and then displaced, human labour.

In the brief account of Chinese and Japanese rice farming given in earlier sections we saw that the technological changes which accompanied the increased investment of human labour allowed an increase not only in the amount of grain produced by a single crop but also in the number of crops the land could produce in a year. However, in spite of the fact that labour productivity in rice may have increased fivefold since the late nineteenth century (see Table 3.1), our comparisons with labour productivity in US agriculture highlight the fact that rice-growing by such labour-intensive methods is a relatively poor source of *income*. The importance of wetland rice cultivation in China and Japan was that it provided a secure supply of surplus food that enabled farmers to develop other income sources from livestock, cash crops and rural industry. This was of primary importance not only for farmers' incomes, but also for the Japanese economy: from the 1870s to the 1920s the share of silk production in the value of agricultural output rose from 5 to 15%, and in the 1920s silk exports accounted for 30% of Japan's foreign exchange earnings. The industrial development of Japan in the twentieth century was accompanied by an increasing location of industry in the countryside and a move towards part-time farming. Thus, by the 1980s, the number and size of farms had changed little from the time of the post-war land reforms, but over 70% of farmers earned more from non-agricultural activities than they did from farming.

Summary

This section has described how Asian industrialisation, principally in Japan, both increased the need to produce more from a fixed land area, and enabled this to be achieved through the use of large applications of cheap fertiliser. Since heavy fertiliser doses caused traditional rice varieties to grow tall and collapse, Japanese researchers produced shorter, stiff-strawed varieties capable of converting increased fertiliser into increased grain yields. This 'biological–chemical' technology of breeding crops to grow with very high levels of mineral nutrients and well-controlled irrigation later became known as 'green revolution' technology when used to increase cereal yields (wheat and rice) in tropical Asia.

4.2   The green revolution in tropical Asia

Our account of the development of wetland rice farming in the twentieth century has so far centred upon the development in Japan of the 'biological–chemical' technology of rice varieties bred to take advantage of large applications of fertiliser. The new varieties were of the *japonica* subspecies (see Section 2.3), and similar varieties were swiftly bred and adopted in other areas growing rice of the *japonica* type: China, Taiwan and Korea. The subsequent success of the new varieties in increasing rice production in these countries prompted moves to seek varieties of the *indica* subspecies that could serve the same purpose in tropical Asia.

Since the characteristics required in the new varieties (short straw, early maturity, flowering insensitive to day-length) were in many respects the opposite of those found in the best *indica* varieties (tall and leafy, flowering and maturity dependent upon day-length), the search for new varieties concentrated on efforts to obtain fertile hybrids between *japonica* and *indica* varieties. The establishment in 1960 of the International Rice Research Institute (IRRI) in the Philippines with funding from the Ford and Rockefeller Foundations provided a focus for these efforts, and in 1962 the Institute produced a short-strawed variety, IR8, with grain type intermediate between *indica* and *japonica*, from a cross between the varieties 'peta' from Indonesia and 'dee-geo-woo-gen' from Taiwan. In the following two decades many such high-yielding varieties (HYVs) were produced at the IRRI and by national research organisations, and were introduced throughout the rice-growing areas of the Indian sub-continent, Indo-China, Malaysia, Indonesia and the Philippines. As a result, and despite consumers' dislike of the culinary properties of the new rice varieties, rice production in these areas increased by 60% and countries which traditionally imported rice, such as the Philippines, Malaysia, India and Indonesia, were able to achieve a large measure of self-sufficiency, or even become exporters of rice. It was this transformation which became known as the **'green revolution'**.

For a number of reasons, however, the effect of the green revolution upon rice-growers' living standards in these countries was more equivocal and has been the subject of much criticism. In particular, in contrast to the radical land reforms implemented in Japan, China and Korea, the European and American administrations which resumed control of their tropical Asian colonies after the end of the Second World War left the influence of rural landlords undiminished. Thus, at the time that the higher-yielding varieties were introduced into south and south-east Asia in the 1960s and 1970s, not only was tenancy a predominant condition for rice farmers, but also the terms of tenancy often placed the landlord in a strong position to gain from increased productivity, because the rent was a proportion of the crop rather than a fixed value. While the details of tenancy arrangements varied greatly between and within countries, the prevalence of share rents has been documented by studies such as that carried out by the United Nations Research Institute for Social Development (UNRISD) (see, for example, Pearse (1980)). These indicate that landlords had the right to as much as 50% of the crop where the tenant paid for inputs such as seed and fertiliser, but that the landlord's share might rise to as much as 75% if he provided inputs.

Against this unpromising background for tenant farmers to increase their income through higher productivity, the early promotion of new rice varieties in the form of a technological 'package' created further difficulties. The 'package' approach meant that farmers were encouraged to use the new varieties only if they also applied the recommended rates of fertiliser and insecticide, for without this agrochemical input the new varieties were unlikely to outyield traditional *indica* varieties. The package at once diminished cultivators' influence over technical decisions in rice-growing. But, more importantly, the need to purchase inputs increased the working capital necessary to grow rice, and so moved control sharply in the landlords' favour because of their greater access to capital, from either their own resources or from institutional (bank) credit secured against their land. Not only were landlords in a position to lend money to tenants for input purchase – debts on which interest could be claimed – but, because the inputs came from outside the rural areas, landlords were also well placed to act as intermediaries in the input distribution chain. Although these mechanisms did not always operate in every situation, all can be found in the UNRISD case studies.

Share tenancy did not persist uniformly in tropical Asia. Land reform implemented in the Philippines in the 1970s converted share tenancies to fixed-rent tenancies with an option to purchase the land over a 15-year period. However, tenant farmers' scope for an improved livelihood continued, even in the 1980s, to be constrained by the scarcity of alternative sources of income, by comparison with small-scale farmers in Japan, Taiwan or Korea in the 1950s. Among the reasons for this were the legacy of rice monoculture inherited in some cases from the expansion of irrigated rice-growing during the colonial era (see Section 3.1), and the relatively greater penetration of industrial manufactures into rural areas in competition with local 'artisanal' manufacture. The lack of high-value agricultural or rural industrial production left the growing rural population the alternatives of migration to the cities in search of work, staying in their villages to engage in unremunerative household-based trading and food-processing activities, or further intensifying their output from rice farming.

On land where irrigation permitted year-round crop production (about 30% of the rice-growing area of tropical Asia in the 1980s), the introduction of the early-maturing HYVs made possible the growing of two crops in one year in tropical Asian countries like the Philippines where it had been impossible with the traditional *indica* rice varieties. Indeed, with irrigation it now became feasible to grow two crops of rice and a third, different, crop (mung bean was a common choice) in the same field in a single year. However, such cropping schemes introduced extreme peaks of labour demand in order to harvest one crop in time to plant another and the 1980s witnessed the rapid adoption of labour-saving technology. In some cases technological change was remarkably simple: the substitution of the sickle for the *ani-ani* harvesting knife in Indonesia reduced the labour required for harvesting. But of no less importance were the use of:

* machinery such as small-scale motorised tractors and portable mechanical threshers originally manufactured for rice-growing in Japan and Korea, and increasingly built in less industrialised countries like the Philippines

* herbicides developed for mechanised rice-growing in the United States.

▲ *Small tractors, developed during the 1950s and 1960s in Japan, and now manufactured in many parts of Asia, allow mechanised tillage in rice fields. (Sichuan Province, China.)*

The second of these was particularly attractive because it not only reduced greatly the work of weeding the crop but also allowed the labour-intensive work of transplanting rice seedlings to the field from a nursery to be abandoned in favour of broadcasting (scattering) seed directly onto the field. Without herbicide, the practice of broadcasting HYV seed produced weed growth of practically unmanageable proportions because the soil could not be kept covered with water while the rice seed germinated, and during this germination period weeds would also germinate. Under these circumstances the large fertiliser applications would feed rampant weed growth, rather than the crop.

It is important to note that whereas the introduction of labour-saving technology into rice-growing in Japan, Taiwan and Korea allowed farming families to take up more productive off-farm employment, in much of tropical Asia labour-saving technology in rice-growing is a means of cutting costs in what may often be the most remunerative form of employment available to farmers. Unless some form of shared ownership of machinery is undertaken, ownership of even small power-tillers will inevitably be concentrated among wealthier farmers, who will be able to hire their equipment to poorer neighbours. The use of herbicide, on the other hand, is more accessible to farmers with little capital because it is more divisible – it need only be purchased in the quantity needed for immediate use. Furthermore the expiry of patents on many commonly used herbicides has resulted in a decline in their cost relative to manual weeding.

The labour so displaced from rice farming in tropical Asia is principally hired labour, those who have little or no access to land, and who, in the absence of alternative remunerative employment, must suffer a decline in living standards.

4.3   Food production, rural development and the environment

In 1987 rice production accounted for 55% of all cereal supplies for the three billion people in Asia. Overall, rice output in Asia increased steadily by between 1.5 and 2% each year from the late 1950s onwards. In tropical Asian countries like the Philippines, Indonesia, Thailand and Malaysia the annual rate of increase during the 1960s and 1970s was much higher, at between 3 and 5%. In earlier sections we have seen how this increase has resulted from technological change in rice farming. At the centre of these changes is the improvement of water control which allows more than one crop per year, and the selection of early-maturing rice varieties that can make use of high fertiliser rates to produce more grain. As we have seen, when this new technology was first introduced, it tended to increase the amount of labour needed to grow rice. By the late 1970s, however, the availability of mechanical and herbicide technology for small-scale rice farming resulted in their use as a substitute for agricultural labour. In this section we will consider how these changes have affected the impact of wet rice cultivation upon the physical environment. The section will end with a consideration of the future development of agricultural systems based on rice-growing.

Wetland hazards

In the intensive rice-growing systems developed in China and Japan up until the twentieth century, the emphasis on maintaining flooded soil conditions meant that 'leakage' of water and plant nutrients was low. The absence of soil channels impeded water drainage from the field soil into ground water and, although several tons of organic fertiliser was commonly applied to each hectare of rice, losses to the ground water could similarly be avoided as long as the fertiliser was incorporated within the oxygen-free anaerobic layer of the flooded soil. Under these circumstances the only movement of material out of the soil, except that into the crop, was the production of gases, particularly methane, which bubbled to the surface and into the atmosphere. Once the soil was drained, however, and oxygen allowed to enter below the top centimetre of the soil depth, the chemical transformation of some nutrients, of which nitrogen was the most important, would result in significant losses when the soil was next flooded (as we saw in Section 2). A further hazard of intermittent flooding should be noted. Water consumption for a rice crop grown under intermittent flooding may be as much as three times that required with continuous flooding, and most of this extra water constitutes drainage into the groundwater. As a result, the level of the water table may rise, and in dry climates with high evaporation rates the minerals dissolved in the groundwater will accumulate as salt at the soil surface. As outlined in Chapter 2, this process of **salinisation** will, within a few years, make it impossible for the soil to support the growth of any crop. Salinisation is an important environmental hazard where rice is grown under irrigation in arid climates – in Pakistan, some parts of India and in many parts of Africa.

More generally, perhaps the greatest environmental hazards in this system of rice cultivation were the increased exposure of farmers to diseases associated with standing water. The most important of these are malaria and schistosomiasis, or bilharziasis. Malaria affects some 200 million people worldwide and is carried by mosquitoes which need

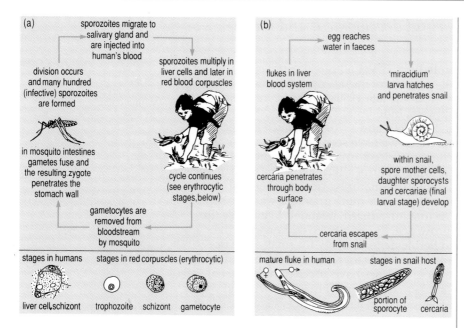

▲   *Figure 3.7   (a) The life-cycle of* Plasmodium vivax *(causing benign tertian malaria). (b) The life-cycle of* Schistosoma mansoni *(causing bilharziasis).*

stagnant water in which to develop their larval stage: see Figure 3.7(a). Schistosomiasis is caused by a range of parasitic worms which live alternately in the human bloodstream or intestines and in aquatic snails: see Figure 3.7(b). It affects 500 million people and is frequently associated with the development of irrigation, but it appears to be less widespread in Asia than in Africa or Latin America.

Further disease risks stemmed from the use of animal and human excrement as fertiliser in wet rice fields. The risk of transmission of pathogens and parasites was eliminated to some extent, however, when effective composting was carried out before application to the fields. The addition of animal manure directly to ponds to encourage the growth of fish for human consumption was potentially more problematic, although the very few studies carried out to assess this risk of pathogen and parasite transmission from animal manure to humans via fish have indicated that the risk may be low.

Fertiliser hazards

The introduction of HYV was coupled first and foremost with the use of heavier applications of nutrients to the soil and, in order to achieve this, recourse was made increasingly to industrially manufactured 'chemical', or inorganic, fertilisers.

Q   Why was this?

A   The main reason was that the concentration of plant nutrients in organic fertilisers is relatively low. The content of important plant nutrients like nitrogen are usually in the range 1–5% in animal manures, and in the range 7–15% in fishmeal. By contrast, nutrient contents of inorganic fertilisers like superphosphate and urea are over

40%. Thus the effort required to transport and apply high doses of nutrients is considerably reduced by the application of inorganic fertilisers.

A further, though somewhat double-edged, advantage of inorganic fertilisers is that all the nutrient content is in soluble form. The advantage of this is that nutrients can be rapidly absorbed by the growing crop and the effect on the crop can be immediate. A disadvantage is that if the crop is unable to absorb all of the nutrients supplied by fertilisers shortly after their addition to the soil, which is frequently the case when the crop plants are small, then nutrients may be wasted through drainage, absorption by weeds or micro-organisms, or transformation into a chemical form which cannot be absorbed by the crop. This disadvantage is particularly important in the case of nitrogen fertilisers, and in order to avoid losses of nutrients the total amount of fertiliser applied to the crop is usually divided into a number of doses applied at different stages of the crop's growth.

In wetland rice farming this presents a problem, because in order to be absorbed by the rice roots the fertiliser nitrogen must be introduced into the subsurface soil layers and – unlike applications of fertiliser to the soil surface in upland crops on freely drained land – the percolation of rainwater through the soil pores cannot do the job. In fact, most recommendations of fertiliser use in flooded rice advise incorporating the first dose into the soil before planting and then lowering the water level in the field at each subsequent dose to allow the fertiliser to come into contact with the soil. In practice, farmers keen to save labour are more inclined simply to scatter fertiliser into the flooded field. In the case of urea, the most commonly used nitrogen fertiliser, this practice results in the formation of ammonia in the surface water, which is then lost to the atmosphere as ammonia gas. Research has indicated that over half the nitrogen nutrient contained in the fertiliser can be wasted in this way. This is significant, for each kilogram of nitrogen represents 2 kg of fossil fuel expended in its manufacture and transport.

Q   With normal nitrogen application rates in the region of 50 kg per hectare of rice field, what is the wastage of fossil fuel per hectare?

A   Of 50 kg of nitrogen up to 25 kg may be lost, and every kilo of nitrogen represents 2 kg of fossil fuel, so that total losses may be equivalent to 50 kg of fossil fuel per hectare.

Pesticide hazards

Although twentieth-century rice growing can be wasteful of high-energy fertiliser inputs, it is the use of pesticides which has caused more immediate concern for environmental consequences. Insecticides were used fairly widely in rice farming in the 1950s and early 1960s, even before the introduction of the high-yielding varieties. In the Philippines 60% of rice was reportedly being sprayed as early as 1965. However, the first HYV introduced into tropical Asia – IR8 – was very susceptible to damage by insect pests such as stem borers (*Chilo suppressalis*). Consequently the use of insecticide was considered a necessary component of the technological 'package' recommended to farmers starting to grow the new variety. Insecticide use was in some cases subsidised by governments anxious to increase rice output. In the Philippines it was effectively supplied free through spraying carried out by the government's Bureau of Plant

Industry, using pesticides provided as part of bilateral aid programmes. In Indonesia, the 1967–70 BIMAS programme to introduce a high-yielding rice 'package' to farmers was contracted out to, among others, the Swiss-based agrochemical manufacturer CIBA. By the late 1960s it was common for insecticide to be routinely applied to every rice crop.

Since many of the insecticides used, such as dieldrin, DDT and endosulfan, are toxic to fish, an immediate consequence of this policy was a drastic reduction in the fish in flooded rice fields. In parts of Malaysia the decline in fish harvests has been estimated at 50–60% (Tait and Napompeth, 1987). Not only does this remove a significant, and in some cases major, source of dietary protein for rice farmers, it also blocks off a potential route to diversification of farmers' sources of income. In addition to the damage to fish production, pesticide use in Asian rice farming posed a direct threat to human health, through the absence of adequate protective clothing for those preparing and applying the chemicals, and the ease with which contaminated water from sprayed rice-fields could drain into watercourses used for livestock and domestic purposes. This danger became even greater as farmers substituted chemicals such as DDT, considered dangerous because of their residual and cumulative effect, by products such as carbufuran, which could kill insects within the plant but which were more immediately toxic to humans. Although many cases of acute pesticide poisoning have been documented among Asian rice farmers, the effects of chronic exposure to lower levels of pesticide over a long period of time are still practically unknown. Risks to farmers and the environment are compounded by the fact that pesticides are often used in the third world long after they have been banned in first world countries because of proven harmful effects.

The final irony of insecticide use on rice was that it was fairly unsuccessful in protecting the crop. Reports from both the Philippines and Malaysia indicate that insecticide spraying in the 1970s to control one pest, the stem borer, was commonly followed by resurgence of another pest, the brown plant-hopper (*Nilaparvata lugens*), necessitating further spraying, and so on, leading to a 'chemical treadmill'. The development of a new variety of rice, IR36, resistant to the two known strains or 'biotypes' of plant-hopper, appeared to provide the answer in the early 1980s, and in 1983 this variety was being grown on about 10 million hectares. By the mid-1980s, however, a third biotype of the plant-hopper, hitherto unknown and against which IR36 had no resistance, was causing considerable damage, despite the repeated spraying of rice fields. Although IRRI's plant breeders quickly released a new variety, IR56, reputed to be resistant to the third plant-hopper biotype, the United Nations Food and Agriculture Organisation (FAO) was experimenting, with some success, with an alternative approach known as **integrated pest management (IPM)** in Indonesia. This strategy concentrated on promoting control of the pest by natural predators such as spiders, and therefore eliminated the use of some 57 insecticides used in the past which were toxic to these predators. Insecticides that have little effect on the predator species are still permitted where plant-hopper damage threatens, but the Indonesian experiment showed that a reduction of insecticide use of about 80% was possible while at the same time increasing yields. As the FAO experiment was extended in 1988 to other countries in tropical Asia, and to China, it raised the prospect of a radical reduction in insecticide use in Asian rice farming by the end of the twentieth century.

Reduction in herbicide use, however, was less likely. As we have seen in the previous section, herbicides are an important means by which farmers can reduce hired labour required for transplanting and weeding, and so increase their income. The long-term consequences of herbicide use

in rice farming have yet to be studied and documented. It is certainly known that some herbicides are toxic to both humans and fish, so that the hazards described above for insecticide use will certainly apply to herbicide use. In addition many herbicides will have negative effects on the blue-green bacteria and *Azolla* plants which are widely planted (or 'inoculated') in rice fields in order to provide a source of nitrogen for the rice crop. FAO estimates that about 1.3 million hectares of rice are cultivated with *Azolla* in China alone. Many herbicides attack only specific plants and it is quite possible that herbicides will be identified which are compatible with both fish and *Azolla*. Until that has been achieved, however, farmers using herbicides on rice risk losing options for agricultural diversification in the pursuit of short-term gains in income from rice.

4.4   Future prospects for Asian rice cultivation

The green revolution in Asian rice farming has exposed sharply both the strengths and weaknesses of twentieth-century science as an instrument for improving livelihoods in poorer, agriculturally based societies. It has demonstrated a capacity to produce enough food for a population believed to have grown perilously beyond the 'carrying capacity' of the fixed amount of land available. It has simultaneously demonstrated that this, though necessary, is not by itself sufficient to improve rural food security or living standards. We have seen that social and institutional factors play an important part in determining the benefits of improved agricultural productivity. Certainly the conditions under which people have (or do not have) access to land is one of those factors, but this operates in a wider context of the alternatives which exist for people to earn a living.

We may take, as an extreme example, the case of Indonesia, where, with some 58% of its population nominally engaged in farming in 1983, rice production was sufficient to cover the food requirements of its people. However, on the densely populated island of Java, rice farmers with land of their own produce as little as 1 kg of rice for each hour of their labour. This is about a quarter of the labour productivity in rice production in Japan and Taiwan (as shown in Table 3.1). But in rural Java, rice growing is one of the most remunerative forms of work available. For the 50% of households who have no land other than that on which their house stands, the productivity of their labour is even lower (White, 1982).

In such situations it is easy to perceive the economic pressures which encourage poor people to have large families. The low productivity of their labour allows parents little opportunity to accumulate savings during their working lives. Large numbers of children – who work rather than go to school – are a means whereby parents try to increase family income and provide security in their old age. (You may wish to refer back to the discussion of economic development and demographic transition in Chapter 1, and in particular to Caldwell's theory of fertility change and inter-generational wealth flows in Section 4.2.)

In considering the future development of Asian rice growing it is necessary to separate the need for rice production from the need for improved livelihoods. As we have seen, many rice-producing countries in Asia have achieved self-sufficiency or are net exporters of rice. By the late 1980s governments in Japan and Taiwan were confronting problems of overproduction of rice, similar to those of overproduction of food in the European Community. As in Europe, political pressure from farmers had resulted in the use of wealth generated by highly productive industry to pay higher prices to rice producers, who in turn used higher levels of inputs to increase rice output further, leading to surpluses of rice too expensive to sell on the world market.

If such conditions apply in richer, industrialised economies, what are the prospects of improved livelihoods for peasant rice-producers in the non-industrialised economies? As in Japan, a half century earlier, the question of *diversification* of rural employment, into industry or high-value agricultural production, is central to improved living standards in Asian rice-growing areas. Some observers have commented that in the case of Japan, Taiwan and Korea, industrialisation was made possible by earlier development of high labour productivity in agriculture founded upon a closely managed and highly developed water control infrastructure. They point to the relatively small proportion (20–30%) of cropland which is irrigated in tropical Asia as an obstacle to agricultural diversification. This in turn has been attributed to 'social barriers to efficient resource allocation to land infrastructure' (Hayami and Ruttan, 1985) which means that the relationship between rich and poor effectively prevents local collaboration to develop and manage irrigation and other works such as soil conservation, which have a public rather than strictly private utility. According to this analysis, the barriers to improving the productivity of agricultural labour in tropical Asia are therefore political rather than physical.

We shall conclude this chapter by briefly considering the fate of one attempt to overcome such barriers: the formation of the communes in China in the late 1950s (see Box 3.1).

The low efficiency of many of the small-scale rural industries established by the communes made them viable only while all alternative supplies for urban large-scale industries were excluded from the rural areas. This was, in fact, the policy carried out by the central state authorities, which prompted one writer to observe 'small and beautiful, if it is not to be wiped out by competition, presupposes the big and bureaucratic' (Kitching, 1989). While future attempts to diversify the income opportunities of rice producers by developing rural industry may be expected to avoid some of the errors of the Chinese communes, in the era of the global market the capacity for any small-scale rural industry to withstand international competition must be open to question.

The Chinese experience in the latter part of the twentieth century emphasises how closely the continued high productivity of land and labour in rice production systems depends on large-scale environmental management. The decline in the maintenance of irrigation infrastructure which has accompanied the break-up of the communes comes at a time when concern is mounting about a more general failure of environmental management in China. Unregulated deforestation of the upper watersheds of the Yangzi and Yellow rivers has caused a marked increase in the siltation rates of these rivers in recent years. Reduced and erratic river flows pose a direct threat to Chinese irrigation, and suggest that in the twenty-first century rice production will require stronger and more wide-ranging social and institutional collaboration than ever before.

Activity 4

Based on what you have read in this chapter, write down ways that the future options of small-scale rice farmers in Asia differ from, or are similar to, those of farmers in other parts of the world.

Box 3.1 The communes in China

As originally conceived by Mao Zedong, the communes were part of a three-tier structure of rural self-management: the communes were subdivided into brigades, which were further divided into work teams. In principle, election of officials at each of the three levels of organisation ensured that the communes represented the interests of their peasant members and acted to defend those interests against coercion by the centralised state bureaucracy.

In Mao's conception, the economic role of the communes was to pursue diversification through rural industrialisation and to increase agricultural productivity through the construction of infrastructure such as irrigation and hydro-eleciric power. The emphasis was to be on local self-sufficiency, and particularly the generation of local savings for local investment.

In practice, the economic role of the communes was severely undermined by the failure of their political function: the authoritarian state and party structure effectively pre-empted local democracy, with the result that commune officials were party nominees and state employees. The development of rural industry, particularly during the 'Great Leap Forward' (1958–1960), was, as a consequence, heavily influenced by the industrialisation policy of the centralised state. This resulted in inefficient and wasteful small-scale heavy industry (epitomised by 'the backyard blast-furnace').

Agricultural diversification was severely constrained by centrally defined priorities emphasising grain self-sufficiency, which translated into pressure for grain monoculture.

The Great Leap was a failure, with widespread famine occurring in the early 1960s, and the role of the commune in economic management was subsequently reduced, with the planning role being taken over by the centralised state and the farm management being carried out at brigade and subsequently work team level. In the 1980s collective farm management was made responsible for agricultural output through a contract system.

Despite the evident failure of the communes to develop high-productivity rural industry, they achieved a very significant investment of human labour in the creation of irrigation and other infrastructure with long-term benefits to the productivity of both land and labour in farming. The magnitude of this achievement is indicated by the fact that 97.3 million new workers were absorbed by the agricultural sector from 1957 to 1975.

The pattern of events which followed the final dissolution of the communes in the 1980s is instructive. In the first five years in which agricultural production was carried out by households on an individual basis, grain output rose, reaching an all-time record level (407 million tons) in 1984. However, in subsequent years grain shortages became common. Two reasons have been given for this decline. Firstly, farmers diversified into higher-value farm products, particularly meat, whose production consumed part of their grain output. Secondly, the dismantling of the communes eliminated the local authority with responsibility for organising the collective effort to maintain the irrigation works, and some deterioration occurred.

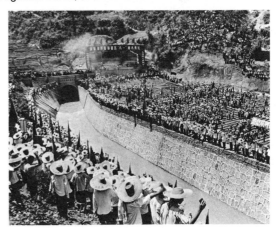

▲ Effective water management requires investment in infrastructure: in China, such investment was achieved in the 1960s and '70s by employing 5% of the rural labour force (17 million people) on irrigation construction projects. Left, digging an irrigation canal near Zibo in Shandong Province, China; and, right, the opening ceremony in 1970 of the Ouyanghai Irrigation Project which brought more than 23 000 hectares of farmland in southern Hunan Province under irrigation.

References

HAYAMI, Y. and RUTTAN, V. (1985) *Agricultural Development* (rev. edn) Baltimore, MD, Johns Hopkins University Press.

KITCHING, G. (1989) *Development and Underdevelopment in Historical Perspective*, London, Routledge and Kegan Paul.

PEARSE, A. (1980) *Seeds of Plenty, Seeds of Want*, London, Oxford University Press.

SARRE, P. and REDDISH, A. (eds) (1996) *Environment and Society*, London, Hodder and Stoughton/The Open University (second edition) (Book One of this series).

SILVERTOWN, J. (1996) 'Ecosystems and populations', Ch. 7 in Sarre, P. and Reddish, A. (eds).

SIMMONS, J. (1990) 'The impact of human societies on their environments', Ch. 2 in Sarre, P. and Reddish, A. (eds).

TAIT, J. and NAPOMPETH, B. (1987) *Management of Pest and Pesticides: farmers' perceptions and practices*, Boulder, CO, Westview Press.

WHITE, D. (1982) 'Population, involution and employment in rural Java', in Harriss, J. (ed.) *Rural Development*, London, Hutchinson.

Further reading

BARKER, R., HERDT, R.W. and ROSE, B. (1985) *The Rice Economy of Asia*, Washington, DC, Resources for the Future.

BRAY, F. (1986) *The Rice Economies: technology and development in Asian societies*, Oxford, Blackwell.

BULL, D. (1982) *A Growing Problem: pesticides and the third world poor*, Oxford, Oxfam.

PEARSE, A. (1980) *Seeds of Plenty, Seeds of Want*, London, Oxford University Press.

SANCHEZ, P. (1976) *Properties and Management of Soils in the Tropics*, New York, John Wiley.

Answers to Activities

Activity 1

Herbivores:

- insects (leaf-eating, sap-sucking)
- fish
- birds (such as ducks and grain-eating species such as *Quelea quelea*)
- small mammals (such as rodents).

Carnivores:

- insects (such as mosquitoes)
- fish
- amphibians (frogs etc.)
- reptiles
- birds (feeding on insects, frogs and fish).

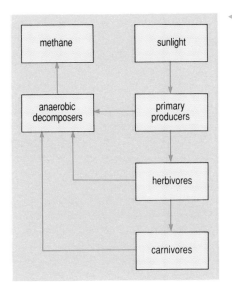

◀ *Diagram of energy flows.*

Activity 2

(a) Examples of social and economic developments that promoted changes in rice cultivation methods:

• prior to the expansion of European influence in the eighteenth and nineteenth centuries, the growth of population and extension of state power in Asia was the main force behind the extension of irrigation to new areas and the search for technologies allowing higher land productivity (for example, double cropping, and the use of fertiliser);

• the relative shortage of capital in this period of Asian history meant that irrigation and rice-farming technology were relatively labour-intensive;

• the absence of large-scale manufacturing presented rice-growers with many opportunities to diversify their economic activity through high(er)-income, household-scale processing (such as silk);

• the growth of European-owned plantation production was responsible for more capital-intensive expansion of irrigation oriented principally towards rice monoculture;

• the industrialisation of Japan led to increased pressure for higher yields of rice and made available larger supplies of fertiliser; subsequently, industrialisation led to pressure for mechanisation of rice production.

(b) Some aspects of rice production techniques which shaped socio-economic relationships:

• the need for close control of water promoted the development of farmers' water management organisations. The need for overall coordination of water use promoted the development of authoritarian, centralised states;

• the labour-intensive production methods for rice promoted high densities of population settlement;

• the intensive management of the rice fields required to manage the water and to carry out operations such as transplanting and manuring at the right time, placed much of the control of the production process in the power of the cultivators and hence increased their bargaining power in relation to land-owners and the state.

Activity 3

Comparisons of rice-growing systems, ranking from highest to lowest:

Labour input	Japan	1729 hr/ha
	Philippines	576
	USA	17
Total energy input	USA	60 646 MJ/ha
	Japan	30 138
	Philippines	7 667
Land productivity	USA	6160 kg/ha
	Japan	4848
	Philippines	1654
Labour productivity	USA	362 kg/hr
	Philippines	2.9
	Japan	2.8
Overall energy efficiency	Philippines	3.3 MJ out/MJ in
	Japan	2.45
	USA	1.55

These comparisons indicate that from the standpoint of total energy efficiency the low-input Philippine system is more efficient than the other two high-input systems. However, from the point of view of land and labour productivity, the US system is most efficient. This is particularly the case for *labour* productivity. It is clear that the assessments of efficiency will depend on which of the factors (land, labour or energy) we consider the most important. It should be noted that the figures for this exercise relate to studies in the 1960s and '70s, and that input and output levels are likely to have changed considerably in the 1980s in both the Philippines, as the result of the green revolution, and in Japan, as a result of fast economic growth.

Activity 4

Future options for small-scale rice farmers in Asia may be considered to be similar to those of farmers in other parts of the world in that rice is now in surplus supply in many Asian countries and, as in North America and Europe, farmers can expect the prices of agricultural goods to decline relative to the prices of manufactured goods. This long-term trend will produce a tendency for agricultural incomes to decline relative to urban incomes, and a pressure to diversify rural incomes from food production.

The situation for Asian rice farmers is different from that of their American and European counterparts in that their existing incomes are lower (due to smaller farms and less mechanisation), they have less access to industrial jobs and, unlike Europe and North America, governments in the less wealthy economies of Asia are unable politically or economically to finance higher incomes for farmers by making urban food consumers pay higher prices for food.

▲ *Plate 1*
Ethiopia

▲ *Plate 2   A poster in China exhorting the population to keep to the 'one couple, one child' policy. Although this campaign met with great difficulties, during the 1970s China's birth rate was halved.*

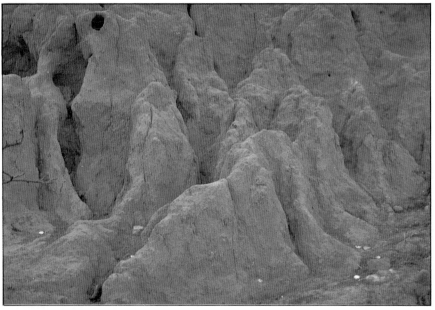

▲ Plate 3a
An irrigated beet field in
Colorado in the United
States blanketed with salt,
a result of poor drainage.

▲ Plate 3b An example of severe soil erosion in the Gambia. Following loss
of protective vegetation cover, water erosion has created deep gullies, making the
land useless for productive purposes.

◄ *Plate 4*
*Masai with flocks of
sheep and goats near Lake
Natron in Kenya. In some
areas desertification is the
end-result of overgrazing
through the keeping of
larger flocks than the area
can support.*

▲ *Plate 5   Terraces in the Philippines: water
management for making flooded rice fields in
upland landscapes.*

▲ *Plate 6   Transplanting rice seedlings into flooded fields
in Yunnan Province, China. Note the earthen 'bunds' or
dykes which ensure that water is adequately distributed in
different parts of the field.*

◀ *Plate 7*
Oger village in the Champagne region of France, showing the prairie-like landscape created by EC policies in the cereal-producing lowlands.

▲ *Plate 8*
Extensive cattle ranching in North America.

▲ Plate 9
Exmoor, looking south-west from Deddy Combe. Coastal heather and gorse is in the foreground, with heather moorland protected by its status as Common Land in the background. In the middle on the right can be seen the 'traditional' in-bye of small fields enclosed by hedgebanks, while larger-scale, post-war moorland reclamation is shown on the left.

◄ Plate 10
Wear and tear along the Pennine Way. Some of the popular parts of this track have become so badly affected by over-use that the surface has had to be artificially covered.

◀ *Plate 11*
The major limestone quarry at Tunstead can dominate visitors' experience of the Peak National Park, even though it is sited just outside the park boundary.

▲ *Plate 12   The Canary Wharf office development, intended as a magnet to draw financial services into London's former Docklands.*

▲ *Plate 13 Bazaar area within the walled city in Lahore, Pakistan. The indigenous design of these historic cities contrasts starkly with the urban environments built by colonial powers.*

▲ *Plate 14   An aerial view of Mexico City, showing the very extensive built-up area and the flat terrain of this former lake-bed.*

◄ *Plate 15*
Air pollution over Mexico City, taken from the Latin American Tower looking east towards Netza, the largest barrio of Mexico City, which itself has a population greater than that of most other Mexican cities.

1   Introduction

As you read this chapter, look out for answers to the following key questions:
- What have been the main changes in temperate agriculture in recent years?
- How have different kinds of temperate agriculture related to their environment?
- How do environmental impacts in the UK differ between extensive and intensive systems?

Temperate agriculture probably exemplifies, as well as any other human activity, the conflict between maintaining economic well-being on the one hand and the quality of the natural environment on the other. Some of the most technologically advanced farming practices are being applied in temperate agriculture; through food trade and aid, the food surpluses generated play a crucial role in sustaining the populations of developed and developing countries alike. But the same farming practices are producing damaging environmental impacts, the scale of which has been fully appreciated only since the early 1980s.

This chapter aims to analyse the changes that have occurred in temperate agriculture in recent years. No part of agriculture has remained untouched, from the inputs purchased for farming, such as fertilisers, machinery and seeds, through the mix of activities and methods of production on each farm, to the ways in which farm produce is processed and distributed between the farmgate and the consumer. We will find that each agricultural change has had its particular environmental impact.

The argument progresses in four stages, each taking us into a more detailed examination of the relationship between agriculture and the environment. Section 2 assembles some of the concepts and empirical evidence from Chapter 2 on farming systems and their locations, to give an overview of temperate agriculture. In particular it develops a conceptualisation of farming as a modified agroecosystem whereby understanding the changing inputs to and outputs from each farming system can help us to understand both the nature of agricultural changes and their environmental impacts.

Section 3 is a detailed examination of recent trends in temperate agriculture, with evidence from the European Union. Temperate farming around the world displays many similar features, often summarised in the term 'the industrialisation of agriculture', involving change in inputs, production trends and marketing of crops and livestock.

The next stage in the argument, in Section 4, is an examination of a number of temperate farming systems to establish the variety of relationships that exist between agriculture and the environment. Limitations of space dictate that just five systems can be selected here for detailed consideration. They are a range of intensive and extensive farming systems, with crop as well as livestock products: extensive arable, intensive arable, extensive grassland (livestock), intensive grassland (livestock) and intensive (non-grassland) livestock. This mixture of 'intensity' with 'land use' forms a common approach in many classifications of farming systems.

Section 5 focuses in detail on two farming systems in the United Kingdom: hill sheep farming and intensive cereal farming. They lie at either end of the environmental impact continuum and illustrate the nature and scale of temperate agriculture's environmental impacts.

2   *Defining and locating temperate farming systems*

In Chapter 2 you met some ecological concepts underlying the
development of farming. In this chapter this idea of agricultural systems
depending on and modifying ecosystems is formalised into the concept of
an agroecosystem.

As you saw in the last chapter in the case of wetland rice, an ecosystem
(environmental unit) can be viewed as an integrated network of energy and
mineral flows in which the major functional components are populations of
plants, animals and micro-organisms. Each system is regulated and
stabilised by the cycling of mineral nutrients for re-use and powered by
energy from the sun. Although a given ecosystem comprises several
thousand species of plants, animals and micro-organisms, each is
continuous to some degree with adjacent systems. (A forest ecosystem, for
example, can merge with an adjacent grassland ecosystem.)

Q   Can you apply the concept of an ecosystem to an agricultural context
    and suggest what might be meant by an **agroecosystem**?

A   An agroecosystem can be defined as an orderly set of interdependent
    and interacting components, or elements, within a defined farming area
    (nation, region, sub-region or farm), none of which can be modified
    without causing a related change elsewhere in the system. Five major
    aspects of any farming system can be identified: its goals (objectives),
    boundaries, activities (components) and their internal relations,
    external relations and the relationship between the internal and
    external relations.

One of the principal differences between natural and managed ecosystems
lies in the simplification of the latter to a relatively few species of crops and
animals. In broad terms, the greater the simplification of the agroecosystem,
the greater the quantity of energy that must be introduced into it, not least
to control pests and diseases, and the greater its potential instability. The
concept of agroecosystems can offer an interdisciplinary approach to
understanding the relationships between agriculture and the environment.
It enables a holistic description of the relationships to be established by
incorporating the socio-economic, political, environmental and
technological elements of each farming system. Equally, the approach
enables the interrelationships between small subsets of components in a
farming system to be isolated for detailed examination.

The practical difficulties of analysing a farming system, especially
measuring the material and energy flows, usually result in farming systems
being defined loosely and used mainly as conceptual devices. Nevertheless,
by focusing on the components of a farming system, and the structure of
flows within it, changes to the farming system can be identified and their
impact on the interacting components evaluated. The impact of changes in
the human subsystem, especially farm inputs, upon the environmental
subsystem is the central concern of this chapter.

Central to a discussion of this subject is the concept of **type of farming**.
Duckham and Masefield's (1971) typology shown in Table 4.1 is very useful
in this respect: although the details have since changed considerably, the
principles remain.

Table 4.1 A typology of farming systems

	Tree crops		Tillage with or without livestock		Alternating tillage with grass, bush or forest		Grassland or grazing of land consistently in 'indigenous' or 'improved' pasture	
	Temperate	Tropical	Temperate	Tropical	Temperate	Tropical	Temperate	Tropical
Very Extensive	Cork collection from Maquis in southern France	Collection from wild trees, e.g. shea butter	—	—	Shifting cultivation in Negev Desert, Israel	Shifting cultivation in Zambia	Reindeer herding in Lapland, Nomadic pastoralism in Afghanistan	Camel-herding in Arabia and Somalia
Extensive	Self-sown or planted blueberries in the north-east of the USA	Self-sown oil palms in West Africa	Cereal-growing in Interior Plains of North America, pampas of South America, in unirrigated areas, e.g. Syria	Unirrigated cereals in central Sudan	—	Shifting cultivation in the more arid parts of Africa	Wool-growing in Australia. **Hill sheep in the UK.** (Sheep in Iceland.) Cattle ranching in the USA.	Nomadic cattle-herding in east and west Africa. Llamas in South America
Semi-Intensive	Cider apple orchards in the UK. Some vineyards in France	Cocoa in West Africa. Coffee in Brazil	Dry cereal farming in Israel or Texas, USA	Continuous cropping in congested areas of Africa. **Rice in south-east Asia**	Cotton or tobacco with livestock in the south-east of the USA. Wheat with leys and sheep in Australia	Shifting cultivation in much of tropical Africa	Upland sheep country in North Island, New Zealand	Cattle and buffaloes in mixed farming in India and Africa
Intensive	Citrus in California or Israel	Rubber in south-east Asia. Tea in India and Sri Lanka	Corn Belt of the USA. **Continuous cereal growing in** the UK	Rice and vegetable growing in south China. Sugar-cane plantations throughout tropics	Irrigated rice and grass beef farms in Australia. Much of the east and south of the UK, the Netherlands, northern France, Denmark, southern Sweden	Experimental stations and scattered settlement schemes	Dairy cattle in parts of the Netherlands, New Zealand and England	Dairy in Kenya and Zimbabwe highlands

Note: Bold type indicates case studies in Chapters 3 and 4.

Source: Duckham, A. N. and Masefield, G. B. (1971) *Farming Systems of the World*, London, Chatto and Windus, p. 16.

Activity 1

Use Duckham and Masefield's typology in Table 4.1 to answer the following questions.

Q   What are the three attributes used to construct the typology?

A   • The input intensity of the system.

    • The dominant land use of the system.

    • The climatic regime within which the system operates.

(Make a note of the farming types that occur under temperate climatic regimes).

Q   What are some of the limitations of the typology for temperate agriculture?

A   The typology omits intensive livestock (pigs and poultry), even though they form an important part of agriculture in many temperate farming systems; it incorporates no direct measure of the level of technology being applied in a farming system; and it contains no indication of the type of social organisation being practised. (By this we mean the mode of production, including the different economic principles by which farming is organised – free market, modified market or central planning.)

The third concept we need to employ is that of the **agricultural region**: that is, the dominant presence of a farming system over an area of space. Figure 2.6 in Chapter 2 shows the set of regions at the world scale proposed by Derwent Whittlesey in 1936 which have withstood the test of time.

Activity 2

Examine Figure 2.6 and revise Chapter 2 to answer the following questions.

Q   What five attributes are used in the typology?

A   (a)   The crop and livestock association (e.g. grain and/or dairy);

    (b)   the methods used to grow crops and produce livestock (e.g. herding or plantations);

    (c)   the intensity of application to the land of labour, capital and organisation, and the output of product which results;

    (d)   the disposal of the products for consumption (used for subsistence on the farm, or sold for cash or other goods – commercial);

    (e)   the ensemble of structures used to house and facilitate farming operations (explicit in small farms and plantations; implicit in, for example, commercial grain or dairy).

Q   Which criteria are found both in Table 4.1 and Figure 2.6?

A   Type of land use, degree of farming intensity and climatic regime.

Q   Which types of agricultural region are not found under the temperate climatic regimes?

A   Types 3 (primitive subsistence), 4 (intensive subsistence) and 5 (plantations).

These classifications imply – and the concept of an agroecosystem explicitly states – that farming systems integrate a variety of social, technical and environmental factors. These can be seen as a hierarchy of constraints. At a general level, features such as soils, climates and topography define the *ecologically feasible* farming systems at any location, from which the farmer can make a choice.

At a second level, human or infrastructure factors are influential, such as farm size, available capital, population density, distance from market, market prices and state intervention in agriculture. Together these features define the *economically feasible* enterprises at any location from which a farmer can make a choice. Such a choice includes artificially extending the range of ecologically possible farming systems by manipulating the agroecosystem; this can be achieved through such practices as irrigation, housing for livestock and glasshouse cultivation.

A third level of constraint is provided by *operational considerations* in farming: for example, steep slopes or small units might constrain the size of machines to be used, seasonality of rainfall might require particular conservation practices or the best seed varieties may be unavailable in a particular area.

A final constraint on the farming system is provided by the *personal behaviour* and *preferences* of individual farmers, for example in relation to features such as profit maximisation versus conservation, the level of information on farming methods or commitment to a particular breed of livestock.

This hierarchy of constraints can explain the choice of farming systems at most locations. As we will see later in this chapter, when profit-maximising behaviour in a farming system is pursued to its logical conclusion, and when governments intervene in agriculture to stimulate production, the consequences for the environment can be damaging, often in unpredictable ways.

3   *The dynamics of temperate agriculture*

3.1   *Introduction*

Before turning to a selection of case studies, we need to gain a more detailed understanding of the general changes that have been taking place within most temperate farming systems. It is the impact of these changes on the rural environment that so concerns society at the present time. To focus the discussion, attention is restricted to capitalist farming systems, thereby excluding temperate agriculture in the countries of the former socialist bloc of eastern Europe and the Soviet Union and in China. Agricultural production in eastern European communist countries shared its basic features with western European agriculture – the industrialisation of agriculture and the effort to produce large quantities of food. However, its damaging environmental effects were even more severe. This was due mainly to two reasons:

(a)   centralisation of agricultural production both at local level (forced collectivisation) and at state level, meaning that centrally made decisions were implemented with no regard to varying local conditions and without any possibility of changing them;

(b) exclusion of economic constraints from the effort to achieve large amounts of production – the plan had to be fulfilled at any costs.

Agricultural models varied from country to country, but this was the prevailing form. The collapse of communism led to adoption of exactly the same model of agricultural production as exists in west European countries. Its effects were social (laying off of a large part of the workforce) as well as environmental (reduction of its negative impact, at least in some areas).

To assist in organising the material in the discussion that follows, we will first consider the inputs to temperate agriculture, then the production process on farms, and finally the marketing of crops and livestock through the food chain to the consumer. Illustrative data are provided on temperate agriculture within the European Union (EU), noting that the EU has grown from six to fifteen countries during the period considered.

3.2   *Inputs to temperate farming systems*

Conventionally, inputs to agriculture, or **factors of production**, are classified under the main headings: land, labour and capital. The term 'land' is used to denote the sum of the components of the natural environment (climate, soil and topography) and is often used in the context of the quality of the physical resources available to farming. Most countries, for example, employ land capability maps on which land is classified into grades according to the physical limitations imposed on agriculture. In the United Kingdom, for instance, only 1.8% of agricultural land falls into the Grade I (best) category, with 33.7% in Grade V (poorest). The quality of farmland can be modified only with the expenditure of capital and energy on items such as organic (animal and green manures) and inorganic (chemical) fertilisers, drainage and irrigation. The quantity of farmland is another input for farming systems. In most countries farmland has been a declining input as land has been transferred to alternative uses such as forestry, urban development and roads, and water catchment. But, at the same time, new farmland has been created by draining wetlands, ploughing moorlands and felling woodland. Each of these strategies impinges on natural ecosystems and the total farmland input is in a constant state of flux. In the EU, for instance, there has been a net loss of farmland in the order of 0.3% per annum for the last decade, although this proportion would have been higher but for the considerable areas of semi-natural habitats that have been 'improved' for agriculture to offset the losses for urban and industrial development.

The term 'labour' includes the managers, owners and tenants of farmland, together with their families, hired workers and those who supply services on contract for drainage, ploughing, harvesting and crop-spraying operations. Like the farmland base, labour has formed a declining input into temperate agriculture for many decades. For example, the twelve-nation European Community (as it was until recently) experienced an average annual reduction of around 3% in 'persons mainly employed in agriculture' over the last two decades. In terms of a fall in actual numbers, between 1970 and 1991, 7.9 million were lost to agriculture from an initial total of 16.3 million. We need to be cautious in accepting these figures at face value; there are real problems in determining who is employed in agriculture, not least because the labour input can be casual, seasonal, spare-time, part-time or full-time in nature. Indeed the proportion of farm-owners with more than one occupation is increasing in most temperate farming systems: in the EU, 30% of farmers also have gainful occupations outside agriculture.

Considerable variation exists in the rate of change amongst the different categories of agricultural labour. For example, until recently the number of full-time, hired farm-workers was falling at a faster rate than the number of owners and tenants of farms. At present hired workers are in the minority and most farms no longer employ full-time workers. Within the 'hired' category, though, full-time workers have been in more rapid decline than seasonal, casual and part-time workers, especially women.

The trends in land and labour are closely related to the third category of input – capital. Capital is now by far the most significant factor of production since it can substitute for both land and labour. On land, capital can be used to 'purchase' additional hectares through inputs of animal feed. For example, in 1987, in the then EC 5.7 million tonnes of cereals were imported to manufacture animal feed; this is equivalent to more than a million hectares of farmland, assuming an average European cereals yield of 5 tonnes per hectare. In addition 21.8 million tonnes of cereal substitutes, such as manioc, and 29.5 million tonnes of protein-rich products, such as soya cake, were imported for animal feed. The productivity of the land resource can be also raised by the application of capital to other inputs such as chemical fertilisers, genetic materials in the form of improved seed varieties and livestock breeds,

Table 4.2 Total consumption of inorganic fertilisers[1] in selected countries, 1956–1985 (000 tonnes; kg/ha in brackets)

Year	West Germany	France	Netherlands	United Kingdom
1956	2114 (148)	1924 (56)	468 (201)	– (–)
1965	2897 (209)	3123 (93)	566 (250)	1555 (79)
1975	3300 (251)	4850 (152)	638 (306)	1800 (95)
1985	3185 (265)	5694 (181)	701 (346)	2524 (135)

Note: [1]Nitrogen, phosphate and potash.

Source: Author's calculations from agricultural statistics (*Eurostat* 1960, 1970, 1988, Brussels).

◀ *Agricultural workers in East Anglia. Although the number of full-time farm workers has declined dramatically, casual, often female, labour is still used on a seasonal basis.*

and agrochemicals such as pesticides, herbicides and fungicides. The volume
of all of these inputs has been rising in temperate agriculture for several
decades. In the case of fertilisers, for example, both the total volume and the
average rate of application per hectare of agricultural land have increased,
albeit at different rates between farming systems, although recently there are
signs of application levels being reduced (Table 4.2). As we will see in more
detail later, excessive applications of fertilisers can have damaging
consequences for groundwater supplies as well as surface water in streams,
rivers and lakes. In the case of agrochemicals, increasing concern has been
expressed about their impact on food health (Figure 4.1).

Activity 3

You can monitor the national and local press to build up your own file
of cases where attention is drawn to the impact on food health of the
use of agrochemicals in agriculture. You should divide the material
between criticisms and defence of the use of agrochemicals.
  Read the extracts in Figure 4.1 taken from newspapers in 1989 and
1990 and then answer the following questions.

Q   What national government departments and agencies in the United
    States and the United Kingdom are mentioned as having a role to play
    in the control of agrochemicals?

A   United States: Environmental Protection Agency. United Kingdom:
    Ministry of Agriculture Fisheries and Food (MAFF); MAFF Advisory
    Committee on Pesticides; National Rivers Authority (NRA);
    Department of the Environment.

Q   Which environmental organisations and pressure groups are
    mentioned as having drawn attention to the dangers of using
    agrochemicals?

A   Natural Resources Defence Council; Friends of the Earth; World Health
    Organisation; Parents for Safe Food.

Turning now to substitutes for the labour input, capital can be used to purchase
a wide range of farm plant and machinery, as well as petroleum for the motive
power. From a position only fifty-five years ago when farming operations
employed mainly labour and horsepower, today almost all are mechanised
to some extent; no temperate farming system has been exempt from the
trend. Modern machinery can plant and harvest most crops, automatically
deliver feed and water to livestock, and wash, grade and process farm
produce. But the mechanisation of agriculture has led to its own environmental
impacts. For example, the use of heavy machinery for successive farming
operations, such as ploughing, seeding, fertilising and harvesting, has
caused soil compaction in some regions; in addition, mechanised trimming
has had a damaging impact on trees and shrubs in hedgerows.
    When the capital needed to purchase increasingly expensive farmland
is added to the costs of fertilisers, machinery and agrochemicals, it is
understandable why temperate agriculture can be described as 'capital-
intensive'. Figure 4.2 shows the full range of inputs for EC farming systems
in 1987, with data for the United Kingdom included for comparison.
Animal feed is by far the most important purchased input (41% of the total
by value), followed by fertilisers (12%) and farm machinery (12%). By
comparison, fossil fuel (petroleum – 10%) and agrochemicals (5.3%) are not

Crop spray link to cancer

James Erlichman, Consumer Affairs Correspondent

Makers of fungicides linked with cancer have drastically reduced their use in the United States, but have not ordered cuts in Britain.

US farmers have been told by Rohm & Haas and three other manufacturers to stop spraying the fungicides on most of their crops. The blacklist, issued last month, covers 70 out of 83 previously permitted crops including apples, beans, cabbage, cherries, lettuces, melons, oats and strawberries.

The fungicides have been widely used in Britain for nearly 40 years on a similar range of crops but no instructions to curb their use have been issued either by the manufactuers or by the Ministry of Agriculture.

Rohm & Haas spokesman, Mr George Bochanski, defended the company's failure to alert British farmers: "We continue to believe that these products are safe and it is not our job to pre-empt the regulatory bodies in the UK."

The company says that residues found in British foods are low enough to meet any restrictions imposed by the US agency.

The Ministry of Agriculture said yesterday that it knew of the US crop withdrawals. Its Advisory Committee on Pesticides is reviewing new data in 'the area of consumer risk'.

The voluntary cuts in the US follow threats by the Environmental Protection Agency to ban or severely restrict the use of mancozeb, maneb and zineb, fungicides in the ethylene bis-dithiocarbamate (EBDC) family.

The US agency believes that continued use of the EBDCs may cause at least 125,000 additional cancer cases among the population of 250 million. Children are believed to be most at risk because of their high fruit consumption, low weight, and the length of time they have to develop cancer.

Scientists are most concerned about ETU, a breakdown chemical found in the fungicides which causes cancer in animals and accumulates most in heat treated foods like tomato paste, ketchup, and apple juice.

Apple spray sales halted

Alar, the apple spray linked to cancer, was withdrawn from sale for food use by its maker, Uniroyal, yesterday after disclosure of more tumours in animal tests.

The Ministry of Agriculture, which had repeatedly said the spray was safe, will now have to revoke the product's licence, most likely tomorrow when the Advisory Committee on Pesticides meets.

Mr Walter Waldrup, at the US Environmental Protection Agency, said yesterday that lung tumours were found in a "low dose" study of mice which was received from Uniroyal at the beginning of October. Liver tumours were found in rats.

Mr Malcolm Tyrell, Uniroyal's European sales manager, said the company could no longer offer "a totally clear data package" and had formally asked the US authorities to revoke Alar's licence.

Concern about Alar and its use in Britain was first disclosed by the Guardian last May following a report from the Natural Resources Defence Council, a US environmental group. It predicted that 5,500 American children would eventually develop cancer directly from Alar-sprayed apples.

Pressure from the NRDC prompted the US Environmental Protection Agency to demand new laboratory studies from Uniroyal. These tests showed "an inescapable and direct correlation" between the use of Alar and "development of life-threatening tumours". The US authorities then instigated a ban on Alar from next year.

But the Ministry of Agriculture, after reviewing the same studies, gave Alar the all clear and refused to halt sales. In June, Uniroyal withdrew Alar from the US market but refused demands from Parents for Safe Food, the pressure group formed by the actress, Ms Pamela Stephenson, to withdraw it in Britain.

Figures compiled in 1983 suggested that about 100 million apples, or about 7 per cent of the British crop, were sprayed with Alar.

Forty-year fertiliser legacy comes home to roost

Nitrates

Excessive amounts of nitrate, a mineral form of the nitrogen naturally present in soil, are toxic, according to the World Health Organisation, whose guideline is 100 milligrams a litre.

Nitrate is converted in the body to nitrite which combines with haemoglobin in the blood to reduce the uptake of oxygen, putting young babies at potential risk of blue baby syndrome. Cases are rare in Britain and water authorities with high nitrate levels have provided bottled water for babies.

Studies in other countries suggest babies could suffer oxygen deficiency without showing symptoms from concentrations around the EEC limit. Friends of the Earth says this should be urgently investigated.

Nitrite is also converted into substances which cause cancers in laboratory animals but there is no clear link with gastric cancer in humans.

Throughout much of east and central England, 40 years of increasingly intensive farming has sent a steady trickle of nitrates down into underground water sources which provide 30–40 per cent of local needs.

When the drinking water directive came into force in 1985, around 50 supplies contained more than the permissible maximum concentration of 50 milligrams a litre. The Government freed suppliers from the obligation to observe these limits on the grounds that there was no health danger with a concentration below 100 mg/l.

A complaint by Friends of the Earth to the Common Market Brussels Commission brought a threat of legal action and last year, the Government withdrew these waivers, or derogations. It was also forced by the EEC to alter the method of testing compliance with the law from an average level over three months to a single result. As a result the number of people receiving supplies over the limit rose from a million to 4 million overnight.

The authorities mainly involved, Severn Trent, and Anglian, are working to meet the limit by closing contaminated sources and blending water.

They have sent their clean-up programmes to the Department of the Environment which will soon announce its strategy for complying with the EEC directive.

Cleaning up supplies may take several years. The main problem, however, is to halt the rise of nitrate levels underground, and in rivers where concentrations are expected to start reaching the limits over the next 10 years.

The Water Bill will allow water authorities and the National Rivers Authority to limit or forbid the use of nitrogenous fertilisers in protection zones around boreholes. Compensation for farmers has been agreed in principle.

▲ *Figure 4.1   The impact of agrochemicals on food and water quality, as reflected in news stories, 1989–90.*

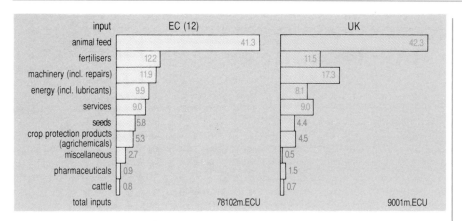

input	EC (12)	UK
animal feed	41.3	42.3
fertilisers	12.2	11.5
machinery (incl. repairs)	11.9	17.3
energy (incl. lubricants)	9.9	8.1
services	9.0	9.0
seeds	5.8	4.4
crop protection products (agrichemicals)	5.3	4.5
miscellaneous	2.7	0.5
pharmaceuticals	0.9	1.5
cattle	0.8	0.7
total inputs	78102m.ECU	9001m.ECU

▲ Figure 4.2   Inputs to UK and European Community farming systems, 1987
(percentage of total inputs).

major inputs by value for the farming systems of the EC, although when
converted into energy equivalents (see Section 4.3 below), high inputs per
hectare and per labour unit are revealed (Simmons, 1989, pp. 239–55). All
these inputs have their origins outside agriculture in the industrial sector of
the economy: as labour has been displaced from farming, so it has been
gained by those industries supplying the inputs. At the same time farming
systems have become interrelated by creating inputs for each other, such as
animal feed for livestock for fattening, and they in turn have become part of
the larger industrial economic system. The most significant feature for the
environment, however, is the increasing volume of purchased inputs into
temperate agriculture and the energy flows that they represent. By 1991,
EC(12) inputs were 87 431 ECU and for the UK, 10 302 ECU.

3.3   Production trends in temperate agriculture

The upward trend in purchased inputs at least until the end of the 1980s,
both per hectare and per person, employed in agriculture, has produced an
increased level of **intensification** in most farming systems. The degree of
intensification still varies from system to system, being most developed in
pig, poultry and horticultural production, and least in grassland systems of
beef and sheep farming, especially in uplands and mountains. There is a
direct relationship between inputs and outputs per hectare from farmland;
consequently the productivity (yield per hectare) of most temperate
farming systems also shows an upward trend. This can be measured in
terms of the per hectare yield of a crop, the density of livestock per hectare,
or the annual output of meat or livestock product (eggs, milk) per animal.
Again using data for the EC, the upward trend in the intensification of
agriculture until recently can be demonstrated through crop yields. There is
naturally some variation in crop yields from year to year, but Figure 4.3
illustrates an overall trend of a general rise in the yield of most crops: low
for oats, rye and barley, quite large for wheat and maize and very large for
soya beans. Taken together with changes in the allocation of land between
different crops and livestock, increased yields have raised farm production
to the greatest extent for fats and oils, wheat and poultrymeat, although
decreases in production have been recorded by a few products, especially
potatoes, fresh fruit and vegetables: see Table 4.3.

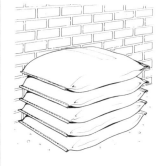

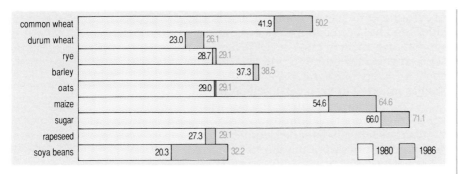

◄ Figure 4.3
Yields of selected crops in the European Community (12), 1980–86 (100 kilogram/hectare).

Table 4.3   The changing volume of agricultural production in the European Community (9), 1992–95 as percentage of 1971–72

Larger increases		Smaller increases		Decreases	
Vegetable fats and oils	197	Sugar (beet)	138	Vegetables	96
Poultrymeat	167	Maize	120	Fresh fruit	93
Wheat (soft and hard)	166	Eggs	113	Potatoes	91
Meat	148	Barley	104		
Wine	143	Milk (raw)	103		

Source: Calculations from agricultural statistics (*Eurostat*, 1976, 1988, 1995, Brussels).

◄ Incorporation of organic material from animal slurry helps to maintain soil structure.

Not surprisingly, many of these production trends have had an adverse impact on the environment. Looking just at one example, in the Netherlands the number of pigs and dairy cows have increased to such an extent that in some regions disposing of their urine and dung now poses a major environmental hazard. The quantity of manure is so great that the soil and field plants cannot absorb all the available nutrients; rainfall washes the nutrients from the manure when it is spread on the fields, resulting in accumulations in soil and water bodies. While some nutrients may be toxic in high concentration in drinking-water, all tend to promote algal growth and hence reduce oxygen in water, sometimes killing fish and plants. These problems have resulted in controls being placed on the quantity, frequency and timing of manure spreading in affected regions of the Netherlands.

While the output of temperate agriculture has been increasing until recently, the domestic demand for food has been rising at a slower rate. This reflects the low level of population growth in most developed countries, and also the tendency to spend a lower proportion of income on food as incomes rise. The outcome for temperate agriculture has been a rising level of surplus production as reflected in the level of self-sufficiency in food production. Table 4.4 demonstrates the problem for the EC, with overproduction for the domestic market evident in all products, but especially milk powder, sugar and cereals. In a free market, equilibrium would be restored through lower product prices, but this has not occurred in temperate agriculture for two reasons. First, individual farmers are locked into a technological 'treadmill'. That is, as new output-increasing technology is developed, innovative farmers adopt that technology to reduce their unit costs of production. For a time they gain an economic advantage over non-adopting farmers. But gradually other farmers are forced into adopting the same technology so as to remain competitive and in business. Eventually output is increased for all those farmers who have not become marginalised and a further round of adopting new technology ensues. Farmers wishing to remain in business have little option but to remain on this technological treadmill.

Secondly, most states have intervened in agriculture to maintain the prices of farm produce which otherwise would be driven downwards in an oversupplied market. (This intervention takes place for a variety of reasons which will be outlined in Chapter 5.) A wide range of support measures has been used in temperate agriculture from import levies, guaranteed prices and support buying by intervention agencies, through direct income supplements and headage subsidies on cattle and sheep, to financial subsidies on the purchase of fertilisers, farm buildings and machinery. Only recently have some governments and the EU turned to reducing price support levels and supporting 'environmentally sensitive' farming practices. By subsidising production, governments in all developed countries have effectively encouraged and supported the increased intensification of agriculture with damaging environmental consequences.

Table 4.4 Self-sufficiency in the European Community, 1975–1993 (% total supply)

Product	1975 (EC = 9)	1980 (EC = 10)	1990 (EC = 12)	1993 (EC = 12)
Wheat (soft and hard)	100	123	126	125
Barley	103	112	124	125
Potatoes	98	100	102	103
Sugar	105	135	129	135
Wine	98	104	105	106
Milk:				
whole milk powder	–	378	334	211
skimmed milk powder	171	128	118	124
Cheese	104	106	107	106
Butter	97	119	133	104
Eggs	100	102	102	102
Beef	101	103	108	104
Pigmeat	99	102	102	102
Poultrymeat	101	110	107	104

Source: Commission of the European Community, The Agricultural Situation in the Community, 1978, 1988 and 1995 Reports, Brussels.

Activity 4

Environmentally Sensitive Areas (ESA) in Britain attempt to relieve the pressures on agriculture for further intensification and specialisation.

Read the extracts from the Countryside Commission's publications in Figure 4.4 about the early days of the scheme and answer the following questions.

Q   How many ESAs had been designated in the UK by December 1987 and where were they located?

A promising start

The Government is to designate six areas of England and Wales as the country's first environmentally sensitive areas (ESAs). It has accepted in full the Commission's proposed boundaries for the Broads, Pennine Dales, Somerset Levels and West Penwith, plus the eastern end of the South Downs and northern and southern sections of the Cambrian Mountains.

The Ministry of Agriculture says it hopes that schemes for incentive payments to farmers will be set up early next year, once the formal designation order has received parliamentary approval.' Meanwhile the management prescriptions for each area and the levels of payment to be offered are still under discussion between ministry officials and farmers' representatives.

Welcoming the Government's move to protect these landscapes threatened by agricultural change. Sir Derek Barber, Chairman of the Commission, said that the chosen areas would provide good experience for the future development of ESAs. He hoped farmers would take up the payments offered in return for joining the management schemes.

"I suspect that the chances of further areas being designated may well depend on the success of these six in attracting farmers' support," he added.

In April, the Commission submitted to ministers a list of 14 'priority' areas which it felt met the criteria for ESA designation, together with management prescriptions and suggestions for the levels of payment appropriate to each. Other areas on that list were Anglesey, Breckland, Clun, the Lleyn Peninsula, North Peak, Radnor, Suffolk River Valleys and the Test Valley in Hampshire.

New ESA options

Cereal farmers in East Anglia and the South Downs are being encouraged to convert arable land back to heathland or grassland, in the new Environmentally Sensitive Area (ESA) schemes announced by the Ministry of Agriculture at the end of November.

Annual payments from £100 in Breckland to £200 in the Suffolk River Valleys are being offered for every hectare of cereal land which reverts.

Breckland farmers also have the option of payments of £300 per hectare to leave six-metre strips of uncropped land at the edge of arable fields, to allow natural regeneration and encourage wildlife. Another option, in return for £100 per hectare, is to limit spraying at the cropped edges of arable fields, creating 'conservation headlands' – an initiative particularly welcomed by the Game Conservancy.

Altogether there are six additions and two extensions to the ESA list in England and Wales. To the six designated in December 1986, are added Breckland, North Peak, Suffolk River Valleys, Test Valley, Shropshire Borders and the Lleyn Peninsula. Extensions have been made to the South Downs and Cambrian Mountains ESAs.

The North Peak scheme is the first to encourage heather moorland conservation. Payments of £10 to £20 per hectare are available to sheep farmers carrying out appropriate moorland management in this area. The higher rate of grant requires the regeneration of at least one hectare of moorland per year.

There are now 1,290 square miles covered by ESA schemes in England, and 560 square miles in Wales.

In addition, Scotland now has five ESAs – Breadalbane, Loch Lomond, the Machair of the Uists and Benbecula, Whitlaw/Eildon and Stewartry – covering 865 square miles. Northern Ireland has two: Mourne and Slieve Croob and the Glens of Antrim, 185 square miles in total.

A   19. (12 in England and Wales, 5 in Scotland and 2 in Northern Ireland.)
      They were widely distributed to cover a range of farming systems and
      landscape types of special merit. (See Figure 5.6 for a map locating the
      ESAs.)

Q   What protective measures for the environment are offered within the
      boundaries of the ESAs?

A   Payments to subsidise approved (traditional) farming practices;
      management schemes under which farmers agree to abide by approved
      farming practices; conversion payments for placing arable land into
      grassland or heathland; conservation headland payments. (Chapter 5
      discusses the development of the ESA concept through to the mid-1990s.)

At the same time as increasing their inputs to agriculture, farmers have
gained economic benefits from simplifying their production systems. The
term **specialisation** can be applied to the process whereby farmers focus their
resources of land, labour and capital on an increasingly narrow range of
crops and livestock. Usually the least profitable enterprises are discarded to
enable economies of scale to be achieved in producing the remaining
products, for example, just cereals with oilseeds or beef with sheep. It is
increasingly common for only one product to be the outcome of a farming
system, for instance pigs, poultry or milk. Of course, new strains of crops and
livestock are always being introduced and these diffuse as an innovation
through a farming system. The spread of oilseed rape and maize through the
United Kingdom in the 1970s and 1980s are two well-known examples. Also,
the economic fortunes of different products change, through alterations in
either domestic demand, supply conditions in the international market, or
government policies. In these cases the enterprise balance in a farming
system, for example between beef and sheep, can be altered. These processes,
therefore, introduce elements of **diversification** into temperate farming
systems that otherwise show increasing specialisation. The trend towards
diversification is gathering pace in the late 1980s as governments try to scale
down the very high costs of supporting the food surpluses from existing
farming systems. In the EU, for example, farmers are now being offered
incentives to withdraw land from the production of crops in surplus ('set-
aside') and to develop a range of farm activities which are less damaging to
the environment. (Chapter 5 discusses these developments.)

An indication of the level of specialisation in agriculture within the EC
can be obtained through the proportion of national agricultural production
accounted for by individual farm products: this is illustrated by Figure 4.5.
In the Netherlands, for example, a high level of specialisation is evident in
milk, pigmeat and vegetables: together these products account for over half
of national farm output by value. In the United Kingdom, 35% of national
farm output comes from milk and beef/veal; lower levels of specialisation
exist in other countries. In Greece, a more recent member of the EU, only
one category – fresh vegetables – exceeds 10% of the output from a
relatively diversified agricultural sector.

As has been noted in Chapters 2 and 3, the trend towards increased
specialisation reduces the stability of agroecosystems. For example, the
genetic variety within each crop and livestock type has been reduced,
thereby limiting the capacity of farming systems to withstand changes in
climate (weather) or attack from pests and diseases. The specialised systems
require increased quantities of pesticides and fungicides, with increased
danger of pollution by residues.

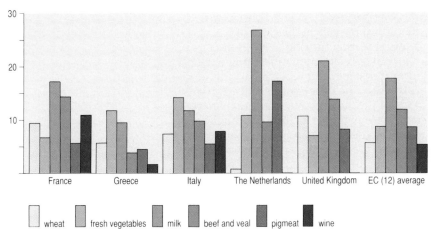

◀ Figure 4.5
Agricultural production in
selected countries and
products, 1987 (% total
value of production). In
general terms the levels of
specialisation have been
maintained into the 1990s.

Chart x-axis labels: France   Greece   Italy   The Netherlands   United Kingdom   EC (12) average

Legend: □ wheat   ■ fresh vegetables   ■ milk   ■ beef and veal   ■ pigmeat   ■ wine

Government subsidy of farm prices has not prevented their real value from being eroded through time by price inflation. Indeed governments have allowed this erosion to accelerate in recent years as a way of reducing the real cost of supporting farm prices. Input costs, on the other hand, have continued to increase, in some cases in real terms and in others at the rate of inflation. Consequently, farmers have been caught in a 'price–cost squeeze' and some have been forced out of farming by bankruptcy. Indeed given the number of family farms, the massive reduction of the labour input could not have been achieved without this process. Under capitalism, small farms are those most likely to be bankrupted, with their land purchased by the remaining, larger farm businesses. The costs of the latter are thereby spread over more units of production, while individual farm families are able to accumulate wealth. Through time, therefore, land becomes owned by fewer, larger farm businesses, a process described generally as **concentration** in agriculture. In some farming systems polarity is evident in the farm-size structure: small numbers of large farms control most of the land resource, while a large number of small farm units occupy the remaining farmland. This degree of concentration is present in the farm-size structure of some regions in the United States, for example California, and in western Europe, for instance the Highlands of Scotland, southern Italy and southern Spain. Elsewhere, however, more medium-sized farms have been maintained despite the increasing dominance of large farms in the control of the land resource. One consequence for the environment of this process of concentration lies in the environmental attitudes of those who occupy the larger farming units. It has been found that occupiers of large farms tend to have less sympathetic attitudes towards the environment and the need for active measures on conservation compared with small farmers.

An increasing concentration is evident in the structure of farm production as well as land occupancy. As farms grow larger and more specialised, so the proportion of output controlled by a few farms increases. We can measure this degree of concentration by the proportion of production controlled by the largest producers. Concentration is greatest for farming systems concerned with intensive livestock (pigs and poultry), and least for extensive farmland with beef, cattle and sheep. In the case of pig production in the EU, for example, the largest 5% of herds controls 63% of all animals. For beef production, the largest 20% of herds contains 61% of all cattle, while in dairying a similar percentage of herds accounts for 50% of milk cows.

When large groupings of farms are considered, we find variations between farming systems and agricultural regions in the rates of intensification, specialisation and concentration (Bowler, 1992). One outcome has been the increased regional specialisation of farming through which differences between agricultural systems and regions have become drawn more rather than less sharply. Also, the processes described in this discussion have introduced methods of production into farming that are increasingly industrial in character. These parallels include the purchase of inputs from outside the farming system, specialisation in the labour and production function, mechanised methods of farming and, as we will now see, the onward transfer of farm produce for specialised processing and packaging. Not surprisingly the term **industrialised farming** is now widely applied to temperate agriculture, although it is a term more appropriate to intensive (pigs, poultry, dairy, cereals, horticulture) rather than extensive farming systems.

3.4   Marketing the outputs of temperate agriculture

It is no longer useful to think of agricultural systems as ending at the farm gate. The farm, as a production unit, is now part of a much broader economic system, sometimes referred to as the **food chain*** , which includes the processing and distribution of food as well as the manufacture of purchased inputs. Only a small proportion of farm production now reaches the consumer without some form of value-added processing, even if this takes the form only of washing, grading and packing the produce. Moreover, the marketing channels whereby food reaches the consumer have become increasingly complex. In temperate agriculture, some farm produce still passes initially through either state-financed intervention boards or producer-controlled marketing boards, both of which act to keep prices artificially high. The Intervention Board for Agricultural Produce, for example, acts in the United Kingdom as the agent of the EU: the Board purchases, stores and subsequently markets surplus farm products, often at considerable cost to taxpayers in the Union. The normal channels of food distribution from the farm, however, are shown in Figure 4.6. The figures are for 1979, since when the relative levels of expenditure will have changed, but the principles remain the same.

*This is not the same usage as in ecology (see, for example, *Silvertown*, 1996a and page 86 above).

Activity 5

Examine the information on the UK food chain in Figure 4.6 and answer the following questions.

Q   What proportion of the value of food (domestically produced and imported) passes through the food manufacturing sector?

A   54.5% ([£3 100m + £2 900m ÷ £5 700m + £5 300] × 100): this demonstrates the key role played by the manufacturing sector in the food chain.

Q   After food manufacturing ([£6 000m ÷ £12 000m] × 100 = 50%), which sector in the food chain adds the greatest proportional value to its inputs?

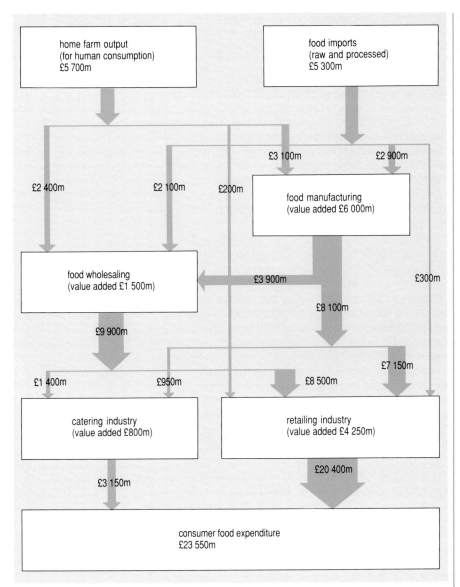

Figure 4.6
The UK food chain (1979 prices).

A   The catering industry (25%); food wholesaling is 15%, and the retailing industry is 21%. The catering industry (restaurant, hotel and institutional trade) serves a fast-growing sector of consumer food expenditure.

Usually food-processing firms, amongst which there is considerable concentration in most countries, hold back from direct ownership of farms, but assure their supplies by placing forward contracts with individual farmers. For example, within France nearly 50% of poultrymeat and 93% of vegetables are marketed under contract and figures for the Netherlands are 90% for each of these products. Forward contracts allow the processor to control the price, quantity and quality of produce delivered from the farm. This degree of control can be so rigid that the farmer becomes little more than a 'landed labourer': the farmer supplies farmland and labour, while capital, materials, farming operations and marketing are controlled by the processor. In these circumstances, the environmental impact of agriculture is determined not so much by the farmer as by decisions taken by the managers of food-processing industries.

Contract farming has been an important catalyst in the development of co-operative marketing groups in some farming systems. By grouping together to offer a product of assured uniform quality and quantity, producers are sometimes able to negotiate more favourable prices with processors. This form of marketing varies in importance between farming systems and countries, for it is culturally as well as economically defined. Pig, vegetable and dairy producers in Denmark, pig producers in France and vegetable producers in the Netherlands, for example, have been particularly amenable to marketing their products co-operatively. Much greater resistance to this form of co-operative behaviour is evident in countries such as Italy and the United Kingdom.

The retailing sector, with its large multiple retailers, is increasingly able to determine the price, quantity and quality of food purchased from the manufacturing and wholesaling sectors or directly from larger farms. The catering industry, especially restaurant and fast-food chains, is also increasing its sales and becoming a significant element in how food reaches the consumer. With all of these developments, the price of food is determined more by the value added beyond the farm gate than on the farm itself (look again at Figure 4.6). This observation has a broader implication. If we become concerned with the efficiency with which temperate farming systems use resources such as energy, then that use needs to be placed within the context of the energy consumed by the whole food chain, including packaging, transportation and distribution. Pimentel and Pimentel (1979), for example, show that only 45% of the energy needed to produce a loaf of bread is consumed on the farm; the figure for a can of sweet corn is 15%. On the other hand, for products where there is little processing, most energy is consumed in farm production – 98% in the case of beef. Concern with the energy-intensive nature of temperate agriculture needs to be extended beyond the farm down the whole food chain.

3.5   Summary

Temperate agriculture is characterised as being energy and capital intensive, with decreasing inputs of land and labour. Individual farming systems show evidence of increasing intensification, specialisation and concentration, although these trends vary within agriculture. Nevertheless, one common outcome of the industrialisation of agriculture is the merging of farming systems with larger industrial systems as regards both purchased inputs and the processing of food outputs. In this way energy and nutrients enter temperate farming systems from the industrial sector, while losses occur when produce (crops, livestock and livestock products) leaves the farm gate to pass down the food chain to the consumer.

Within the food chain, decisions reached on the marketing of food in the retail and manufacturing sectors are passed back to individual farms: these decisions influence the volume, variety and methods of food production in temperate agriculture, with direct consequences for the environment. Not all these impacts need be negative. Since the late 1980s, for example, consumers have expressed a demand for 'organically produced' food, that is food produced without the use of chemical fertilisers, agrochemicals such as pesticides and added hormones in livestock. Large multiple retailers, realising the market potential of this demand, are placing contracts for 'organic' farm produce and thereby stimulating a more environmentally sensitive from of agriculture. Nevertheless, the environmentally damaging consequences of modern temperate agriculture tend to predominate and the next section looks at a number of temperate farming systems to examine these impacts in more detail.

Activity 6

One way of focusing your attention onto the environmental impacts of temperate agriculture is to speculate on the form that an 'environmentally friendly' farming system would take. List the features you would favour in such a system:

- What would be the pattern of inputs?
- What farming practices would be included/excluded?
- What energy flows would take place within your model farming system?
- What outputs would be produced?
- What would be the environmental impacts of your system?

4   Case studies of temperate farming systems

4.1   Introduction

The previous section looked at some of the general trends to be found in temperate agriculture and began the task of identifying their associated environmental impacts. However, Section 2 drew our attention to the wide variety of farming systems within temperate agriculture and it would be surprising if evironmental impacts, as well as their severity, were similar in all the systems. In this section, therefore, we examine five temperate farming systems in greater detail: firstly, to expose the varying agriculture– environment relationships to be found in each system; secondly, to estimate the relative seriousness of the impacts; and, thirdly, to begin an assessment of the need for remedial action. The five case studies are: extensive arable farming; intensive arable farming; extensive grassland farming; intensive grassland farming; and intensive livestock farming. These farming systems range over a variety of crop and livestock types, include both intensive and extensive farming practices, and cover different types of physical environment. Attention is focused on the degree of ecosystem manipulation that each represents.

4.2   Extensive arable farming systems

Extensive arable (crop) farming takes place mainly at the *extensive margins of cultivation*, where commercial crops are at their ecological limits because of deficiencies in rainfall or temperature (and therefore growing season) or both factors acting together. Historically the agroecosystem had been simplified to continuous cereal cultivation, mainly wheat, punctuated by fallows of between one and three years' duration depending on soil moisture conditions. But in recent decades rotational cropping has been introduced, including grain, sorghum and grass leys for beef cattle and sheep. Despite the adoption of high-yielding crop varieties and the periodic application of chemical fertilisers, outputs per hectare remain characteristically low and variable, whereas farming operations are highly mechanised so as to produce high crop yields per worker. In order to maintain the economic viability of farm businesses, farm sizes must be relatively large with production costs per hectare kept to a minimum. One farm practice in North America, for example, is to rely on peripatetic contractors to harvest the cereal crop, rather than having to employ farm labour all year and own the harvesting equipment. The following are examples of large-scale, specialised grain-farming regions: the spring wheat region of the Dakotas and eastern Montana, together with the adjacent part of the Canadian prairies; the winter wheat belt of Kansas, Colorado and Oklahoma; the Palouse country in Washington; the Cordoba–Santa Fé–Bahia Blanca region of Argentina; the spring wheat region of western Siberia; and the Victoria–South Australia–Queensland wheat belt of Australia (type 7 on Figure 2.6).

With their location at the semi-arid margins of cultivation, extensive arable farming systems are very vulnerable to climatic and economic fluctuations. Taking the Great Plains of the United States as an example, 'boom' periods of cereal farming can be identified in the initial period of permanent settlement following the 1862 Homestead Act, the first two decades of the twentieth century, the years immediately following the Second World War and, more briefly, the late 1960s. Conversely, 'bust' periods have been experienced in the 1890s, 1930s, 1950s and 1980s. However, it is not easy to distinguish between the effects of climatic and economic fluctuations on some of these cyclical movements. The run of dry years which produced the infamous Dust Bowl and land abandonment of the 1930s, for instance, coincided with a downturn in the world economy leading to the Great Depression. Similarly, the 1940s witnessed a coincidence between relatively high rainfall years and an increasing demand for wheat in war-torn Europe. On the other hand, the economically depressed conditions of the 1980s can only be accounted for by a change in government policy towards agriculture: price supports for cereals were severely and sharply reduced, leading to a downturn in the incomes of extensive grain producers. Table 4.5 gives data for a county in Colorado on the Great Plains which show the continuing decline in the number of farms as well as the rural population, but fluctuations in the number of hired workers, reflecting the contemporary state of 'boom or bust' in the cereal economy.

Specialised, large-scale grain production has tended to replace natural grasslands formed under semi-arid continental climates. Farmers have sought to cope with this marginal agricultural environment by manipulating the agroecosystem using a variety of dry-farming techniques:

- clean tillage where stubble is ploughed in during the autumn using a mouldboard plough

- conventional tillage where stubble is surface disc ploughed in the autumn or spring

- conservation tillage where stubble is undercut in the spring to remain on the surface as a mulch

- minimum tillage, as above but with less ploughing of mulch and weeds
- chemical tillage, again as for conservation tillage but with weeds controlled by herbicides.

In each case the objective is the exercise of farm-level control over the **soil moisture budget** by a number of methods: reducing weed growth to a minimum between harvest and planting; keeping wheat stubble upright for as long as possible in order to increase snow accumulation and reduce wind velocities; preserving a straw mulch until planting time to reduce wind impact, evaporation and run-off; maintaining a rough field surface with large clods of earth during fallow and after planting so as to reduce wind velocities and trap snowfall; and returning plant residues to the soil so as to maintain the nutrient cycle (Figure 4.7).

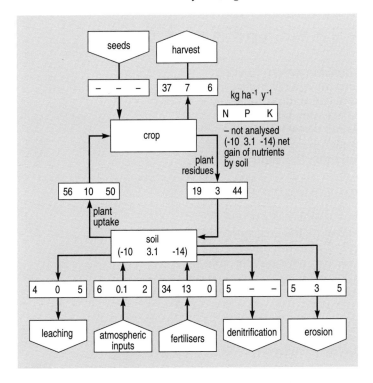

Figure 4.7   The nutrient cycle of a wheat farm in central Kansas, USA.

Source: Briggs, (D. J. and Contney, F. M. (1985) *Agriculture and the Environment: the physical geography of temperate agricultural systems*, London, Longman Group.

Table 4.5 Socio-economic indicators in Sedgwick County, Colorado. 1910–82

Year	Total population	Hired labour	Number of farms	Average farm size (ha)
1910	3061	–	448	144
1920	4207	399	487	195
1930	5580	–	560	222
1940	5294	213	505	232
1950	5095	145	474	268
1960	4242	279[1]	376[1]	335[1]
1970	3405	571[2]	327[2]	428[2]
1980	3266	390[3]	253[3]	520[3]

Notes: [1]1959; [2]1969; [3]1982.

Source: Spath; H. J. (1987) in Turner, B. L. and Brush, S. B. (eds) *Comparative Farming Systems*. New York, Guildford Press, p. 319.

When farmers make inappropriate modifications to the agroecosystem in these semi-arid environments, the consequences can be disastrous. For example, clean and conventional tillage, which leave no protective stubble on the field surface after harvest, proved catastrophic for parts of the Great Plains during the sequence of dry years in the 1930s. Although this farming practice has nearly been eliminated, wind erosion of the soil remains a significant problem in particularly dry years, for example 1975: see Table 4.6.

Activity 7

Examine Tables 4.6 and 4.7, then answer the following questions.

Q   What features of chemical tillage recommend the farming practice to profit-maximising farmers? (Table 4.7.)

A   Farmers are able to trade off reduced machine operations (discing, ploughing, weeding), with lower energy inputs, against higher inputs of chemicals. The loss of soil moisture is reduced (fallow efficiency rises), cereal yields are raised and the ratio between total energy outputs and inputs increases. You should note that while the dependence of dry-farming on purchased energy inputs is reduced, chemical tillage suppresses both weeds and other flora and fauna.

Q   Are there any underlying trends in the three different wind-erosion conditions in Colorado between 1969 and 1983? (Look at Table 4.6.)

A   Despite the well-known problem of wind erosion, the area suffering cropland damage appears to be increasing in Colorado; the area of crops destroyed shows little long-term reduction, although the area under emergency tillage is falling.

▲   Extensive arable farming: combine harvesters in the Czech Republic.

▲   *The morning after a windstorm in the Texas Panhandle, one of the United States'
most productive cotton-growing regions. During a critical time between harvest and
regrowth while the soil was bare, the wind lifted the silt and left the sand.*

Table 4.6 Wind erosion in Colorado, 1969–83 (ha)

Year	Cropland damage	Crops destroyed	Emergency tillage[1]
1969	10 988	56 356	29 525
1972	29 525	20 011	68 396
1975	261 737	388 189	535 681
1978	168 083	76 184	50 791
1981	925 830	19 213	35 843
1983	147 187	31 003	14 345

Note: [1]By producing large soil clods, deep tillage can serve as emergency tillage for wind
erosion control.

Source: Spath, H. J. (1987) in Turner, B.L. and Brush, S.B. (eds) *Comparative Farming Systems*.
New York, Guildford Press, p. 320.

Table 4.7 Tillage systems and their agro-ecological features (north-east Colorado)

Tillage system	Number of mechanical operations[1]	Fallow efficiency (%)	Yield (kg/ha)	Energy output/input ratio
Clean tillage	5–8	20–24	1680–2352	5–15
Conventional tillage	4–6	24–27	2016–2688	10–15
Conservation tillage	4–5	30–33	2352–2688	15–20
Minimum tillage	2–4	30–40	2688–3360	20–30
Chemical tillage	1–3	33–45	3024–3696	20–35

Note: [1]Ploughing, harrowing, discing, spraying, weeding, sowing, harvesting.

Source: Spath, H. J. (1987) in Turner, B.L. and Brush, S. B. (eds) *Comparative Farming
Systems*. New York, Guildford Press, p. 330.

4.3   *Intensive arable farming systems*

Intensive arable farming is characterised by crop rotations and high inputs
producing high yields per hectare. The crops found in these farming
systems include cereals (wheat, barley, oats, maize), oilseeds (sunflower,
soya bean, colza), roots (sugar beet, turnips) and vegetables (cabbage,
carrots, potatoes and so on). Whereas extensive arable farming systems aim
to regulate the soil moisture budget, intensive systems are designed
primarily to manipulate the nutrient cycle. Intensive arable systems employ
high throughputs of nutrients but relatively simple and minor internal
circulations.

Each annual crop cycle terminates with the removal of nutrients in the
harvested crop for commercial sale, the removals being replenished by
organic or purchased chemical fertilisers. In this sense the farming system
uses energy to act as a converter of mainly chemical inputs (nitrogen,
potassium and phosphorus) into useful food products. Figures 4.8(a) and
(b) show the nutrient cycle for an intensive arable farm in the Netherlands
growing potatoes, wheat and sugar beet in rotation. Under the rotation,
each crop makes slightly different demands on the soil in terms of
nutrients; in addition the farmer can exercise greater control over pests and
diseases while enjoying the economic benefits of spreading risk over several
crops. Figures 4.8(a) and (b) can be compared to gain an appreciation of the
importance of fertiliser inputs as well as the management of crop residues.
While the latter return important soil nutrients to the system, especially
nitrogen and potassium, unless carefully managed, crop residues can also
harbour pests and diseases for subsequent crops, together with any toxicity
remaining after the spraying of agrochemicals. Figures 4.7 and 4.8 also
permit a comparison of the nutrient cycle under extensive and intensive
arable systems: it shows the relatively high level of fertiliser applications
needed to maintain intensive systems, together with the high losses of
nutrients from leaching, denitrification and harvesting. As we will find
when examining intensive cereal farming in the United Kingdom, the high
losses of nutrients through leaching pose a major environmental hazard
both for supplies of drinking water and for wetland and river habitats.

Like extensive agriculture, intensive arable farming is found
throughout the world. When intensive crops (often under irrigation) are
combined with tree fruits (olives and citrus) and rain-fed wheat, a
characteristic Mediterranean agricultural system can be recognised (type 6
in Figure 2.6). When crops are combined with livestock in rotation, the term
'mixed farming' is commonly employed (type 8 in Figure 2.6): it is a
particular form of agriculture in both eastern and western Europe, eastern
North America and parts of other regions of European settlement in the
Argentine pampas, south-east Australia, South Africa and New Zealand.
Even so, large areas largely devoid of livestock can be found within these
regions, for example in the intensive cereal-growing areas of eastern
England and north-central France.

Dorel (1987) offers us a detailed view of one such intensive cereal-
growing area on the chalky plains of Champagne, located to the east of the
Paris basin. From being one of the poorer agricultural regions of France,
with small farms and much land in either forest or fallow, the region has
been transformed since the Second World War. Forests were cleared and
land drained; farms were amalgamated to create large businesses, and
intensive cereal production (wheat) set within a crop rotation of wheat/
barley/sugar beet/wheat/peas/wheat was rapidly developed. The
transformation was so rapid – only twenty years – that Dorel remarks on

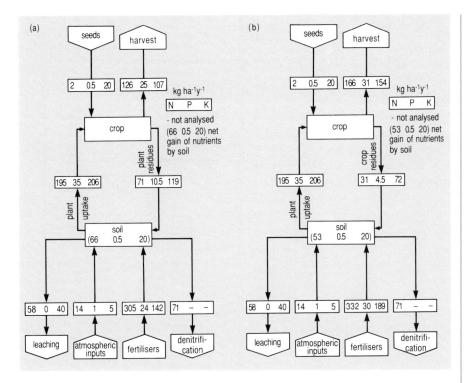

◀ Figure 4.8
*The nutrient cycle of an
intensive arable farm with
potatoes–wheat–sugar beet
rotation on clay soil in the
Netherlands.*
*(a) Crop residues retained
on the land.*
*(b) Crop residues removed
from the land.*

Source: Briggs, D. J. and
Courtney, F. M. (1985)
*Agriculture and the
Environment: the physical
geography of temperate
agricultural systems*, London,
Longman Group.

'the uniform, regional farm patterns reminiscent of the Great Plains of the United States – the general flatness, the small number of trees, the large cultivated fields, the scattered homesteads'. The principal causes of this transformation, through the intensive applications of fertilisers, agrochemicals and mechanisation, have been identified already: they were the financial incentives offered by the price support system of the Common Agricultural Policy and the penetration of agriculture by agri-inputs and food-processing firms. The outcome, however, has been a transformed landscape: enlarged fields to accommodate the powerful tractors and combine harvesters of modern cereal farming; hedgerows and woodland swept away, together with their habitats for flora and fauna; modernised farm buildings on those holdings still in production, with decaying or converted buildings on those farmsteads no longer actively engaged in agriculture. (See Plate 7.)

The energy-intensive nature of modernised arable farming is caused by successive management operations which include seedbed preparation, sowing, fertiliser application, pesticide spraying, harvesting and residue disposal. Each operation requires the expenditure of energy through the use of machinery and other materials produced off the farm. In the 1970s, concern over the apparent 'energy dependency' of modern farming systems spawned a wide range of studies, each attempting to measure the elusive concept of 'energy efficiency'. Our discussion at the beginning of this chapter provides a background to this analysis. Figure 4.9, using data for 1971–72, shows the principles involved in calculating an energy ratio for an individual (arable) farm. Clearly the calculated energy ratio between outputs and inputs of 2.1 means little on its own; the figure only gains value when set alongside similar calculations for other farming systems. However, researchers in this area are unable to agree on the details of the computations: they differ on the range of inputs to be included, their

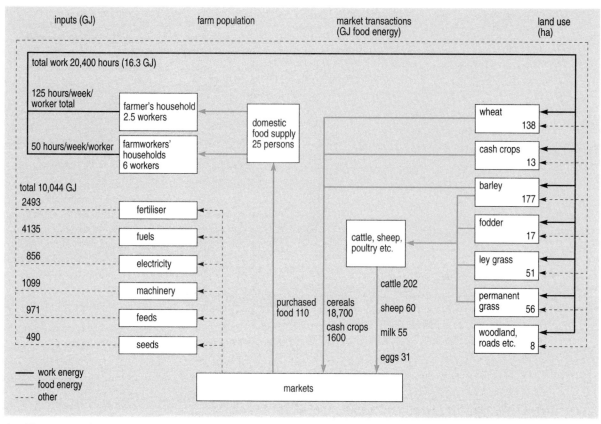

Figure 4.9 Average pattern of energy flows on a 460 hectare arable farm in southern England, 1971–72.

$$\text{Energy ratio} = \frac{\text{Crops sold} + \text{animal products sold}}{\text{Work input} + \text{purchased inputs}} = \frac{20\,300 + 348}{16.3 + 10\,044} \text{ GJ} = 2.1$$

imputed energy costs and the crop and animal yields to be assumed.
Varying assumptions are also employed on the labour input to different
farming operations: Spath (1987, p. 342), for example, excludes harvesting,
drying and transportation operations from the energy budget since the
energy inputs occur after the production process in the field is complete.
Other researchers, however, include these inputs, arguing that the totality
of the production process should be examined. There is also some dispute
over which index of energy use should be employed: Bayliss-Smith (1982),
for example, offers a choice between energy productivity, surplus energy
income and energy yield, in addition to the energy ratio itself.

Pimentel and Pimentel (1979) provide us with the most comprehensive
review of energy budgeting. Their calculations produce comparable
statistics for crops and livestock farming systems in the United States. Their
procedure was to divide the fossil energy input by the energy value of the
protein output (not the total food energy) for various crops and livestock
products. In Table 4.8 their results are inverted to give an energy ratio of
the same type as in Figure 4.9. The ratios obtained in this case are all less
than one (more fossil fuel input than protein output). They clearly
demonstrate the energy advantages of crop over livestock farming, and of
extensive over intensive systems. For each unit of fossil fuel input applied,
the protein output obtained is much greater for vegetable crops than for
livestock products (with rice the only exception, for US production

Table 4.8   Energy ratio for crop and livestock products in the United States (energy value of protein output/fossil energy input)

Crops		Livestock	
Soya beans	0.49	Beef (rangeland)	0.099
Oats	0.39	Eggs	0.076
Wheat	0.29	Lamb (rangeland)	0.062
Brussel sprouts	0.28	Broilers (chicken)	0.045
Maize	0.27	Pork	0.028
Potatoes	0.24	Milk	0.028
Rice	0.10	Beef (feed-lot)	0.013

Source: Adapted from Pimentel, D. and Pimental, M. (1979) *Food, Energy and Society*, London, Edward Arnold, pp. 56–9.

◄   *A cereal field showing the monoculture of modern industrialised agriculture.*

◄   *Crop spraying is favoured in intensive arable farming to achieve high yields.*

systems), and for extensive beef, eggs and lamb production than for
intensive systems.

Even so, most temperate farming methods are inefficient in the
conversion of energy in the sense that only a small proportion of total
energy inputs is ultimately consumed as food. But energy accounting is
simply a tool for describing a farming system. It does not prescribe a
healthy diet, nor quantify animal welfare, amenity or landscape aesthetics,
nor value the amount of food capable of being produced by intensive
farming methods. In most parts of the world, the main concern with
farming systems is their ability to generate profits and/or deliver food to a
population rather than their energy efficiencies.

4.4   Extensive grassland farming systems

Extensive grassland farming systems (ranching) tend to occupy locations
beyond the margins of cultivation in either excessively arid, cold or
mountainous environments. Large farm units predominate to exploit
previously natural, but now modified, grasslands (rangelands) using low
input-output farming methods for the production of beef cattle or sheep. The
major ranching areas lie on the prairie rangeland of western United States,
the llanos of Venezuela, the sertão of Brazil, the pampa of Uruguay, the chaco
of Patagonia, the karoo of South Africa, the arid interior of Australia and part
of South Island, New Zealand (type 2 in Figure 2.6). (See Plate 8.)

Ranching emerged as a major agricultural system only in the second
half of the nineteenth century, the main factor being the growth in demand
for beef and wool in the urbanised areas of North America and western
Europe. Consequently, the initial marketing infrastructure of these farming
systems tended to be export-oriented, with transport networks and meat-
processing plants focused on ports so as to serve distant markets. In recent
decades other farming systems have been able to produce beef, wool and
sheepmeat at competitive prices; consequently, ranching has had to adopt
the same process of intensification as found in other farming systems in
order to survive. In this way the rangeland and environment have come
under increasing pressure from extensive grassland farming.

The central feature of rangeland management is the livestock-carrying
capacity of the native grasses, herbaceous plants and shrubs. In the sense
that the very act of grazing transforms herbage composition, little 'natural'
grassland remains. For example, in Australia the widespread adoption of
sheep-grazing led to the removal of kangaroo grass, a predominantly
summer-growing species, and its replacement by essentially winter-
growing species such as *Danthonia* and *Stipa*. Nevertheless, a distinction can
be drawn between those pastures that receive little management and those
that are subject to improvement through rotational grazing, reseeding,
controlled burning and fertilising. Improving the grassland by these
methods fundamentally changes its character and composition, but allows
higher densities to be achieved. Grassland improvement has been
introduced along the more humid margins of rangeland areas, including
the cultivation of some fodder crops. Such developments have moved the
farming system towards a mixed agricultural economy but led to a
continuing reduction in the extent of semi-natural grassland. Even on the
'unmanaged' pastures, the pressure to increase farm output has tended to
lead to higher stocking densities, overgrazing and soil erosion by both wind
and water. Excessive grazing depletes the vegetational cover, reduces the
reincorporation of organic matter into the soil, compacts the soil and

initiates soil displacement on sloping areas. Indeed the first signs of
overgrazing can be seen in the down-slope movement of soil caused by
'trampling displacement'. The terraced livestock paths leave the topsoil
vulnerable to rill and gulley erosion.

Figure 4.10 shows the structure of a grassland system of farming. It
highlights the interrelationships between soil, grassland and livestock. In
ranching, grazing is the chief manipulation of the environment. The stock
are replenished by breeding and only a regulated number of animals leave
the system each year. Adjusting the number of cattle or sheep to rangeland
conditions on an annual or even monthly basis is central to good management
practice. In the highly variable semi-arid environments of extensive grassland
farming, a balance has to be struck between the forage consumption rate of
different types or ages of stock, grass yield (biomass), the species composition
of different rangeland areas, and the need to leave approximately half of
total forage growth for the pasture to recover for the following year.

Similar considerations apply to extensive grazing systems in upland
and mountain areas located within otherwise intensive farming regions.
Cattle and sheep farming on the moorlands of the Scottish Highlands and
the Cambrian Mountains of Wales, as well as the alpine meadows of
western Europe, are cases in point. Nevertheless, from the perspective of
the llanos or the sertão, with carrying capacities in the range of 2–6 hectares
per sheep, hill sheep farming in these areas appears to be a relatively
intensive farming system at 0.8–4 hectares per sheep. A more detailed
consideration of upland and hill sheep farming in the United Kingdom is
provided in Section 5.2

Judged in relation to extensive and intensive arable-farming systems,
significant environmental damage from extensive grassland farming
appears to be relatively limited and localised. This observation suggests the
existence of a continuum of environmental impacts: from farming systems
with relatively low to those with high and damaging consequences for the

▼ *Figure 4.10   An
extensive grassland farming
system.*

Source: Briggs, D. J. and
Courtney, F. M. (1985)
*Agriculture and the
Environment: the physical
geography of temperate
agricultural systems*, London,
Longman Group.

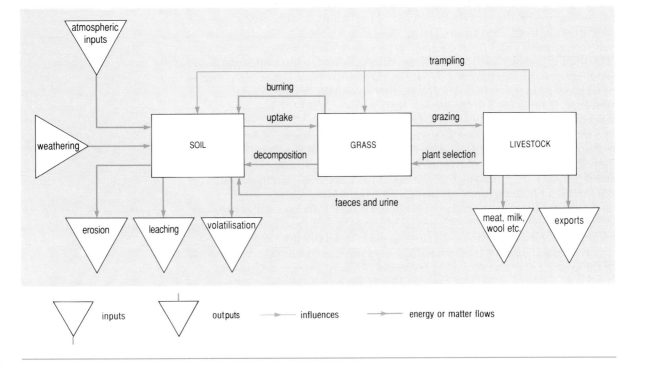

environment. With their localised occurrences of over-grazing, soil trampling and induced soil erosion, the extensive grazing of cattle and sheep can be placed towards the low-impact end of the continuum. But an alternative view can be taken on some grazing systems. For example, throughout the Mediterranean basin over-grazing by goats is widely viewed as a major contributor to the serious problem of soil erosion, especially along watersheds. In the western USA poisonous burroweed has become common and many woody species have invaded the grassland, for example mesquite, sagebrush and juniper. In Tasmania the overgrazing of pasture has encouraged the spread of unpalatable tussock grass and thorn bush.

4.5   Intensive grassland farming systems

Intensive grassland farming systems produce meat (veal, beef, lamb and mutton) as well as livestock products such as milk, hides and wool. Both cattle and sheep graze artificially grown pasture of three main types: permanent pasture, rotation pasture (leys) and temporary pasture. In general, rotation and temporary pastures contain more nutritious grasses – ryegrass, timothy and clover – with higher yields of forage. Turning to Figure 4.11, intensive grassland farming regulates the components of soil, grassland and livestock to a greater extent than under ranching. With rotation and temporary pasture, for example, the sward is periodically ploughed and reseeded, while nitrogenous fertilisers are applied to promote leaf growth. With so much nitrogen removed by harvest and grazing, as well as that lost in drainage waters, replenishment is needed to maintain biomass yield. Nevertheless, as with fertilisers used in arable systems, careful management is needed to reach the optimum rate of application per hectare. If the forage is too nitrogen-rich, for example, cattle can be in danger of nitrate poisoning.

Rotation pasture is usually renewed in a five- to ten-year cycle, whereas temporary grass, being part of an arable rotation, usually has a life of only one or two years' duration before being ploughed in. In both cases, grass-seed mixture is selected so as to produce a high forage yield throughout the year. Permanent pasture, in contrast, is ploughed only infrequently and is refreshed by surface treatment of lime, nitrogenous fertiliser and grass-seed. There is a long-standing debate in agriculture over the relative merits of temporary as compared with permanent grass; if managed well, however, there is little to choose between the two, and both can achieve stocking rates of between 3 and 4 cattle per hectare.

Intensive grassland farming is found mainly in association with areas of European settlement in western Europe, North America, North Island of New Zealand and parts of coastal south Australia. Farm size tends to be small but, as with ranching, the success of the intensive system depends on the quality of management supplied by the individual farmer. Three particular management practices should be noted. First, grass is conserved either as hay or silage to feed livestock during the winter months when the growth of the pasture ceases. Removing grass from the field in this way involves major losses of nutrients. Consequently, and secondly, the growth of the pasture is promoted using fertilisers. Thirdly, the grazing of the pasture is controlled using a variety of techniques. Under a *'free range' system*, for example, livestock are allowed to graze at will over the whole area of available grassland. Under *set stocking*, however, the stock are rotated from field to field on a daily or weekly basis so as to exploit the full potential of the available fodder. *Paddock grazing* is an elaboration of this

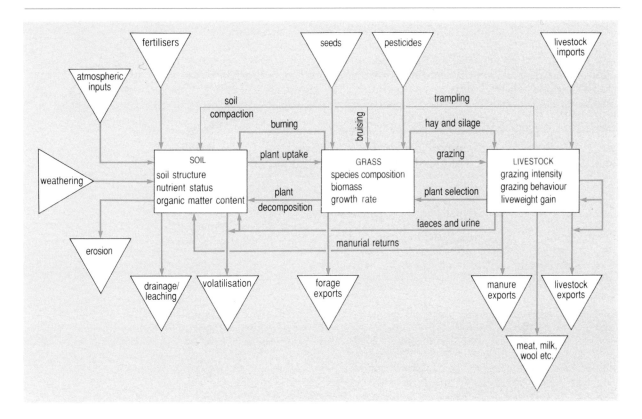

▲ *Figure 4.11*
An intensive grassland farming system.

Source: Briggs, D. J. and Courtney, F. M. (1985) *Agriculture and the Environment: the physical geography of temperate agricultural systems*, London, Longman Group.

technique: fields are subdivided into small, fenced areas (paddocks) and the stock are grazed intensively on each in turn. *Strip grazing* introduces a yet more intensive practice: an electric fence is moved daily or hourly in stages across the pasture; at each stage livestock have access to only a narrow strip of grass which is grazed to the full. Finally, under *zero-grazing* the grass is cut daily and taken from the field to the stock which are housed in stalls or open yards. This system minimises the waste of forage involved in trampling and fouling under field-grazing methods, while the stock expend little energy in obtaining their food. On the other hand, the nutrient losses are significant and there is no recycling of nutrients through animal faeces. Indeed, the problem of disposing of the slurry effluent created by cattle kept at high densities has been mentioned already, while accidental leakages and spillage from storage areas (tanks or lagoons) can have a devastating effect on the ecology of local streams and rivers. Indeed so toxic is the effluent that in many countries – including the UK – farmers can be fined for leakages into water courses, especially where public water supplies are involved. In Denmark recently enacted environmental laws have been applied to the storage and application of manure and slurry so as to control the pollution of groundwater. Farms with more than twenty animals have to be able to store a minimum of six months' production of slurry and manure; strict construction standards have been applied to slurry pits and lagoons; animal housing must be placed at certain distances away from domestic water supplies; and restrictions have been placed on the use of nitrogen on pastures.

The manipulation of the agroecosystem is most developed in dairy farming (type 9 in Figure 2.6) with high levels of stock density, capital and labour inputs achieving high milk yields per hectare and per cow. Livestock

▲   *The containment and disposal of slurry is an enormous problem. In England and Wales 2.2 million dairy cows produced 23 million cubic metres in 1992. (Above) A slurry store in a sleeper-walled compound; (below) good practice in solid and liquid slurry stores: surface rainwater on the muck soon drains out when it stops raining.*

diversity is minimised: Holstein and Friesian dairy breeds are dominant, with some regional specialisation in breeds such as Normandy, Guernsey and Ayrshire. Successively lower intensities of inputs and outputs are found in cattle and sheep production but, with the increasing economic pressure on agriculture, even these systems are beginning to adopt more intensive methods of management. In sheep farming, for example, winter housing has been introduced, together with supplementary feeding of concentrates during the winter and spring months, especially around lambing time.

A wide variety of farming practices can be found within these broad features. Dairy farmers in North Island, New Zealand, for example, benefit from the growth of grass all year; there is no need for energy-intensive

▲   *A modern milking parlour*

grass conservation or winter housing for the stock. In beef and sheep
production, some farms breed, rear and fatten stock, for example in the
Welsh borderland of the United Kingdom, whereas only one or two of the
three stages of livestock production are practised elsewhere. To give one
example: the permanent pastures of south-east Leicestershire have been
used traditionally to fatten, but not breed or rear, beef cattle and sheep.

   To some degree, all intensive grassland systems are supported by state
farm policies. The farming system employs a relatively large number of
farmers and hired workers, but its products tend to be in over-supply on
domestic and world markets. So as to yield a socially acceptable farm
income, most states offer price supports and market intervention for
products such as milk, beef and sheepmeat. The milk quota scheme
operated under the Common Agricultural Policy of the European Union is
a typical example of such state intervention.

4.6   *Intensive livestock farming systems*

Intensive livestock farming is among the most regulated of the
agroecosystems, and comparable with intensive horticultural production
under glass. The system has been applied to a number of livestock products
including stall-fed veal, feed-lot beef and zero-grazed dairy cows.
However, the term 'intensive livestock' is most commonly applied to the
production of pigs and poultry.

   There are a number of distinctive features in the farming system.
Firstly, the stock are housed in buildings with closely regulated
temperature and lighting conditions. Poultry and pigs are kept at high
densities in cages and pens respectively, with feeding, watering, egg
collection and slurry removal carried out by mechanical means. Economies
of scale are obtained through the large number of stock handled by each
production unit, and not surprisingly the system is widely described as

▲   *Intensive livestock rearing, in this case for veal. High inputs of energy are needed to maintain a controlled environment, but a high value output is produced.*

'factory farming'. These farming practices have attracted increasing criticism from the aspect of animal welfare, while the food health standards of beef and poultry products have also been questioned.

Secondly, intensive livestock farming is primarily a method of converting cereals into a food product such as bacon, pork, eggs and poultrymeat. The feed conversion efficiency of the livestock determines the profitability of the enterprise. In a sense, therefore, land is incidental as cereals are usually purchased as the output of another farming system. Since a significant proportion of the cereal inputs come from developing countries, the 'land' supporting intensive livestock farming is international in its distribution.

Thirdly, rather than having a large number of independent family-owned farms, intensive livestock farming is concentrated into a small number of very large agribusinesses in which vertical integration is commonplace, with feed mills, egg hatcheries, production units and processing plants all placed under one ownership.

Fourthly, compared with other systems, intensive livestock farming is characterised by the absence of state regulation. Of course, there are exceptions to this generalisation. The incidence of salmonella led to the compulsory testing of poultry in the UK. In Canada the poultry sector functions under supply management (i.e. production quotas), but across the international border, in the United States, the market is not regulated. As a result, the family structure of poultry production has been retained in Canada, whereas agribusinesses are dominant in the United States.

As with intensive grassland farming systems, the main environmental problem associated with intensive livestock systems is the disposal of animal effluents. When intensive farming units are dispersed amongst other farming systems, the effluents (slurry and manure) can provide useful sources of nitrogen, phosphorus and potassium when spread on neighbouring farmland. But when large numbers of such units are

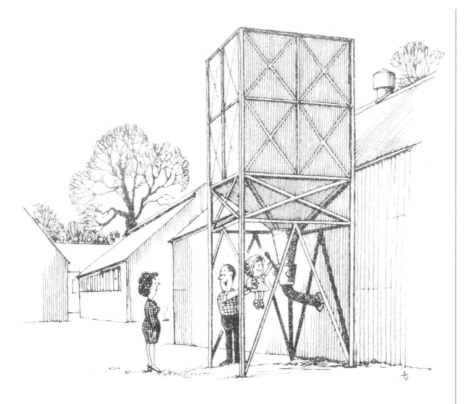

'She loves feeding the animals.'

concentrated in a relatively small area, for example, in northern parts of
Brittany (France), the Po Valley (Italy), the Republic of Ireland and East and
West Flanders, including the Kempen area (Belgium), disposal of the
effluent creates environmental hazards in terms of ground and surface
water pollution and soil toxicity.

4.7   Summary

This chapter has discussed only a few of the very many farming systems
found in temperate agriculture. Nevertheless, the case studies reveal
differences in the internal structure of the systems and the varying extent to
which the agroecosystem is manipulated. This understanding should now
lead us to approach generalisations about temperate agriculture with some
caution, although the farming systems tend to have one feature in common:
a dependency on energy inputs purchased off the farm. Nevertheless, the
farming systems vary in the efficiency with which that energy is converted
into useful food products, intensive livestock farming being relatively
inefficient in this respect. The discussion has also identified some of the
environmental effects of the different farming systems, together with the
need to place each system along a continuum of low–high impacts. The
next section takes an even more detailed view of these agriculture–
environment relationships by focusing on just two types of agriculture in
the United Kingdom: hill sheep and cereal farming.

5   The environmental impact of temperate agriculture

5.1   Introduction

Concern over the damaging environmental consequences of modernised – some would say industrialised – agricultural systems was first expressed in the United States as part of the evolving environmental movement. Rachel Carson's *Silent Spring* (1963) was amongst the first popular expressions of anxiety over the damaging impact of pesticide residues on the ecosystem, an anxiety that later spread to fertilisers, farm effluent, soil erosion and the loss of habitats. Transferred to a British context, official concern was expressed as early as 1970 about the impact of modern farming on soil structure, and later on the wider environment, while the environmental debate was given a higher profile in agricultural circles following Marion Shoard's book, *Theft of the Countryside* (1980), and debate on the Wildlife and Countryside Act 1981. Today 'agriculture and the environment' is high on the political agenda in most countries. Indeed a considerable literature has been developed around the topic which we can deal with here in only a partial way. Rather than attempt a summary, attention is focused on two farming systems in the United Kingdom that lie at each end of the environmental impact continuum. Hill sheep farming has been selected to represent a relatively 'low impact' system, while intensive cereal growing acts as an example of 'high impact' farming.

5.2   Hill sheep farming in the United Kingdom

With upland areas accounting for approximately 40% of Britain (7.7 million hectares), the land resource occupied by hill farming has considerable ecological significance. In particular, the moors and heaths of the uplands act as a refuge for many threatened species of flora and fauna, including several rare species such as the red kite. Three main environmental impacts can be identified: moorland reclamation, grassland species composition and bracken incursion. In the wider perspective it can be argued that none of these impacts is particularly severe; nevertheless, all three have the effect of reducing both the diversity of natural and semi-natural habitats and the visual amenity (landscape value) of hill and mountain areas.

   Moorlands and heathlands in Scotland, Wales, the Pennines, North York Moors, Dartmoor and Exmoor were largely created by woodland clearance. Contemporary improvement, commonly known as *moorland reclamation*, therefore, continues a historic trend by bringing such land into cultivation. In some cases the pre-existing vegetation (for example *Calluna*, *Vaccinium* and *Pteridium* species) can be cleared by light surface cultivation, burning or spraying with herbicides; the soil can then be prepared by applying lime and chemical fertilisers, followed by the sowing of preferred grass species such as ryegrass, meadow grass and clover. Where soil conditions are very wet, however, deep ploughing and drainage may be necessary before surface cultivation can commence. Quite marked reductions in the area of moors and heaths have been recorded in the 1950s

and 1960s in regions with hill sheep farming. On Exmoor, for example, over 4000 ha of moorland (17% of the total area) were reclaimed between 1947 and 1979, while similar figures have been recorded for the North York Moors between 1950 and 1963 and the Brecon Beacons from 1948 to 1975. Most reclamation and enclosure occurs on the fringes of the open moors and heaths, and considerable areas are secondary reclamations of land that fell out of cultivation in the early part of this century, especially during the 1930s. Much upland remains under threat from any further intensification of hill farming.

The motives for land reclamation are relatively straightforward. As hill sheep farms have come under economic pressure, so their owners have attempted to raise farm output by increasing the number of hill ewes and cattle. Reclaiming hill pastures raises the carrying capacity of a farm and enables more stock to be fed. But the high capital costs of moorland reclamation, when faced by the relatively low prices of lambs and cattle for fattening, render the practice economically marginal. Consequently, successive governments in the United Kingdom have provided both grant aid to farmers per hectare of reclaimed land, and support for their incomes by headage payments on cattle and sheep. Although the grant aid has now been discontinued, the headage payments remain as Hill Land Compensatory Allowances (HLCA) under the Common Agricultural Policy. By 1991 these had increased in value up to £63.30 per hill cow and £8.75 per hill ewe. All the hill and upland areas of the United Kingdom are now eligible for these direct income supplements under the European Union's Less Favoured Areas Directive (see Figure 5.1).

The ecological effects of reclamation are disputed. On the one hand, reclamation clearly changes the character and composition of upland vegetation from heather and coarse-grass species – cotton grass, bent-grass, purple moor-grass – to agricultural grasses. In addition the diversity of habitats is reduced, while the buffer zone between intensive agriculture and open moorland is removed. Such changes impinge on the animal species which are dependent on the moorland habitat for food, cover and nesting ground. But other research has shown that drainage, fertilisers and fencing can increase rather than decrease populations of insects, wood mice, pygmy shrew and field voles, as well as their predators – foxes and owls. Consequently the ploughing and fencing of open moorland has been a contentious issue, with Exmoor providing a particular cause célèbre. A long-running battle has been fought between conservation and farming interests, with origins that can be traced back to the 1960s. A formal enquiry was held in the early 1970s, under Lord Porchester, to investigate the problems surrounding moorland reclamation; it resulted in financial compensation for some of the land-owners who gave up their plans to plough the moorland in exchange for management agreements. These financially expensive arrangements were later formalised under the Wildlife and Countryside Act 1981 and have served to preserve limited areas of moorland with high ecological and amenity value. Nevertheless, the remaining area of moorland is now fragmented and vulnerable to further piecemeal reclamation on a farm-by-farm basis, especially in the dark areas shown in Figure 4.12. (See also Plate 9.)

Turning now to the second major environmental impact, *species composition*, the activities of burning and grazing have considerable impacts on the species found in upland pastures. Heather burning is carried out mainly in the interest of grouse for gameshooting as discussed by *Silvertown* (1996b, Section 4.2). Figure 4.13 summarises the successional transitions that occur in moorland habitats under different burning and grazing regimes.

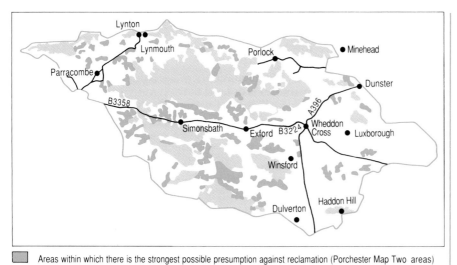

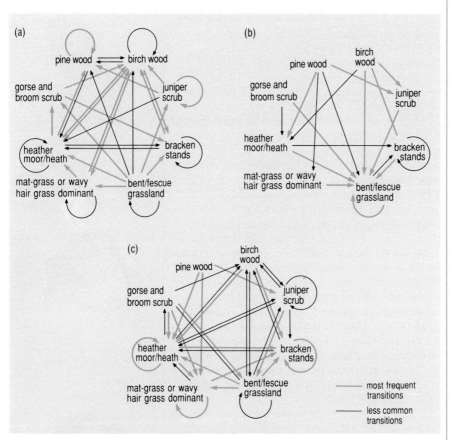

◀ Figure 4.12
Exmoor moorland, 1980.

◀ Figure 4.13
Successional transitions
between common types of
semi-natural vegetation
under different grazing
pressures on well-drained
acid mineral soils.
(a) Low grazing pressures
(<1 sheep equivalent per
hectare per year) and
without burning.
(b) High grazing pressures
(>2–3 sheep equivalent per
hectare per year) and
frequent burning.
(c) Intermediate levels of
grazing (1–2 sheep
equivalent per hectare per
year) and burning.

Activity 8

Examine Figure 4.13 and answer the following questions.

Q   What effects do high grazing pressures and frequent burning have on
the transition between vegetation types?

A   With the exception of bracken stands, all types of vegetation are
succeeded by bent/fescue grassland which eventually becomes
dominant. Burning and grazing together prevent regeneration of a
species-rich habitat.

Q   What are the main differences in the successional transition between
low grazing pressure without burning and intermediate levels of
grazing and burning?

A   Low grazing pressure without burning allows pine and birch wood to
regenerate, bracken is less dominant, and there are more frequent
transitions from mat-grass and bent/fescue grasslands to juniper scrub
and birch-wood habitats. Intermediate pressure prevents woodland
regeneration and encourages a variety of types including heather,
bracken and mixed grasses.

Research has revealed the differential impact on species composition of
treating moor-grass-dominated grassland with lime and chemical
fertilisers. Under rotational grazing, but after only five years, lime-treated
grassland becomes dominated by sheep's fescue, whereas fertiliser-treated
land has mainly bent-grasses. Thus, in the case of hill sheep, moderate
densities tend to enhance species diversity, while low and high grazing
pressure reduce diversity. With no effective upper limit (up to 6 ewes per
hectare) on the number of sheep for which HLCA could be paid, there have
been periodic claims of over-grazing through 'farming for the subsidy'.
Sheep numbers have undoubtedly increased over the period 1974 to 1993
(see Table 4.9), so there is now some hard evidence of this problem (see
Chapter 5).

As Table 4.9 also indicates, beef cattle have declined in importance
within the upland farming system. This change in farming practice is
thought to be one of the main reasons for the extension of bracken
(*Pteridium aquilinum*) into both grassland and heathland, though the decline
in numbers has slowed since the mid-1980s. Cattle graze less selectively

Table 4.9   Changes in breeding ewes and beef cows in Wales, 1974–93

County	Rough grazing as % of total agricultural land (1993)	Ewes 1974	Ewes 1993	Change %	Beef cows 1974	Beef cows 1993	Change %
Clwyd	22	452 289	731 988	62	23 328	20 771	−11
Dyfed	19	596 911	1 100 227	84	52 304	53 574	2
Gwent	17	159 725	237 477	49	10 016	9 480	−5
Gwynedd	45	716 293	1 065 875	49	41 753	36 579	−12
Mid Glamorgan	37	141 628	201 547	42	8 856	7 100	−20
Powys	35	1 252 806	1 884 105	50	84 456	75 283	−11
South Glamorgan	4	1 946	34 508	1673	1 885	2 748	45.8
West Glamorgan	41	6 212	10 184	1529	6 256	6 623	6
Wales (total)	29	3 401 238	5 356 961	58	228 854	2 122 158	−7

Source: *Welsh Agricultural Statistics*, 1974, 1986 and 1995, Welsh Office.

than sheep, although their trampling effect is thought to suppress the growth of bracken. In addition, bracken is no longer cut to the same extent as in the past for livestock bedding. Taylor (1980) has estimated an annual rate of expansion of bracken in Wales of 2072 ha between 1936 and 1966, and 10 360 ha for the United Kingdom. Ecologically, bracken suppresses other plants and tends to grow in association with a limited range of shade-tolerant species. There is also some evidence that bracken acts as a carcinogen in livestock, with a similar effect on humans who consume their milk and dairy products over a long period of time.

5.3   Intensive cereal farming in the United Kingdom

The environmental impact of intensive cereal farming can be usefully examined under five headings: field size, chemical pollution, soil structure, pesticides/herbicides and habitat destruction.

The economics of agriculture have moved in favour of farm machinery which requires large fields for maximum operating efficiency. Increases in *field size* reduce the time spent turning machinery at rows' ends, raise the speed of equipment across the land and enable wider implements to be employed. The result has been the removal of small woods and hedgerows, a trend particularly evident in the intensive cereal-growing areas of eastern England where hedges are no longer needed to control livestock. In addition, farmers have been relieved of the costs of trimming and managing woodland and hedgerows, while small gains have been made in the area of productive farmland. Concern about the rate of the hedgerow removal began to be expressed in the 1970s, although most studies have shown that the highest rates of removal occurred in the 1960s. Between 1945 and 1970, for example, hedgerows were destroyed at a national average rate of 8000 km a year, with rates possibly ten times above the national average in cereal-growing areas. Since the 1960s the rates of hedgerow removal appear to have fallen back to more acceptable but still ecologically damaging levels.

The removal of woods and hedges has three important environmental impacts. First, it reduces habitat diversity. Work by the (then) Nature Conservancy Council has clearly demonstrated a marked reduction in the number of mammals, birds and Lepidoptera in areas where hedges have been removed or replaced by wire fences. The felling of woodland reduces roosting and nesting sites for birds and small mammals, while the removal of hedgerows destroys the habitat corridor that links those sites with foraging areas in the fields. Secondly, hedgerow removal detracts from the visual amenity of the landscape by destroying a characteristic element of the traditional British countryside. Thirdly, a number of studies have shown that wind erosion of the soil is more prevalent in those areas where hedgerows have been removed. Wind velocity is reduced leeward of a hedgerow in proportion to its height and density.

Turning now to *chemical pollution*, we have seen already that the application of fertilisers represents a significant modification of the agroecosystem. Unfortunately, not all of the fertiliser is taken up by the field crops in intensive cereal cultivation: the excess is leached from the soil, especially by winter rainfall, and ultimately enters the water courses, including ponds and lakes. The accumulation of nitrates and phosphates in water bodies creates algal blooms and eutrophication and leads to the collapse of the ecosystem. Figure 4.14 shows the problem to be most severe in central and eastern England, with the Broads of Norfolk and Suffolk being particularly adversely affected. It should be noted, however, that some of

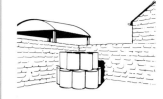

Farmers have been responsible for removing thousands of kilometres of hedgerows, changing the landscape and affecting the ecology of the countryside.

groundwater vulnerable to nitrate contamination

Figure 4.14   The impact of agrochemicals on the environment in 1994: fertilisers.

the pollution is also caused by sewage outfalls. A rising level of nutrients in the Broads for the thirty years up to 1990 has been related to the progressive decline of reed-swamp and the reversion to open water. Under anaerobic water conditions, the plants appear to be susceptible to root damage, while the number of freshwater fish and amphibians has been similarly depleted. In addition, rising nitrate levels have been recorded in groundwater supplies, thereby posing a hazard for human drinking-water.

The impact of arable farming on *soil structure* is also well documented with many researchers concluding that continuous cereal growing can lead to structural damage. The decline in structural stability and microporosity of the soil can be relatively rapid, a process that is exacerbated by the loss of decomposable organic compounds through crop removal and the burning of cereal stubble (now banned). Compaction of the soil surface by

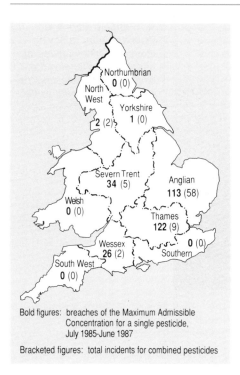

◀ Figure 4.15
The impact of agrochemicals on the environment: pesticides.

machinery can also be problematic when tillage takes place during unsuitable soil moisture conditions. Some researchers point out that, with so many separate machine operations under cereal farming, the compaction of soil over large parts of any field is almost inevitable. However, the impact of changes in soil structure on crop yields, especially through compaction, is difficult to detect, although it has been of continuing concern over the last twenty-five years.

Fourthly, inorganic *pesticides and herbicides* have played a large part in modern agriculture. They vary in toxicity, specificity and persistence. Ideally pesticides should be effective against specific organisms without having damaging effects on other insects and micro-organisms. Unfortunately too many of the early pesticides were broad-spectrum poisons that killed pest and predator insect alike and many were persistent. More recent organophosphate pesticides are less persistent but more toxic and specific. Nevertheless, there is growing concern over the side-effects of toxic residues in human food, while resistances in insect populations require ever more powerful pesticides. The incorrect application of sprays can also be problematic, with wind-drift taking toxic chemicals onto adjacent and inappropriate crops and ecosystems, as well as areas of human settlement. The use of herbicides to kill weeds in cereal crops is open to the same objections. More recently, evidence has emerged on the presence of pesticide residues in drinking-water. Figure 4.15 shows that reported breaches of Maximum Admissible Concentrations are greatest in the vegetable- and cereal-growing areas of eastern England.

Finally, lowland *habitats* have not been immune from the processes of 'reclamation' and 'improvement' that we identified in the uplands in relation to moors and heaths. In the lowland areas, associated mainly with intensive cereal farming, woodland and wetland habitats have been destroyed over the last three decades, the scale of which has only recently been quantified by bodies such as the Institute for Terrestrial Ecology and

◀   *The Halvergate Marshes: arable farming or intensively managed grass required a lower water table so large ditches were dug in an 'improvement scheme'. However, objections to this led to the Broads Grazing Marsh Conservation Scheme in 1985, a blueprint for the idea of Environmentally Sensitive Areas announced in 1987.*

the Nature Conservancy Council (now English Nature). Their research for the United Kingdom has revealed 40% losses of lowland heaths, 50% losses of lowland fens and up to 50% losses of ancient lowland woods. Rates of loss tend to be most severe in southern and eastern England, with less damage in western and northern regions of the United Kingdom. Just as moorland reclamation on Exmoor became a major issue in the late 1970s, so the wetlands of the Halvergate Marshes assumed a similar role in relation to lowland cereal production in the early 1980s. The marshes, forming a 3000 ha triangle of land between the Lower Bure and Yare rivers in eastern Norfolk, in the heart of the Broads, yield a fertile soil when drained, and this potential became increasingly attractive to the land-owners under the cereal price support system of the Common Agricultural Policy. There was a complex and protracted conflict between the land-owners, conservation

pressure groups, Countryside Commission and the (then) Nature
Conservancy Council (O'Riordan, 1986; Blunden and Curry, 1996) but, as in
the Exmoor case, the final outcome yielded the loss of some habitat, on this
occasion to draining and ploughing for cereals, while other areas became
protected by management agreements and financial subsidies to land-
owners to continue traditional livestock grazing (The Broads Grazing
Marsh Conservation Scheme).

5.4   Summary

The two cases have underlined the varying environmental impacts of
temperate agriculture and caution us against making generalisations on
temperate farming systems. In the case of hill sheep farming, stocking
densities have been shown to be the critical factor in habitat modification,
as well as for the economic well-being of the farming community. In
reviewing the evidence on environmental impact we may conclude that the
hill sheep farming system in Britain maintains rather than damages the
ecological diversity of the upland areas. Indeed, other non-farm uses of hill
areas appear to have a greater environmental impact, for example forestry,
water catchment and the recreational use of hill land. Moreover, in an era of
falling profitability for agriculture, including the withdrawal of state
subsidies, the alternative uses of hill land are likely to assume a more
prominent role, including their impact on the upland environment.

Intensive cereal farming lies at the other extreme of the environmental
impact continuum in Britain. The farming system demonstrates a wide
range of environmental consequences, many of them undesirable. While the
landscape impact is the most dramatic visually, the more insidious
consequences of nitrate and agrochemical pollution probably pose greater
threats to the long-term stability of the agroecosystem.

6   Conclusion

There is nothing inevitable or beyond influence in the environmental
impacts that have been identified: farming systems throughout the world
can be developed that are more sensitive to the environment. What
determines developments are the economic conditions of agriculture, the
structure of state policies on farming, and ultimately the attitudes and
behaviour of those who own and farm the land. Of these features, we must
recognise the central role played by the state in:

 - financing the type of research and development that has produced the
new, high output farming technology
 - supplying the price incentives to farmers to adopt that technology
 - until very recently, failing to adapt farm policies once it became clear
that modern farming was having undesirable environmental impacts.

However, individual farmers must also accept a share of the responsibility.
Their profit-maximising behaviour, albeit encouraged through subsidies by
governments in many countries, has yielded products surplus to domestic

demand, at a cost above that prevailing on world markets, and with environmental consequences the scale of which we are only now beginning to appreciate.

Of course, not all impacts yield the same level of severity for the environment. Establishing that level, however, is fraught with difficulty. On the one hand are value judgements on the nature and quality of landscape, and the importance of preserving different types of habitat, including the diversity of flora and fauna. On the other hand there are scientific judgements on the damage to human health of polluting water and food with agrochemicals, including fertilisers. Opinion is by no means unanimous on where the critical levels of pollution lie. But recent years have seen the weight of opinion slowly turning against the excesses of a modernised, industrialised temperate agriculture and remedial policies and developments are slowly taking shape.

A wide range of policies is being implemented to limit the damaging consequences of modern agriculture and these are discussed in more detail in the next chapter. One option of current interest, however, is low input-output farming, of which organic farming is a variety. This new agriculture would recreate the diversity of natural ecosystems by integrating crops and livestock, would maintain varied crop rotations, eliminate the use of agrochemicals and limit the use of fossil fuels. Critics doubt if such an agriculture could produce the volume of food necessary to feed the urban population, but supporters argue that if the same quantity of research money is expended on low input-output farming as has been committed to intensive agriculture, then such problems would be overcome. Until recently, governments have avoided direct involvement with these issues, but intervention to promote low input-output farming is now appearing in agricultural policy-making.

As matters stand, however, and looking just at the United Kingdom, the actual approach being applied is largely to persuade farmers to adopt voluntarily a more sympathetic attitude to the environment by planting woodland, caring for hedgerows, reducing the use of agrochemicals, managing ponds and wetlands for their habitats, and continuing traditional farming practices where they contribute to the maintenance of favoured landscapes and habitats. Financial subsidies from the state are available to promote these practices, backed up by the zoning of selected areas for special protection like the Environmentally Sensitive Areas. Overall, though, a voluntary rather than mandatory approach is being taken to resolve the agriculture–environment conflict.

Turning back finally to the world scene where this chapter began, temperate agriculture cannot be divorced in its impact from farming systems in developing countries. On the one hand, the surpluses generated by temperate agriculture provide the food aid that increasingly underpins the supply of agricultural products in many developing countries. This applies both to emergency situations, for example the droughts which periodically afflict parts of Africa, as well as annual food imports. From the perspective of many developing countries, the continuation of food surpluses in temperate agriculture has a higher priority than the resolution of the environmental impacts we have been discussing. On the other hand, the protective farm policies in temperate agriculture have excluded some of the agricultural exports of other developing countries – although not animal feeds and oilseeds – and so contributed to their economic problems. As noted in the opening paragraph to this chapter, temperate agriculture probably exemplifies, as well as any other human activity, the conflict between maintaining economic well-being and the quality of the natural environment.

References

BAYLISS-SMITH, T. P. (1982) *The Ecology of Agricultural Systems*, Cambridge, Cambridge University Press.

BLUNDEN, J. and CURRY, N. (1996) 'Analysing amenity and scientific problems: Broadland, England', in Sloep, P. and Blowers, A. (eds) *Environmental Policy in an International Context, 2: Conflicts*, London, Edward Arnold.

BOWLER, I. R. (ed.) (1979), *The Geography of Agriculture in Developed Market Economies*, London, Longman.

CARSON, R. (1963) *Silent Spring*, London, Hamish Hamilton.

DOREL, G. (1987) 'High-tech farming systems in Champagne, France: change in response to agribusiness and international controls', in Turner, B. L. and Brush, S. B. (eds), pp. 405–23.

DUCKHAM, A. N. and MASEFIELD, G. B. (1971) *Farming Systems of the World*, London, Chatto and Windus.

O'RIORDAN, T. (1986) 'Moorland preservation in Exmoor' and 'Ploughing into the Halvergate marshes', pp. 191–208 and 265–99 in Lowe, P. *et al.*, *Countryside Conflicts: the politics of farming, forestry and conservation*, Aldershot, Gower.

PIMENTEL, D. and PIMENTEL, M. (1979) *Food, Energy and Society*, London, Edward Arnold.

SARRE, P. and REDDISH, A. (eds) (1996) *Environment and Society*, London, Hodder and Stoughton/The Open University (second edition) (Book One of this series).

SHOARD, M. (1980) *The Theft of the Countryside*, London, Temple Smith.

SILVERTOWN, J. (1996a) 'Ecosystems and populations', Ch. 7 in Sarre, P. and Reddish, A. (eds).

SILVERTOWN, J. (1996b) 'Inhabitants of the biosphere', Ch. 6 in Sarre, P. and Reddish, A. (eds).

SIMMONS, I. G. (1989) *Changing the Face of the Earth: culture, environment, history*, Oxford, Blackwell.

SPATH, H. J. (1987) 'Dryland wheat farming in the Central Great Plains', in Turner, B. L. and Brush, S.B. (eds), pp. 313–44.

TAYLOR, J. A. (1980) 'Bracken – an increasing problem and a threat to health', *Outlook on Agriculture*, Vol. 10, pp. 290–304.

TURNER, B. L. and BRUSH, S. B. (eds) *Comparative Farming Systems*, New York, Guildford Press.

Further reading

HAINES, M. (1982) *Introduction to Farming Systems*, London, Longman.

SIMMONS, I. G. (1980) 'Ecological-functional approaches to agriculture in geographical contexts', *Geography*, Vol. 65, pp. 305–16.

WOLMAN, M. G. and FOURNIER, F. G. (eds) (1987) *Land Transformation in Agriculture*, Chichester, John Wiley.

Chapter 5 Competing demands in the countryside: a United Kingdom case study

1 Introduction

This chapter examines the principal competing demands on the countryside of the United Kingdom and their environmental impacts. It does so for two related reasons. First, the use of rural land became an urgent issue of practical politics in the 1980s as the perceived need for agricultural production diminished and the way was opened for other uses. This issue has also arisen in other European countries and in the United States. Second, policy debates about the use of rural land have had to recognise that separate policies for different sectors – agriculture, forestry, industry and so on – have often failed because of the interaction between sectors. This problem has been a difficult one when only balancing economic motives with social aims; it has become even more complex now that environmental criteria are also being seriously considered. Nevertheless, there have been some attempts at integrated rural management and these are worth evaluating as possible guides to policy improvements in future.

As you read this chapter, look out for answers to the following key questions:
- What are the principal competing demands in the countryside of the UK?
- How have different policies between sectors contributed to environmental problems?
- How did agricultural policy come to dominate, and how is policy responding to reduced demand?
- What are the prospects for integrated rural development?

2 The ascendancy of agricultural interests

For most of the post-war period, agriculture has enjoyed a special status as the favoured use of rural land and the most subsidised sector of the economy. This raises two questions: how was this special status achieved, and why has it now come under question? Answers are to be found in the next two subsections.

2.1 A forty-year perspective

To provide answers to these questions, it is necessary to go back to the Second World War. Then, after a period of agricultural depression which had existed for most of the inter-war years, a coalition government, committed to the maximum output from home farming, decided to address the likely shape of agriculture once peace returned. A committee under Lord Justice Scott was set up for this purpose and it ultimately put forward plans that were to set agriculture firmly in an expansionist phase.

The Report of the Scott Committee made recommendations that formed the basis of the Agriculture Act 1947. This Act guaranteed markets for farmers, a stable and efficient industry, home production levels which would ensure food supplies for national needs, and reasonable incomes to farmers through a system of guaranteed commodity prices. The philosophy contained in this legislation manifested itself through two powerful policy mechanisms. Firstly, a series of financial incentives was made available to improve the infrastructure of farms to allow them to become more 'efficient'. Secondly, to protect these farmers while structural improvements – such as land drainage, field and farm amalgamation schemes, and new and/or improved buildings – were put in place, prices would be supported to allow a reasonable income to farmers. The burden of price support was not to fall on the consumer, but would be paid for from the Exchequer.

During the 1950s and 1960s, however, dependence on price supports grew so that by the time the United Kingdom joined the European Economic Community in 1972, what had been planned as a short-term measure had become the overwhelming proportion of support payments to agriculture. Grants for structural improvements were costing the government only a quarter of that spent on keeping domestic agricultural prices way above those of the world market, mainly through price support.

Throughout this expansionist phase it should be remembered that agricultural interests were not merely the passive beneficiaries of policies which happened to suit government. Under the terms of the 1947 Agricultural Act farmers had been given a seat at the table for the annual negotiation of commodity prices through the National Farmers' Union (NFU). By this alone they had become 'insiders' insofar as they not only had direct access to the Minister of Agriculture, but through him, to Cabinet. The NFU became an increasingly powerful lobby, assisted by the fact that successive governments were well stocked with farmers or land-owners with a vested interest in state support for agriculture. Not only did the NFU have the financial clout which enabled it to put over to local and national politicians a point of view which favoured the continuance of a high level of government assistance, but it also employed full-time liaison officers in the House of Commons and serviced an all-party committee on agriculture. Its public relations department became second to none, issuing thousands of press releases and, in the early 1980s, claiming that over thirty hours of broadcasting each week was devoted to information provided by it. It is not surprising that one commentator (Newby, 1985) concluded that 'in the post-war history of agriculture, it is not only the government of the day, but the NFU which has been responsible for guiding and shaping the destiny of British farmers.'

As agriculture in the United Kingdom became subject to the European Common Agricultural Policy (CAP), the structure of agricultural assistance changed little, although the consumer came to bear a much greater proportion of the direct costs of food production through prices in the shops rather than Exchequer support. Commodity prices were now set annually in Brussels, with differential support for farm products. This increasingly led in the direction of monoculture as farmers tended to concentrate on the most profitable products. Structural support was provided through European Directives or policy instructions to member countries. The most important of these covered farm modernisation, early retirement and socio-economic advice. A fourth directive on 'Less Favoured Areas' applied where remoteness, altitude or otherwise poor agricultural conditions posed a problem for farmers (see Figure 5.1).

By the late 1970s the CAP had attracted wide and popular criticism on three main fronts. First, it was hugely expensive, costing at its peak over 75% of the whole European Community budget. For the United Kingdom

▲ Figure 5.1 Less Favoured Areas. The Less Favoured Areas Directive, introduced in 1972, was extended in 1984 so that it covered 53% of the United Kingdom. Apart from providing more favourable support for agriculture, these areas are now eligible for special assistance for the development of tourism and craft industries.

alone farm support amounted to more than £1100 million by 1980 with support prices averaging 30% of farm incomes. Secondly, it was producing massive food surpluses; butter, beef and cereals 'mountains', wine 'lakes' and so on became accepted European jargon. The third criticism was that all of this very expensive over-production was having a disastrous impact on the environment, an aspect of central concern to this chapter.

However, it was not until 1992 that the reforms to the CAP developed by the then Agricultural Commissioner, Raymond McSharry, offered an opportunity to tackle some of these problems. Certainly, by the middle of the decade these were beginning to have some impact, but the agri-environment policies that are part of these reforms are a very small part indeed of the CAP budget. Lowering the levels of price supports has reduced food surpluses somewhat, although other policies such as agricultural set-aside, under which farmers are paid to leave land uncultivated, have had little effect. Thus the cost of the CAP remains problematic. Although farmers receive less in terms of price support, a series of compensation payments has been introduced with the result that, at the time of writing (1995), the whole cost of the policy is greater in real terms than it was in 1992!

Activity 1

Go back over Section 2.1 and make a list of the main economic and political factors that have contributed to the post-war prosperity of agriculture.

2.2   Environmental impacts of agriculture

The Scott Committee had noted that inter-war agricultural depression had led to much rural dereliction and assumed that a more prosperous agriculture would produce a more attractive landscape. Unfortunately, the structural supports offered to farmers provided incentives to remove features such as hedges and marshland. At that time the Committee was not able to foresee the impacts that large-scale mechanisation and the intensive use of chemicals were to have.

After the United Kingdom joined the EEC it became more profitable to produce cereals than livestock. Grants available for farm modernisation and structural change allowed farmers to claim 50% of the costs of drainage and ploughing. As a result, large areas of former grassland and smaller areas of woodland, marshland and health were converted to intensive cereal production, especially in those lowland areas of England where topography and climate are most suitable.

Activity 2

Re-read Chapter 4, Section 5.3 to revise the principal environmental impacts of intensive cereal production.

Q   Can you, from your own observation, think of at least one other major impact *not* mentioned in Chapter 4?

A   Certainly another important impact on the rural landscape resulting
     from changing agricultural practices concerns the visual effect of new
     farm buildings.

Farm capital grants have encouraged the erection of large farm structures,
further enabling the intensification trend of modern farming in livestock
rearing as well as cereal farming. Increased production generally has
enhanced the demand for a variety of buildings for storage – for example
large silage towers and pits, grain silos and barns. The greater use of ever
larger machinery requires larger buildings to house them and has made
older barns redundant.

As well as increases in the size and number of buildings, there has also
been a considerable change in the style of their construction. Buildings
represent the largest element of fixed capital on the farm and it is therefore
a rational decision to make them as cheap as possible. This has led to an
increase in the standardisation and mass production of buildings, with two
significant results. Firstly, buildings tend to be larger and more uniform
across the countryside and, secondly, they make less use of local materials
and traditional vernacular building styles and so are more out of character
with the local landscape. But unless such buildings have been constructed
within 366 metres of a road, as with all other forms of change in the
countryside motivated by agriculture, no planning consent from the local
authority has been needed. Indeed, the exemption of agriculture from the
provisions of the Town and Country Planning Act 1947 has allowed
farmers a freedom to do what they liked with their land not enjoyed by
other industrial sectors or members of the community. Finally, where
buildings are concerned, farmers were not subject to local rates. The
incentive was therefore to put up large new low-cost buildings rather than
to repair and adapt the old.

Most of these environmental impacts have been concentrated in the

'Nobody will ever build on this land while I'm farming it.'

south and east of England, though they have occurred to a lesser extent everywhere. They amount to a radical change in the visual quality of the landscape, a sharp reduction in the area and diversity of habitats for wildlife and pose a threat of chemical pollution of food and water. Given these environmental costs, it is particularly unfortunate that agricultural policy has been far from an economic success.

2.3   A declining rural economy

The gross economic inefficiency of agricultural support in the UK has already been hinted at in this chapter. It is this kind of economic impoverishment by agriculture that catches the public eye – consumer food prices have been high, surplus agricultural commodities cost hundreds of millions of pounds to store, and the vulnerability of both price and structural support mechanisms to abuse has been considerable. In the view of many economic analysts, such a situation is an inevitable consequence of an industry maintained in hothouse conditions and fuelled by public money, but outside the disciplines of market forces. This system of support has led not only to problems for the rural environment, but also to considerable problems in the rural economy itself.

In order to guarantee a reasonable income to farmers, the system of indirect supports through guaranteed prices to farmers was established. Generally, these prices were set so as to allow marginal farmers to stay in business, but invariably this has meant that farmers who were not marginal were able to earn quite considerable incomes. Indeed, until the late 1980s the City saw such farming prosperity, especially in the south and east of England where it was reflected in high land prices, as a sound investment proposition for pension and other long-term funds.

But what of the agricultural workers? Well, they have done particularly badly out of modern agricultural policy. This is because government grants have been made available for *capital inputs* to land drainage systems, farm machinery, fertilisers and so on, but never to *labour inputs* to agriculture. This has had the effect of making capital relatively much cheaper than labour and as a consequence has caused much labour to be shed and substituted by capital wherever possible. In 1981 there were nearly 100 000 fewer people working in agriculture than in 1971. Moreover, the Transport and General Workers Union has estimated that between 1985 and 1987 there had been a further drop of around 20% in the full-time agricultural labour force, and a further reduction of 30% is anticipated by the turn of the century. Unfortunately for the farm workers, they have not benefited from corresponding higher wages, a factor which in itself has increased the drift from the land.

These kinds of labour losses have had repercussions in rural society as a whole. At one time, agriculture was the principal employer in the rural economy and many rural communities were dependent on agriculture for their livelihood. As the agricultural workforce has declined there has been a drift from the land and a resultant break-up of the local community in many areas. Tales of declining rural services, such as shops, transport, schools and health care, declining rural incomes, and in the more remote areas an ageing population, have provided the focus of rural community concern from the 1970s through to the 1990s. This out-migration of agricultural labour is counterbalanced in some rural areas by an influx of those who wish to live in the countryside but work in the towns, although their presence has only exacerbated the story of decline. For these are the people with their own transport, only too willing to use their cars to travel to group practice surgeries in the next village, or to shop at the nearest supermarket rather than in the village store. This has given a further twist to the downward spiral of provision of local transport and other local services, on which the less advantaged who live and may still work in the village must rely.

Thus the rural economy has experienced vast capital investment in agriculture and a considerable loss of labour, both effects of agricultural support policies. The workings of the rural economy are little understood, but despite the huge levels of subsidy going into agriculture it seems that the overall contribution of this particular sector to the local economy has been slim: even in agriculturally intensive areas such as rural East Anglia, farm output accounts for less than 10% of the region's Gross Domestic Product, and in more marginal areas its contribution could be much less than this. Such large inputs of capital could certainly have been used much more efficiently in other sectors of the economy.

In distributional terms, agricultural support policies have thus often had an invidiuous effect. The huge financial support given to agriculture may provide a reasonable income to some farmers, but does nothing for other workers in the rural economy. This sets up artificial disparities in income in rural areas where, despite farmers' earnings, wages even today remain well below those in urban areas. These distributional effects are at their worst in the remotest areas, where Less Favoured Area supplementary payments to farmers further emphasise the income differences between them and other workers in the local economy.

2.4 Summary

From the days of the Second World War the agricultural production imperative has spawned a very expensive industrialised sector in the countryside as the overwhelming and relatively unfettered land-user. The attainment of production goals has had its costs in the unfairness and inefficiency within the rural economy and in the quality of the rural landscape and ecosystem. The principal impacts on the rural landscape have been: the ploughing of permanent pasture; agricultural intensification which has been manifest in increased field size, the loss of hedgerows and a generally improverished ecology; and a whole range of new industrial farm buildings. The causes of such tangible expressions of rural change can be found in the attitudes to agriculture of the first post-war government and its successors and those of the EU. However, of no small importance has been the effectiveness of the agricultural lobby with the general public, in Whitehall and in Brussels, thus maintaining a dominant role for agricultural support policies long after there was an objective rationale for them. Even the 1992 policy reforms have ensured, through compensation payments, that considerable income support to farmers remains.

Activity 3

Before moving on to look in much more detail at other land-use interests in the countryside, much of Section 2 can be reviewed by answering the following question.

Q   Who have been the beneficiaries from the agricultural policies outlined here and who have been the losers? Make two lists and then compare yours with those given at the end of the chapter.

3   Conservation on an ebb tide

3.1   Amenity conservation and the countryside

Conservation, if it can be called a land use, has provided the sharpest 'on the ground' challenge to agriculture. Interestingly, policies for the conservation of the countryside found favour with the same wartime government that had been responsible for reconstructing national agricultural policy. Even in the nineteenth century there had been a wide range of conservation movements for the countryside, deriving from a number of intellectual origins, but very broadly speaking their parliamentary successes came in two distinct but related spheres – those of amenity conservation and scientific conservation (discussed in the next section).

Amenity conservationists had been essentially concerned with landscape conservation and, for this reason, their objectives had been to ensure greater public access to the countryside. Organisations such as the National Trust, the (then) Commons, Open Spaces and Footpaths Preservation Society and the Council for National Parks did much to ensure that landscape conservation designations such as **National Parks** and Conservation Areas (later to be termed **Areas of Outstanding Natural Beauty (AONBs)**) were incorporated into the National Parks and Access to the Countryside Act 1949. This was one of a series of laws that included the 1947 Agriculture and Town Planning Acts and indeed the New Towns Act 1946, that were to provide the thrust for post-war reconstruction.

These land-use designations for amenity conservation were to offer the principal statutory means of protecting the landscape, but they applied only to England, Scotland and Wales. In Northern Ireland moves were made early on to parallel the British legislation, but nothing was done until the Amenity Lands Act 1965. Even by the early 1990s, though, there were no National Parks, in this respect resembling Scotland. The situation in both countries has been perhaps a reflection of the less intense pressures on landscape, compared with England and Wales, as much as that of the powers of the large land-owners! However, the situation is changing with the question of National Parks back on the agenda in Scotland: a report emerged from the Countryside Commission for Scotland in 1990 offering strong support for them. In Northern Ireland there are now eight AONBs, along with 38 in England and Wales, while Scotland has its own equivalent **Areas of Outstanding Scenic Interest (AOSIs)**. These remain the single common feature of landscape protection in all the countries of the United Kingdom, but they have always been without the more complex administrative arrangements afforded to National Parks. Indeed, over and above normal planning constraints, the designation of AONBs has remained little more than a recognition of a need to safeguard such areas.

In addition to National Parks and AONBs, the concept of **Heritage Coasts** was introduced in England and Wales in the 1970s with the same principal purpose – that of protecting high-value landscapes. But in their role of providing a framework for combined voluntary and local authority management of vulnerable coastal areas they were also to be non-statutory. By the 1990s these amenity conservation designations (shown in Figure 5.2), together with those for scientific conservation (which will be considered

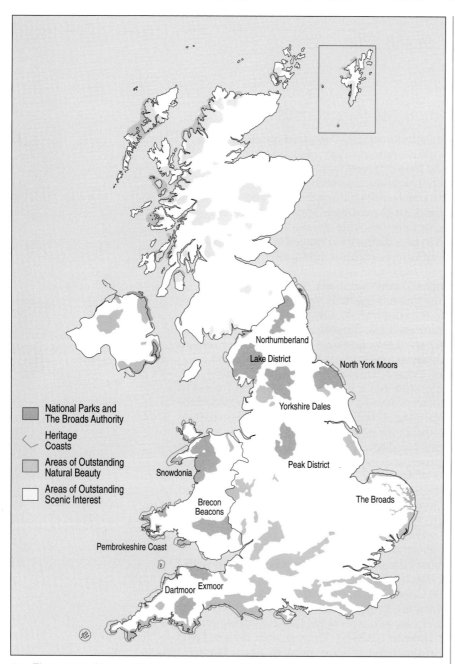

▲ *Figure 5.2 National Parks in England and Wales, Areas of Outstanding Natural Beauty in England, Wales and Northern Ireland; Heritage Coasts in England and Wales; and Areas of Outstanding Scenic Interest in Scotland, 1995. National Parks cover nearly 10% of England and Wales; AONBs cover almost 13% of England and Wales and 18% of Northern Ireland; and AOSIs cover 13% of Scotland.*

below), accounted for a quarter of the total land area in England and Wales – a proportion that in itself can be considered a conservation success.

But how effective have these designations been at protecting the landscape, particularly against the increasing modernisation of agriculture already discussed? In truth, they have had substantial shortcomings in this regard. At the time of the 1949 Act, it was considered that most of the protection of National Parks and AONBs could be adequately carried out with the use of land-use planning control powers introduced in the Town and Country Planning Act 1947. In effect, until the introduction of management plans in the 1970s and stronger legislation in the 1980s (which will be considered in Section 5) there was little more protection for the landscape in these areas than that provided by the normal development control functions of the local planning authority.

Most important in this respect, of course, was the fact that nearly all of the agriculture – and indeed forestry – was exempt from land-use planning controls, even in National Parks and AONBs, so that landscape protection was not made easier in these designated areas. Such voluntary measures as were introduced for farmers in these areas counted for nothing without financial support, because it was the financial incentives of agricultural policy that were the principal spur to agricultural landscape changes. There was thus one government department, now the Department of the Environment (though it has changed its name several times since the war), trying to conserve our more valued landscapes with very limited physical planning controls and another, the Ministry of Agriculture, Fisheries and Food (MAFF), indirectly dismantling the landscape in the ways outlined.

Ironically, most National Parks are in remoter upland areas which, by the 1970s, were precisely the areas designated by the EC as 'Less Favoured Areas' where livestock headage subsidies and grant aid above the norm was available for farm modernisation. For a time in the 1970s it was even possible to get improvement grants for farms on land that MAFF itself classed as unimprovable! The tide began to turn a little in the 1970s with much argument about the ploughing up of moorland in Exmoor National Park (as was mentioned in Chapter 4, Section 5.2). Between 1954, when the 68 632 hectares of Exmoor were designated, and 1975, 20% of its land was enclosed and subject to 'agricultural improvement'. This led to a government Report, produced by a committee chaired by Lord Porchester, which recommended stronger powers over agricultural activities in National Parks. These powers, which were eventually introduced for areas other than just National Parks, are considered more fully in Section 5.

As for AONBs, it is now commonly recognised that the landscape conservation powers relating to them have been and continue to be somewhat ineffectual. Research has indicated that even the development control function in AONBs is often little different from other areas and other efforts towards landscape conservation are very much at the whim of the constituent local authorities. Although at the end of the 1970s the Countryside Commission, the government's advisory body on landscape conservation and countryside recreation, was debating whether to abolish AONBs altogether, a committee which it had sponsored, chaired by Kenneth Himsworth, reported that the mere fact of their designation had at least helped in the proper ordering of priorities in them. On the strength of this comment, the Commission decided to continue with them and by the end of the 1980s the largest of them all – the North Pennines – was designated. Perhaps the prevailing view about AONB designation is best summed up by the then Council for the Protection of Rural England: 'AONBs actually mean very little, but we're awfully glad that we've got them.'

The Chilterns AONB – a rare exception to the general rule. Thanks to an effective AONB management committee and a vigilant environment pressure group, the Chilterns Society, it is still a relatively intimate landscape of woodland and small fields. This has been achieved in spite of pressures from agriculture and urban development – London is little more than 56 km away.

Q   What have been the main weaknesses in terms of landscape conservation in National Parks and AONBs in the forty-five years following their inception?

A   In both, agriculture and forestry were largely exempt from planning control and there was no special protection exercised by the Department of the Environment against the forces promoting agricultural development through MAFF. If National Parks were designed positively to promote conservation and enjoyment objectives inside their administrative structures, for AONBs there were to be no such arrangements – only a presumption in favour of landscape protection, a factor which could merely be taken into account in any planning decision.

   Many pressure groups wanted the 1949 Act to make it mandatory for the land-use planning function of local authorities in a National Park to be carried out by a single body known as a joint board, where the Park covered more than one county. Although there were obvious advantages in such a cohesive approach, in the end the joint board issue was left to be decided locally. Only two such joint boards currently exist – for the Lake District and the Peak Park – and these National Parks have proved to be the most successfully administered. However, the Environment Act 1995 contains a proposal to create independent boards for all the National Parks.

3.2   Scientific conservation and the countryside

If the amenity conservation provisions for the countryside since the War have been widely considered to be less than fully effective, **scientific conservation** has had some noted successes. The key to this was, initially at least, in the use of the word 'science'. Pressure groups prior to the 1949 Act, such as the British Correlating Committee representing a range of scientific interest groups, and the Society for the Promotion of Nature Reserves, gained much of their momentum from an increasing growth of and interest in a new strand of the natural sciences in the twentieth century – **ecology**. This study of complete natural systems, rather than just individual components of them, led to calls for the conservation of the fauna and flora of the countryside, and not just its landscape, for reasons of scientific research. The scientific 'mystique' of this kind of conservation led to most of the provisions for nature conservation passing through Parliament into the National Parks and Access to the Countryside Act 1949 with hardly any discussion.

In part, scientific conservation – or **nature conservation**, as it is commonly termed – was given stronger powers than amenity conservation although they were to be applied over a much smaller area. The (then) Nature Conservancy (later to become the Nature Conservancy Council, at least until 1991) was to give scientific advice on the control of fauna and flora and was to have a significant role in the provision of a national research programme for nature conservation. Importantly, though, this government agency was also given powers for the purchase and management of specific areas of land as nature reserves.

Thus, some of the more effective land-use designations, in terms of their ability to resist agricultural intensification, were introduced – including **National Nature Reserves (NNRs)**. These were either managed under a nature reserve agreement with the owner or owned by the Nature Conservancy and were thus overwhelmingly disposed towards the conservation and study of our countryside fauna and flora. Other nature conservation designations were also instituted. Areas considered important for nature conservation but not of the national significance of NNRs were introduced as *local nature reserves*, set up by local authorities but managed by wildlife trusts. More specific sites could be covered by a designation known as a **Site of Special Scientific Interest**. This was essentially an area about which the Nature Conservancy could notify the local planning authority of its scientific worth in the context of development proposals. Again, however, some of the most damaging operations in SSSIs were to come from agriculture and forestry, where the local planning authority had little or no control. The Nature Conservancy had to negotiate as best it could with the Forestry Commission and the Ministry of Agriculture separately in order to have its interests better respected.

In terms of land-use control, then, scientific conservationists were given some quite effective powers in National Nature Reserves in which they have been able to resist agricultural intensification reasonably well. Protection of local nature reserves and SSSIs was less effective up to the beginning of the 1980s, because their management and ultimately their ownership have required that scientific conservation has been much more a matter of negotiation. Indeed, at that time the Nature Conservancy Council reported that some 13% of SSSIs suffered damage to their wildlife interest each year. And importantly for the competing demands for land in the countryside overall, all these designations represent only pockets of land in a broader countryside where the impoverishment of a rural ecology has been taking place in the manner described in Section 2.3. But in the 1980s changes in government policy also shifted the land-use emphasis of scientific or nature conservation. This is discussed further in Section 5.

3.3   Disparate views of the conservation purpose

Undoubtedly, the fact that agriculture dominated conservation as a competing demand in the countryside was principally due to the drive to increase food production aided by a powerful unified agricultural lobby. But a number of commentators have suggested that the development of the conservation movement for the countryside has in part been inhibited by the fact that there are so many disparate views about how and why the countryside side should be conserved. These are often conflicting and as a result their political force is weakened. Certainly conservation interests have not had any of the 'insider' influence of the kind enjoyed by the NFU and, unlike the farmers' lobby, they do not speak for any specific interest other than society at large.

It has already been mentioned that before 1949 countryside conservation was broadly split into the amenity and scientific camps. But even within these there was a broad spectrum of (overlapping) views. This situation has become even more complex in the years since the Second World War, and countryside conservation has been characterised by a wide range of different pressure groups all championing their own causes. The Council for the Preservation (now Protection) of Rural England, for example, has been concerned with both scientific and amenity conservation and it has been particularly effective, not so much in conservation management, but in political lobbying. The Royal Society for Nature Conservation, on the other hand, has developed an excellent voluntary management framework for scientific conservation in the countryside through the County Wildlife Trusts, but this has been at the expense of its political influence. The Royal Society for the Protection of Birds has a good political profile but has a conservation emphasis that is narrower than some. The National Trust is very much an 'establishment' pressure group, but its early interest in nature conservation was considered by many to be misdirected. Although the National Trust Act 1907 specified that it should acquire and manage sites of special interest to naturalists, it was slow to secure these and did so in a random fashion with little regard to their national importance. The Society for the Promotion of Nature Reserves was set up in 1913 precisely to rectify this omission. Although organisations such as these are quick to form affiliations in arguing a common cause for countryside conservation, their overall profile inevitably remains somewhat fragmented, particularly when set against the lobbying powers of the agriculture industry.

The clarity of the countryside conservation purpose has been weakened further by the 'alternative life-styles' revolution of the 1970s which spawned more global environmental groups such as Friends of the Earth and Greenpeace, and the 'green' revolution of the 1980s which has seen environmental issues move to the top of the political agenda. Ironically, the growth in environmentalism that these developments have brought about has done little to raise the profile of countryside conservation, quite simply because very few politicians make any clear link between the conservation of the ozone layer and the diminution of the greenhouse effect on the one hand and the sustenance of the diversified countryside ecosystem on the other. For example, in 1989 at the same time as the government was backing a global reduction in the use of chlorofluorocarbons (CFCs) in refrigerators and so on, it was cutting the Nature Conservancy Council's budget in real terms and, somewhat contentiously, proposing to amalgamate it with the Countryside Commission in Wales and Scotland. These new arrangements would break the mould of forty years' experience relating to the separate existence of these bodies and their predecessors, although at the time when amalgamation was first discussed, the government was arguing in Part VII of its proposed Environmental Protection Bill for a more sensitive and accountable framework for conservation as it 'continues to rise in importance on the public agenda'. The new combined country agencies would be able, it said, to tailor the delivery of conservation more closely to regional and local needs. This was not how most of the major wildlife and conservation organisations saw it, however, and they expressed considerable reservations, not least because, in their view, the British Isles form a bio-geographical entity which is best served by a co-ordinated scientific approach. Such criticisms were, however, only partly mollified by the decision to set up a joint committee to handle matters of common interest in England, Scotland and Wales, including scientific standards, and the commissioning and support of research which transcends country boundaries.

Nevertheless, in spite of the claims by government during the progress of the Bill in both Houses of Parliament in 1990 that these proposals – like the 1949 Act – were 'a milestone in the promotion and appreciation of our natural heritage', misgivings continued to be voice. These largely concerned the likely funding of the combined organisations and the general failure by government to achieve consensus on the proposed reorganisation; the all-important non-government agencies continued to argue that conservation in practice would be considerably weakened by these changes and that Part VII of the Bill should be withdrawn pending proper consultations. Moreover, as the Countryside Commission in England pointed out, not only was the emphasis of the proposed joint committee on nature conservancy rather than wider countryside policy, but the Commission would not itself have a place as of right on that committee. Its fears were seen to be well founded when the Bill passed into law in 1991, an Act which amalgamated the Countryside Commission in Wales with the activities of the Nature Conservancy Council there to form the Countryside Council for Wales. A similar marriage in Scotland established Scottish National Heritage.

Q   Can you identify any links between countryside conservation and global issues?

A   Both are being affected by the industrialisation and intensification of agriculture.

3.4   Summary

The principal designations for amenity conservation have been detailed and their shortcomings examined, concluding that, until the 1970s and 1980s, they offered little more protection than that provided by the Town and Country Planning Act 1947 and even this did not apply to agriculture and forestry. In addition there were the policy conflicts between the Department of the Environment with its landscape protection interests and those of MAFF, often working in the opposite direction. In contrast, scientific conservation has fared better, gaining stronger powers especially in the designation of NNRs which have often been able to resist agricultural intensification. An examination of the conservation lobby has revealed its basic weaknesses compared with that for agriculture; that its interests have suffered as a result is beyond doubt.

4   Increasing pressures from other land uses

Since the War tensions between agriculture and conservation in the countryside have been perhaps the most obvious because of their widespread visible results. But during this period a whole range of other rural land uses has been growing, making the competing demands for the countryside more intense. This section examines a number of these before placing them all into a specific geographical context by assessing competing land-use pressures in a National Park.

4.1   The growth of tourism and leisure

There have been two main types of policy for countryside tourism and leisure since the War. The first, social policies, have been concerned to increase the opportunity for the enjoyment of the countryside by the public at large. The second, land-use policies, have focused on the more legalistic mechanisms relating to how public access to private land might be achieved. Despite the first set of policies being the principal thrust for government action, it is this second set that has been the most controversial since they have attempted to procure from land-owners access rights to areas that have invariably been given over to competing land-uses.

Up to the National Parks and Access to the Countryside Act, there had been active pressure on the government both to define the network of public footpaths in England and Wales more clearly and to allow greater access to open land, particularly in the mountains. Provision was made in the Act for the proper identification of all footpaths to which the public had a right of way through the production by each local authority of a Definitive Map. Although a fine notion, this has been fraught with difficulties and the work is still not complete! In the Act, provision for access to open land was not very satisfactory. In the designation of Access Agreements, local authorities had to negotiate payment to the land-owner in exchange for access – a system that was never widely adopted. Such a state of affairs created problems in the 1950s and 1960s, when the popularity of countryside recreation and tourism began to grow

◀ *Many pathways in the countryside are ploughed up, as in this example, or remain obstructed or unmarked, in spite of the provisions of the National Parks and Access to the Countryside Act 1949. A key preoccupation for the Countryside Commission in the 1990s is to see that this situation is remedied.*

considerably. Increasing car-ownership, leisure time and paid holidays led many people to fear a 'recreation explosion' where the countryside would be flooded with urban populations at play.

During the 1960s the Labour government was keen to develop further public opportunities for access and introduced a whole new range of essentially local authority facilities in the Countryside Act 1968. Country parks, picnic sites and transit caravan sites were designed to cater for the growing popularity of rural leisure, but local authorities implemented their designation in a rather different spirit. Literally hundreds of country parks and picnic sites were designated during the 1970s, but primarily as a means of *containing* the recreation 'explosion' that they felt might overwhelm National Parks and AONBs, rather than with the main aim of increasing opportunity.

These areas, however, provided yet another set of rural land-use designations, in addition to those for conservation that were considered in the previous section, that allowed some protection from agricultural development. Since the 1970s, though, there have been problems with these designations and the reasons behind them. Firstly, it has never been clearly established that the recreation explosion ever really happened in the way that most leisure planners had feared. Visits to the countryside seem to be much more susceptible to the vagaries of the weather and the day of the week than to indicators of material affluence. Furthermore, there has never been any comprehensive study of whether a large number of visits actually harms the physical and ecological environments of the countryside anyway, although some isolated examples (such as footpath erosion on the Pennine Way – see Plate 10) are evident enough. Generally, most people seem reluctant to stray more than 100 metres or so from their cars.

Secondly, surveys in the 1980s have shown that people do not actually like country parks and picnic sites much – they prefer the 'real' countryside. This has put the pressure from recreation in the countryside squarely back onto the public footpath network, reviving the frictions that have existed for over 200 years between the rambler and the land-owner. The attitude of the land-owner has changed slightly in the later 1980s and into the 1990s, as rural recreation and particularly tourism have been reassessed for their potential in agricultural diversification, a term used to describe attempts made by farmers to find ways of earning income other than from the production of food commodities which are in surplus. (This is discussed more fully in Section 5.)

The problem of using recreation in particular for diversification, though, is that much countryside recreation is available freely and as of right to the public and it is therefore difficult to generate much farm income from it. Quite naturally, this has led to a keenness for the development of farm tourism rather than just countryside recreation because it is the overnight stay that offers income potential. There are inherent problems here, however, in competing with the cost and the climate of foreign holidays and in developing an economic base that will ever be anything more than seasonal. Besides which, over-provision of accommodation can easily arise, as has already happened in recent summers in Devon and Cornwall.

Farm diversification, however, is encouraging more active forms of recreation in the countryside, since this is income-generating, and it is in this quickly growing sector that recreation impacts on the rural environment are likely to be the greatest. In the late 1980s and early 1990s, literally hundreds of proposals for golf-courses reached district planning authorities each week. New sites for hang-gliding, clay-pigeon shooting, motor-cycle scrambling and so on were increasingly being set up on farm land and the number of set-piece war games that could be witnessed in the countryside on a weekend became considerable.

4.2   *Expanding timber production*

The expansion of timber production in the United Kingdom has many
parallels with that of agriculture. Although more cyclical, when expansion has
taken place it has been rapid. It has similarly had a number of detrimental
environmental effects which have offended public opinion and prompted a
policy change: in 1988 tax concessions that had attracted the very wealthy
were removed from afforestation schemes. This followed the controversy
created by a scheme for planting one of Britain's few remaining primaeval
landscapes, the Flow Country of Sutherland. The largest controversy
surrounding timber production, however, is its economic viability.

The spur to timber expansion has essentially been the perceived need to
hold a strategic reserve of timber. After the First World War timber
depletion was so large that the Forestry Commission was set up under the
Forestry Act 1919 to undertake an aggressive programme of planting to
double the national area of the forest estate. A similar programme was
instituted after the Second World War under the Forestry Act 1945, but as
well as public planting, a complex series of grants and fiscal concessions
was instituted under the Forestry Act 1947 for the private forester, the
principles of which remain with us today.

By the 1970s and 1980s the argument for expanding timber production
had become a very difficult one. Since the early 1970s the Treasury has
indicated that home timber production is not economic since its capital
investment offers a very low rate of return indeed, calculated to be within
the range of 3 to 5%. This was well below the rate which the Treasury found
acceptable for public investment. An attempt to measure the effectiveness
of investment programmes which takes into account social benefits such as
recreation and employment is made by cost-benefit analysis.

Q   What kind of methodological problems can you envisage arising in
trying to assess the return on investment from forestry?

A   (a)   Trees take a long time to mature. It is very difficult to forecast
likely timber prices in, say, 50 years' time.
(b)   If strategic or balance-of-payments assumptions are made, it is not
remotely possible to envisage the likely role of these with respect to
timber production half a century or so hence.
(c)   Forest planting means new jobs, but it is difficult to calculate the
cost of their creation, or, indeed, their value in social terms. Such forests
are often in remote areas, where alternative employment is very scarce.
(d)   People use forests for recreation, but some of these attract very few
and some many hundreds of thousands a year and each visitor may
spend anything from a few minutes to several hours there. Therefore the
measurement of recreational benefits of a forest is problematic.
(e)   The capacity of recreation sites to absorb visitors varies greatly. It
is hard to know just how many can be absorbed by a mature forest
compared with other recreational sites.

However, many economists do not accept the conclusions of cost-benefit
analysis for the above reasons and because they say that the scope of the
analysis is too narrow, and that the whole wood-processing industry and
its allied services which are dependent on United Kingdom forest produce
should be taken into account. For example, the net subsidy per job in the
industry as a whole is much less than if only those employed within the
forest are considered. Thus although cost-benefit analysis *seems* to offer
some precision in determining return on investment, its conclusions with
respect to forestry are dependent on a whole range of initial assumptions.

▲   Commercial timber developments were, until the late 1980s, spurred on by tax concessions for the rich. But they have also proved controversial, especially if it can mean the destruction of one of Britain's few remaining primaeval landscapes, the Flow Country of Sutherland. The lower photo shows the land drained and channelled ready for planting.

Not surprisingly it is the nature of these assumptions that still bedevils reports on forestry. In 1986 the National Audit Office, an independent body set up to review the economic efficiency of public organisations, questioned the real worth of some of the Forestry Commission's claims about the public benefits of forestry. These included employment creation and recreation potential, as well as balance-of-payments and strategic supply arguments. A year later, the House of Commons Public Accounts Committee, the all-party watch-dog on public expenditure, claimed that too many of the Forestry Commission's activities remained unquantified. While conceding that a low rate of return by forestry might be justified on job creation grounds – especially in remote areas – its reasons for making such a contention needed much better supporting evidence.

It is, as with agriculture, however, the landscape and ecological impacts of modern forestry practice that have attracted the most criticism of this second-largest user of rural land. Conifers are fast-growing trees and offer a reasonably quick return on investment by forestry standards. But the size of the plantations required to maximise profits, their geometric boundaries, and the regimented rows they contain, can be a significant eyesore. Some of the more recent attempts to screen these huge blocks of timber with deciduous planting have been considered to be little more than crudely cosmetic. In ecological terms, too, such large-scale monoculture drastically reduces the species diversity of an area, particularly amongst insects, and efficient woodland management reduces the number of decaying and rotting trees upon which much wildlife depends. Coniferous plantations can also lead to soil acidification, thus adding to the effects of acid rain. Moreover, some have also argued that large-scale timber production could lead to an increased demand for more inland water storage reservoirs because forests, relative to pasture land, inhibit surface run-off by about 25%.

Finally, it is worth noting that recent attempts (1994) by government to privatise the Forestry Commission were not approved by the advisory body set up to examine its alleged benefits. However, a number of concerns remain about the scale of the disposal of publicly owned forestry land and the possibility that public access to it might become more restricted as a result.

◀ *Regimented conifers, a Forestry Commission practice that gave way in the 1980s to a less geometric approach and a reversion to more a traditional form of woodland planting, also shown in this example.*

4.3 *Water collection in the countryside*

The most apparent effect of water collection is the large river-regulating reservoirs which drown large tracts of farmland and dwellings and also, in some extreme cases, entire communities. Such reservoirs may mean the loss of fertile agricultural lowlands, as in the case of Rutland Water in the county of Rutland, but more commonly they flood the valley bottoms of upland areas. This is devastating for the local economy since valley bottoms are not only the most agriculturally productive areas but also where settlements and communications networks predominate.

This impact could be minimised by using alternative sources of water collection such as underground aquifers or by employing sea-water desalination plant, but river-regulating reservoirs have traditionally been the cheapest means of storing water and have the added advantage of allowing downstream flood control. Any kind of water collection can obviously only be justified if the water is actually needed. However, it has increasingly been argued that in many cases river valleys should not be flooded either because the water is not actually in demand or because it would be more cost-effective to manage water consumption.

It has also been suggested that, in a water industry dominated by engineers, promotion and prestige are more easily associated with large new reservoir construction projects than with the development of demand management policies. In addition, there has always been a tendency by the water sector to produce too much water because the consequences of shortages are severe, but those of surpluses are negligible. This is because water has historically been cheap to store and distribute and the way that people pay for it – through rates rather than related to consumption – hides any wastage. In addition, the way in which planning permission for reservoirs is obtained always puts national need criteria above those of the locality. Thus local opposition on agricultural and conservation grounds can be quite forcefully overridden by claims of a national need for water.

Since the 1960s proposals for the construction of new reservoirs in rural areas have been hotly contested because the need for their water has never been satisfactorily proven. Certainly, this was the case over the development of the Kielder Reservoir in Northumberland. Kielder was to become the largest reservoir in Europe, built with large contributions from the European Commission's Regional Fund and justified by the requirements of a growing regional economy during the late 1960s and early 1970s. Subsequent recession made Kielder into the reservoir that nobody actually needed. Many people have concluded, with the benefit of hindsight, that the money would have been much better spent on renewing the water pipeline infrastructure of the region, which, originally Victorian, had reached the end of its useful life. By the early 1990s more than a quarter of the total drinkable water consumption in the region could be accounted for through leakages, a figure close to the national average for such losses. More recently, however, the figure has dropped to less than 17%, well below the national average, whilst the recent development of a modern electronics industry in the north-east and the avoidance of water-supply restrictions in the region could suggest that Kielder is at last needed.

The advantages of reservoir water collection as a rural land use lie chiefly in the water-based recreation facilities that they offer. These constitute one of the fastest-growing sports sectors and provide important 'honeypots' to allow people to congregate in the countryside in relatively large numbers. Nevertheless their construction is, as in the case of Kielder, usually resisted by the public on environmental grounds. If the new water companies, following water privatisation in 1989 (except in Scotland), still favour reservoirs as a means of increasing supply, it is because of the high cost of the repair of the distribution system compared with the development of a reservoir and because of the personal proclivities of water engineers. However, if the problem of new water storage sites becomes too acute because of strong opposition, undoubtedly the alternative for them will be the aggressive pursuance of policies of demand management, almost certainly involving the widespread use of water meters.

4.4   *Mineral extraction in the countryside*

For much of this century, iron ore and coal were, in volume, the largest of the extractive industries. Having been long established, they, along with tin-mining, had created an urban countryside where they had been won. The mining villages of County Durham, the Welsh valleys and, for tin, the Cornish coast, had long developed into distinctive physical and social entities in the countryside. In the post-war context their environmental impact became first one of dereliction and decay, and then later one of reclamation and restoration, sometimes back to the agriculture that had preceded them.

In a more localised way, specific minerals have had a very significant impact on particular countryside locations in the post-war period. The winning of fluorspar and the problems associated with waste disposal in the Peak District, the potash mining in the North York Moors National Park, the quarrying of slate and mining for gold in Snowdonia and the extraction of china clay in Cornwall are all good examples of this. (See Plate 11.)

But since the War it is perhaps the increased extraction of building materials that has had the most pervasive effects on the countryside environment at large. The exploitation of sands and gravels for the construction industry has been widespread, for their workings can be found in many of the lower reaches of river valleys in Britain. The demand for them has grown phenomenally in response to increased infrastructural, housing and industrial requirements in a period of post-war reconstruction and economic growth. As shall be seen in much greater detail in Book Three of this series (Blunden and Reddish (eds), 1996), the impact of aggregates extraction on the countryside is significant because of its broad geographical spread and falls into two main phases – the extraction phase and the after-use phase. Fairly soon after the Second World War, legislation was introduced to tighten environmental controls over both of these phases. The Minerals Act 1951 introduced in particular an obligation on the part of the minerals companies to restore mineral workings to an acceptable state for after-use. But this statute was not backdated and so all mineral workings that were started before 1951 – and most of these have been abandoned only since the 1980s – had no such requirements placed upon them. The environmental deterioration in these areas remains a significant problem in the countryside today. The problem of environmental impact has also been exacerbated by the fact that workings begun before 1957 can be exploited as long as their owners think fit. The compensation for closing such works has rarely been regarded as a viable proposition.

The problems of the minerals sector, particularly in terms of its environmental effects on the countryside, were examined closely in the 1970s by three investigating teams set up by the Department of the Environment, largely as a counterbalance to new legislation (Minerals Exploration Act 1971) aimed at providing financial help in the costly exercise of finding and developing fresh mineral resources. Two of these reports called for a more effective planning controls system for minerals, and the national co-ordination of aggregates exploitation in order to reduce environmental impact. The third put forward specific proposals to diminish the environmental impact of all minerals extraction other than aggregate materials.

Despite the valuable work carried out by the three teams, little has changed legislatively to improve the environmental impact of minerals extraction in the countryside. The Town and Country Planning (Minerals) Act 1981 has extended the 1951 Act's minerals after-use requirements to include what is now termed 'aftercare', whereby the minerals companies must now sustain an on-going management regime after minerals have been worked out, rather than just a once-and-for-all restoration programme. But this has done little to help local authority planning departments (usually county councils) with control over permission to extract minerals in the first place. Although planning permission is required for minerals extraction, if it is refused by a county council, minerals companies may appeal to the Secretary of State for the Environment. The problem here is that the Secretary of State may use different criteria for judging the application from those used by the county. Whereas the county will try to balance environmental considerations against local minerals needs, the Secretary of State may balance them against national needs and find in favour of an application that could not be justified in terms of local needs alone. Thus a decision about minerals extraction and the impact such a working can have on the countryside may be taken out of the hands of the local planning authority altogether.

Nevertheless it remains an impact that can be considerable. The whole of the development of the sand and gravel workings now occupied by the Cotswold Water Park in the Thames Valley in Wiltshire and Gloucestershire, for example, which has completely changed the face of many hundreds of hectares of agricultural land, was exploited most intensively as the result of the national need to build the M4 motorway and to develop Swindon. But, unlike this area, many of the 'holes in the ground' have no after-use requirements placed on them at all since they were begun before 1951.

Whether the effect of minerals extraction on the countryside of the United Kingdom is likely to diminish or increase is a complex question. Certainly the demand for minerals will continue to rise and if alternative approaches to their extraction and their use, especially aggregates, are not adopted, then areas of the countryside that so far have remained unexploited in this way will become affected.

One alternative, that of the super-quarry, is now back on the agenda in the 1990s, having first been suggested twenty years earlier. However, the notion of aggregates production on a very large scale at a remote coastal location in Scotland, subsuming the role of many other smaller operations and taking advantage of cheap sea transport to major areas of demand, is not without its own environmental critics. Indeed, such a proposal for Harris has been roundly rejected by its inhabitants following a public inquiry in 1994.

4.5   Military training and other national needs

If the control over minerals development in the countryside suffers from the problem that national need can often overrule local interest, this is also true of a number of other large-scale land uses. Principal amongst these is national defence. Throughout the countryside there are areas set aside for military training, perhaps the best known being Salisbury Plain and large

◄   *Figure 5.3*
Major military training areas in
Britain. Principal training areas are
associated with a wide range of
training facilities together with
permanent camps and bases;
intermediate training areas have
more than 40 hectares devoted to
live firing and/or dry and adventure
training; specialist training areas
concentrate on single combat units,
for example tank training at
Castlemartin.

parts of Northumberland and Dartmoor National Parks. Such areas are
scattered throughout the countryside particularly in its wilder, more
attractive parts and on the coastline – precisely those areas that are most
popular for the purposes of rural leisure, as Figure 5.3 indicates. Indeed, the
National Parks have suffered especially from the demands of the Ministry of
Defence: while Dartmoor and the Pembrokeshire Coast have around 5% of
their land area owned or leased by the MoD, in Northumberland the figure
rises to 22%. Although low-flying aircraft can create disturbance over a wide
area, perhaps more serious are the physical dangers that training exercises
on firing ranges may cause for the straying rambler. Use by the MoD can
cause great archaeological damage and also removes the possibility of most
other rural land uses, particularly of a more productive kind. However,
military training can have its positive side since evidence suggests that the
restriction of public access and many forms of agricultural activity which
such a use imposes can favour the maintenance of great ecological diversity.

In addition to military training there are other large-scale intrusions into
the landscape concerned with defence. Undoubtedly the best-known of these
is the Fylingdales early warning system in the North York Moors National
Park, which comprises a huge truncated pyramid sitting at the top of an open
moorland landscape (installed without planning permission). More
commonly, large radiotelescopes – massive satellite dishes – are to be found
in many, often flat, parts of the countryside where they can be seen for miles.

Over these and many other intrusions in the rural landscape (including
those discussed above) the government has the final say, often through Acts
of Parliament, public inquiries or even Royal Commissions. Thus, ultimately, it
was central government that made the decision to allow the building of a huge
oil terminal in the Pembrokeshire Coast National Park and the now defunct
nuclear power station at Trawsfynydd in Snowdonia National Park.

The building of nuclear power stations at numerous coastal locations –
including Sellafield, Sizewell and Berkeley – since the War has, at the end of
the day, always been a central government decision, for they have been
considered not only controversial as visual instrusions into the landscape
but also environmental hazards. These, along with nuclear waste
reprocessing and weapons plants, can affect adjacent rural land-users. The

▲   Fylingdales Early Warning System – an example of an intrusive development in a National Park – North York Moors.

◄   Designed so as not to break the skyline, Trawsfynydd nuclear power station remains a massive intrusion into Snowdonia National Park. Although it reached the end of its useful life in 1994, concern exists about dismantling procedures. The high levels of radiation present mean that this will have to done in stages, spread over 100 years.

effects of radio-active discharges on livestock and plant life even at low levels have been known for some time, but more recently there seemed to be evidence of increased child cancer around such installations. However, it now seems possible that this is due to the exposure of workers within the plants, leading to the development of serious health problems in their offspring.

Even in terms of 'soft' energy options, in the 1990s decisions are being made about a number of 'wind energy' farms to be sited again in remoter, more beautiful (but windier) parts of the countryside, such as the North Pennines and mid Wales, where the size and density of windmills is already causing a significant visual disamenity.

And it is central government that makes decisions, too, about large-scale infrastructure developments which may have impacts on the countryside. These can be as diverse as the siting of a by-pass through Twyford Down outside Winchester and around the congested town of Newbury, and the most appropriate route for the channel tunnel rail link. Here, trade-offs are constantly being sought between economic expediency and the environmental impact on the countryside. Beyond roads, the effects of developments such as those involved in increasing the capacity of London's airports and in constructing a barrage across the Severn, cause perennial debates about this environmental/economic trade-off.

Finally, in terms of large-scale developments, there is the ever-present threat to the countryside from new towns and new villages, leaving aside constant applications for individual rural dwellings. Thus the post-war new town developments at Milton Keynes, Livingston, Runcorn, Antrim and so on have eaten up many hectares of the lowland countryside of the United Kingdom. But while these and other urban types of development in the countryside have been perceived as having considerably escalated the loss of rural land between 1957 and 1971, they have only affected some 2% of the total land area. More recently, in the wake of a relaxation of planning constraints in rural areas from 1987 to 1990 a new phenomenon appeared in south-east England – the purpose-built country town to be constructed on greenfield sites, each occupying between 360 and 400 hectares. At the height of the enthusiasm for such schemes nearly one hundred of these were proposed, the motivation being, according to the developers, that this would reduce the pressures for infill, urban sprawl and piecemeal development in the countryside, as well as providing a high-quality built environment. However, they were vociferously opposed by those living in the areas affected. Indeed these proposals evoked more heated argument than the development of the (much larger) new towns ever did. Today, such schemes are far fewer in number and are mainly for new villages, such as the proposal for the ecologically friendly Cambourne in Cambridgeshire, with its 3300 homes and all the appropriate social infrastructure (see Figure 5.4).

Certainly, from 1990 the downturn in the economy and a renewed interest in planning as a positive activity in the countryside which found expression in the Town and Country Planning Act 1990 has relieved development pressures in rural areas to some degree but more particularly in the south and east of England. It is not just the new country towns that now have not come to pass but a whole range of other extensive leisure developments.

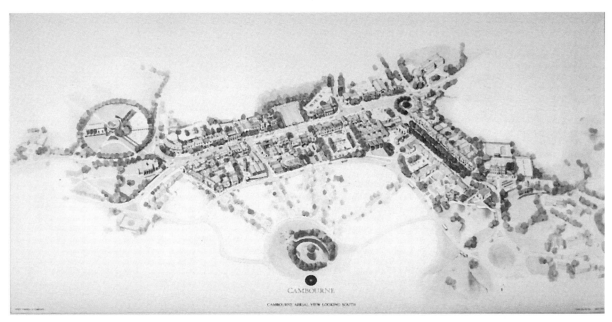

▲   *Figure 5.4   Designing a new community. The primary aim for the development of Cambourne in Cambridgeshire is environmental sustainability. Energy efficiency, of both journeys and buildings, conservation of ecology and recycling of waste materials are some of the objectives underlying the planning of the new community (by Terry Farrell & Partners). The architect's impression shows the proposed new centre, which will serve three 'villages' of residential areas; it is expected to be completed by 2010.*

4.6   Summarising land-use pressures in valued landscapes: a case study

So far this chapter has examined how the drive to increase food output after the Second World War did much to damage the landscape, ecology and indeed the economy of the countryside. Although much of this development was directly at the expense of rural conservation in a number of forms, the competing demands of other countryside land-users – leisure, forestry, water, minerals, other large-scale developments and so on – have created a complex web of land-use policies and land-use interactions, all with their particular environmental consequences. This subsection will look at one specific example to give an indication of how these complexities are manifest in a single area and to summarise many of the points made in this section. Later, the chapter will examine more closely some novel attempts to overcome these complexities.

As with much of the countryside of the United Kingdom, the Peak National Park is predominantly an agricultural area where, because of its physical characteristics, livestock production has been pre-eminent since the War. More recently there have been increasing pressures to plough up land for cereals production. Until the 1980s the National Park Authority could do little formally to prevent this because of the exemption of most of agriculture from planning controls. Incentives to intensify both cereals and livestock production were enhanced in the 1970s by the designation of

*'We're very lucky, when you think about it,
working in such beautiful surroundings.'*

much of the National Park as a 'Less Favoured Area' with higher grant-aid
availability for modernisation. A first clear tension in this area, then, lies in
having an expanding agricultural industry in a National Park, the statutory
objectives of which are to conserve the landscape of the area, as well as to
provide opportunities for public enjoyment. In 1987 changes in government
policy at least made it mandatory for alterations in the use of land for
agricultural purposes or for an extension of woodland to be the subject of
consultation with the National Park authorities, an extension of the
notification provisions contained in the Wildlife and Countryside Act 1981
that we consider further in Section 5.1 below.

On the public enjoyment front, the National Park also faces land-use
problems too. It was in the Park area that the conflicts between
recreationists and land-owners refusing to allow access to their land came
to a head in 1932 when 800 ramblers from Manchester organised a mass
trespass on Kinder Scout. It was as a result of this that the National Parks
and Access to the Countryside Act 1949 introduced Access Agreements.
Nearly all of these Access Agreements have been designated within the
Peak National Park and bring a range of associated management duties as
well as costs.

These alone have not resolved recreation pressures in the Park,
however, and a whole new series of recreation management experiments –
public transport, park-and-ride schemes, cycle hire schemes and the
development of interpretation centres – have all had a role to play in
balancing visitor numbers with the quality of the recreation experience in
the Park. Nowhere is this balance more delicate than in those areas
dedicated to nature conservation. In several National Nature Reserves
particular care has to be taken with the management of public access lest
the effects destroy the very scientific worth of the Reserves themselves.

Conservation and recreation, however, provide only the tip of the iceberg

◀ *In the Derwent Valley area of the Peak District, the National Park authority and the National Trust have combined with forestry and water interests to resolve the problems caused by tourist pressure on the area. (Above) planting schemes have been undertaken to improve the area; (below) keeping out cars has greatly enhanced its recreational appeal and made its conservation much easier.*

in rural land-use conflicts in the Peak National Park. As an upland area at the upstream end of the Severn–Trent water system and close to large industrial populations, historically there have been great pressures within the Park for the construction of water storage reservoirs. Subsequent to their construction, certain of these such as the Ladybower and Derwent have generated their own recreation problems, while during the construction of the Derwent reservoir a whole village (after which the reservoir was named) was drowned. In close association with water developments have been those of forestry. Again, the terrain and soil quality of the area have made the Park susceptible to large-scale coniferous timber expansion with no recourse to planning controls on the part of the Park Authority.

It has perhaps been over or in connection with the extraction of minerals, however, that land-use conflicts in the Peak National Park have been most intense. In addition to the mining of fluorspar, pressure to extract limestone for both the construction and the chemicals industries has been very great indeed. Although applications for minerals extraction have been resisted with some successes by the Peak Park Joint Planning Board, a large number of quarries are worked and associated activities such as cement works provide significant intrusions into the landscape. Even where development has been diverted outside the Park, it has been of sufficient scale, such as the Tunstead ICI quarry, still to have a significant visual impact from within the Park. (See Plate 11.)

Thus reconciling rural land-use conflicts is a paramount purpose of the work of the Peak National Park Authority. Although this often entails significant compromises, sometimes to the National Park objectives themselves, it has sometimes brought forth new initiatives. One of these, in the field of integrated rural development, will be considered more closely, in Section 6 below since it relates to the Peak National Park.

Activity 4

The exercise just carried out in relation to the Peak National Park has strong similarities, if in a more limited way, with the approach to Cumbria in Chapter 1 of Book One (Sarre and Reddish (eds), 1996). You should therefore now be ready to consider part or all of a National Park or an AONB that you know well and try to sort out the different land-use demands made upon it, and then consider the complex web of interactions that exist between these. You might find it useful to do this as a flow diagram.

Check your answer with the example given at the end of the chapter. The diagram there gives you a simplified version of such an exercise, but for a lowland area – the Norfolk Broads, now enjoying a designation similar to that of a National Park. This provides a counterbalance to the upland setting used above.

5   Agricultural diversification: conservation on a flow tide?

5.1   New conservation controls in the countryside

By the late 1970s the damage caused to the countryside by unfettered agricultural expansion had become untenable. The Porchester Report (referred to in Chapter 4 and in Section 3.1) formally proposed new measures for the control of agriculture in high-value landscapes. Both the scientific and amenity conservation bodies put their weight behind these proposals. By the 1980s, budgetary crises with the Common Agricultural Policy and continuing food surpluses led to formal proposals in Europe for the curbing of food output. But in the United Kingdom it was new conservation measures against the environmental damage caused by agriculture that began to bite even before policies to curb food over-production started to take effect. The issue of solutions to check food over-production will therefore be returned to in the following section.

By 1981 the Wildlife and Countryside Act had been passed requiring farmers to notify local authorities of intentions to undertake potentially damaging agricultural operations: objections could lead an authority into negotiating a management agreement with the farmer to desist from such activities. However, in such a case a farmer would be entitled to compensation payments equivalent to the value of output forgone as a result of not carrying out improvements, or in extreme cases could result in the purchase of the land.

Safeguards against damaging operations in National Parks and SSSIs were also introduced in the Act. The latter required SSSIs to be renotified by the then Nature Conservancy Council. This meant that the 30 000 or so owners and occupiers of such sites, designated originally under the National Parks and Access to the Countryside Act 1949, had to be given full details of the site and the kind of operations which, if carried out there, could damage it. The same information also had to be passed on to the local authority, the water authority and the Secretary of State for the Environment.

The sheer cost of these management agreements, particularly when there was no certainty that a farmer would have ever carried out the operations he had threatened, was the subject of much criticism. In an attempt to overcome this in one particularly valuable wetland area that was under threat of drainage, Halvergate Marshes in the Norfolk Broads, the Countryside Commission proposed in 1985 rather than compensation payments, a set of payments for the positive management of agricultural land, but in an environmentally sensitive way. This was known as the 'Broads Grazing Marsh Conservation Scheme'. From these beginnings, **Environmentally Sensitive Areas (ESAs)** emerged and are to be found throughout Europe. Indeed, they have been formally enshrined in United Kingdom legislation in the Agriculture Act 1986.

Some other loopholes of the Wildlife and Countryside Act 1981 were closed in the Wildlife and Countryside (Amendment) Act 1985, particularly in relation to notification procedures for both management agreements and SSSIs, but the principal conservation successes in relation to agriculture have been in relation to ESAs. These were to cover areas whose national environmental significance was threatened by agricultural change, but which could be conserved through the adoption or maintenance of particular forms of farming practice. The areas were to represent a discrete and coherent unit of environmental interest, to permit the economical administration of appropriate conservation aids.

ESA schemes, which are run by MAFF, are voluntary and require the farmer to adhere to particular management regimes in exchange for financial assistance. The particular management system varies from area to area but may include, for example, restrictions on fertiliser use and reductions in stocking densities. The use of chemicals is also commonly discouraged, and traditional features of the landscape retained and maintained. Clearly, much of the purpose of ESA management is to redress directly the worst ravages of agricultural intensification summarised earlier in this chapter.

By mid-1995, 6250 agreements with farmers had been either reached or were under discussion, covering over 360 000 hectares, or 15% of the land area of the UK. The government's budget for ESAs is expected to rise to £63 million by 1996–7. As a result, the ESA scheme has been considered a clear

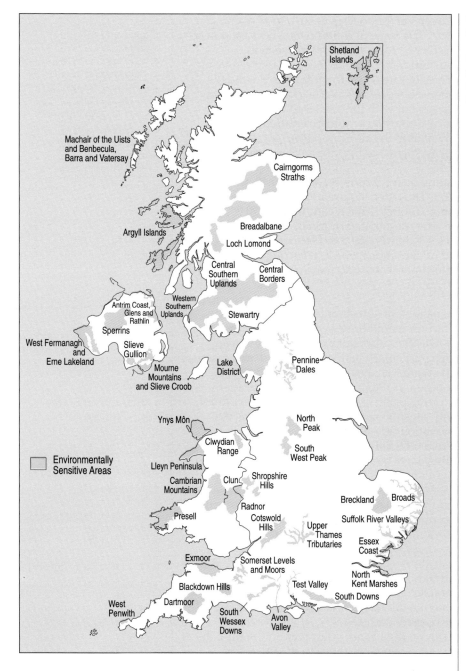

Figure 5.5 The ESAs of the UK. Those designated by 1994 have increased the variety of our protected landscapes and provide a much more effective approach to conservation.

step forward in mitigating the worst ravages of agricultural practice in a climate of expansion. But, at the end of the day, these areas, important as they are, still only cover a small part of the agricultural land of the United Kingdom (see Figure 5.5).

More recently, a broader range of conservation instruments has supplemented the ESA scheme, inspired in many cases by ESA successes. The Countryside Stewardship Scheme has been the most pervasive of these, offering grants and annual payments to manage important landscapes in traditional ways. By 1995 the scheme covered 80 000 hectares with payments of around £100 per hectare. But there are other programmes: the Countryside

◀ *The Somerset Levels –
an example of an
Environmentally Sensitive
Area and one landscape of
high quality that could easily
have been destroyed without
ESA status, now recognised
by the European Community
as a means of reconciling
agriculture with
conservation.*

Premium Scheme, the Habitat Improvement Scheme, the Hedgerow Scheme,
the Farm Woodland Scheme and the Woodland Grant Scheme. All of these
schemes, largely of a voluntary nature, have been introduced by various
government departments between 1990 and 1995 to allow farmers, through
grant aid, to farm in more environmentally friendly ways.

Pollution control measures have also been improved in the 1990s.
Influenced by the Government White Paper of 1990, *This Common Inheritance*,
pollution control measures have become more pervasive. For agriculture
these have included Nitrate Sensitive Areas and Nitrate Vulnerable Zones,
but new regulations and Ministry of Agriculture codes of practice to enhance
the quality of soil, water and air have been directed at farmers.

However, one of the many schemes which are part of the agri-
environment programme and still in the pipe-line at the time of writing
(June, 1995) is particularly important and is aimed at moorland farmers who
over-graze their livestock thus damaging unique upland habitats. The idea
will be to pay farmers operating outside ESAs a higher rate per head for a
smaller number of ewes per hectare. Even so, MAFF has considered serious
over-grazing problems on a case-by-case basis for some time. The Ministry of
Agriculture has also issued guidelines pertaining to the conservation of
natural resources such as energy and water. These have found ready
acceptance by farmers as they square with their commercial concerns.

In general terms, these conservation causes have been helped by
various policies that have been introduced in the 1980s and 1990s, and in
particular the McSharry reforms to the CAP of 1992 aimed at curbing food
over-production. It is to these considerations that we now turn.

5.2   Food over-production solutions

There is a logic that suggests that if much of the damaging impact of
agriculture can be attributed to agricultural expansion, then curbing
expansion will have an ameliorating effect on the environment. This has
been the case in varying degrees, and this section reviews a number of
policies that have been instituted or proposed since the start of the 1980s
which were designed to curb food over-production, paying particular

attention to their actual or likely environmental impact and effects on the rural economy.

The European Commission and most economists readily agree that since it is the high prices of agricultural commodities that encourage farmers to produce more than is required, with attendant environmental consequences, then reducing these prices affords the only long-term solution to the food surplus problem. For political reasons and for the long-term stability of the agriculture industry, this has taken a long time to come into effect. The McSharry reforms of the CAP of 1992 aimed at moving prices back towards those prevailing in the world in general have, however, needed to be applied in conjunction with compensation payments if farmers are to remain solvent. At the time of writing it is too soon to tell what effect these price reductions *per se* will have on the environment; at present we can only speculate.

Taken on their own, they may encourage further intensification in certain areas as farmers seek to sustain their gross incomes in a regime of falling prices. In other areas, farmers may find that the marginal costs of additional production may be greater than output values and will therefore reduce output levels. These choices in turn will depend on the cost of inputs. Price restraint policies could thus have quite unpredictable effects on the environment and land use in the countryside, although since they have been introduced in tandem with the agri-environment policies discussed in the previous section, it is likely that environmental impacts will be positively mitigated.

In addition, it is instructive to look at the impacts of other kinds of restraint measures that are either being introduced or contemplated, since it is with these that price restraint policies are likely to operate in parallel. In 1984 **quotas** were imposed on milk production in the United Kingdom as part of a European Commission policy to reduce output. This simply set a limit on the amount of milk that a farmer could produce before being penalised by levies. In principle, they could also be used to put a ceiling on the amount that a farmer could produce that would qualify for the kinds of price support measures discussed in Section 2 of this chapter. This is less painful than price reductions since the latter reduces the level of subsidy on all, rather than only marginal levels of, output.

The introduction of quotas provoked a quick response from farmers who culled cows and reduced their use of high-protein feed concentrates to get their levels of production down. Some farmers, with heavy borrowing commitments to finance plant modernisation or extensive herd replacement schemes, went out of business. Other farmers wishing to reduce overheads and maintain profitability have purchased milk quotas from those changing the nature of their agricultural enterprise or going out of business. In many cases, dairy farm net incomes have actually risen as a result of this, so the effects on the rural economy, within agriculture at least, have not been completely adverse. They do, however, restrain the most efficient producers. For these reasons and the bureaucratic inefficiencies and loopholes that have been experienced in the implementation of milk quotas, they are unlikely to be extended to other sectors.

Environmentally, too, the effect of quotas is uncertain, but it may not be large. On the one hand, it may lead to lower input, lower output farming with the result that the landscape may appear as less intensively cultivated. On the other hand, it could drive certain farmers out of dairying altogether and into more intensive enterprises such as cereals.

A second means of reducing output is to place adjustable taxes on all outputs of a certain kind. These are called **co-responsibility levies** and were introduced into the United Kingdom for wheat in 1986. These levies are designed not only to reduce production levels, but are also considered to be useful in helping to pay for the storage of wheat surpluses. In practice, co-responsibility levies are little different in their effects to a cut in the prices that farmers receive, except that smaller producers can be exempted from the levy. Environmental and economic impacts are therefore likely to be similar if somewhat smaller than those of a reduction in prices.

Where co-responsibility levies have been introduced in Europe they have generally been very expensive to administer and monitor, and in the main have been too small in size to have any effect on output levels. They thus tend to be considered rather cosmetic as a means of reducing food surpluses. Two further means of curbing food over-production are likely to have greater impacts on both the rural environment and the rural economy than those considered so far. These are extensification and 'set-aside'.

Extensification is a broad notion concerned to lower the intensity of output in agriculture, essentially by reducing the inputs. Clearly, lower inputs to agriculture, as have been seen in the case of introduction of milk quotas, will lead to a less industrial farm landscape without any necessary loss in net farm incomes.

Extensification can take a number of different forms. It could, for example, entail a reduction in farm inputs such as the fertiliser, nitrogen. In fact, a number of people have proposed the introduction of nitrogen taxes as a simple way of reducing output. This is also a very flexible means of controlling production, since output is very sensitive to nitrogen levels and varying the tax would correspondingly vary output. Nitrogen quotas also have been proposed and these would have a similar though more selective effect. However, the introduction of Nitrate Sensitive Areas and Nitrate Vulnerable Zones has reduced nitrogen inputs in particular targeted areas (see Chapter 2, Section 4.4). Reducing farm modernisation subsidies to capital equipment would also decrease the scale of mechanisation being applied to agricultural production.

In the context of extensification, the movement towards organic farming is gaining momentum as part of a general but quite fundamental change in attitudes by the public to food consumption and the environment. This relies on 'natural' inputs to agriculture which remove not only large amounts of nitrogenous fertiliser from the production process, but also pesticides and other manufactured chemicals. Although this appears to reduce the cost of inputs per hectare, this is offset by the increased input to the cultivation process in order to reduce the problem of weeds. Outputs per hectare also fall. But there is a corresponding market premium for organically grown foods so that ultimately revenues per hectare need not be any lower than in conventional farming.

Extensification would be likely to have clear environmental benefits to the countryside. Landscapes would be less industrial in appearance and more akin to the 'patchwork quilt' of the nineteenth century. Some of the principal benefits, however, would come with an enriched rural ecology. Reductions in all forms of manufactured chemicals would go a long way towards reversing the ecological impoverishment process that was charted in Section 2 of this chapter. The rural economy may also experience benefits. With a reduction in capital inputs to agriculture there may well be an increase in labour inputs as a substitute. Indeed, some research has suggested that if organic farming became the prevailing food system in the United Kingdom it could create an extra 1.3 million jobs directly in agriculture. The trickle-down effect on the rural economy of such an increase in rural employment would be considerable. However, contrasting evidence from some organic practitioners indicates that human and mechanical inputs can remain much the same as for other more orthodox systems of production.

Set-aside is an over-production solution concerned specifically with taking land out of agriculture. At the levels of food production which prevailed when the policy was introduced under the Agriculture Act 1986, it seemed that it might be necessary to remove two million hectares from agriculture in order to remove food surpluses in the United Kingdom. Thus farmers were given compensation payments for taking a minimum of 20% of their land out of production. A number of criticisms of such a policy soon emerged, not least that farmers who removed, say, 20% of their land under crops might then use grants to intensify production on the remaining 80%. Thus the farmer might end up producing the same amount of food, whilst receiving compensation payments for setting land aside. Some of the criticisms have, however, been addressed by subsequent modifications. The most recent of these, resulting from the McSharry reforms of 1992, make the payment of any subsidies for grain production (which offers acreage payments rather than payments by volume) mandatory on the set-aside of productive land. Inside such an incentive, set-aside can follow one of two routes: either a total of 15% of land can be set aside on a rotation scheme for six years, or 18% can be set aside for five years, providing the tract of land in question remains the same.

However, the impact of set-aside on the rural economy is slight since it creates no jobs and need affect farm incomes little. Environmentally, there are pockets of land that are now no longer intensively farmed, but with no positive management these are quickly taking on the appearance of derelict land. This has become such a problem that in 1989 the Countryside Commission introduced an experimental scheme in East Anglia that provided payments for the positive management of set-aside land in some non-food-producing use, such as development for recreation.

Activity 5

Tabulate the impacts of the solutions to food over-production discussed in this section with the proposed solution on the left and adjacent to it on the right, favourable environmental impacts and unfavourable environmental impacts.

Your table should show four solutions. Each one will have a favourable environmental impact but *two* will also show possibilities for unfavourable environmental outcomes.

The prospect of large areas of surplus land in the countryside deriving from attempts to find solutions to the over-production of food is, of course, central to this chapter. What should be done with it? Does such a policy mean that the competing demands on the countryside of the United Kingdom are likely to diminish? Well, not exactly. At the same time as the government has been developing policies to stem food over-production, it has been introducing policies to help farmers diversify out of agriculture altogether. These have been policies attempting to establish 'alternative land uses in rural economy' which is precisely the name, shortened to **ALURE**, of the policy package. The environmental and land-use implications of ALURE are now briefly examined.

5.3 Alternative land uses for the countryside

As well as introducing Environmentally Sensitive Areas, the Agriculture Act 1986, based on the ALURE package, introduced a number of measures for diversifying the rural economy. One of these was generally to encourage non-farm developments on agricultural land which was also given force by a Department of Environment Circular of 1987, *Development Involving Agricultural Land*. This was aimed at relaxing planning controls over agricultural land in all areas except Green Belts and National Parks and led, quite quickly, to a spate of planning applications for houses, golf-courses and even whole country towns (as was mentioned in Section 4). As a general policy theme, this relaxation of development controls has probably been the most significant measure in recent years that has increased competing demands on the countryside – and understandably caused much consternation amongst environmental pressure groups and, in specific areas, around the then proposed green-field sites for new villages, amongst a wider public.

A 'Farm Woodland Scheme' provided a second strand to the ALURE package. In the early 1980s the Forestry Commission had already introduced a Broadleaved Woodland Scheme to provide planting and management grants for non-coniferous woodland schemes. The Farm Woodland Scheme was introduced by the MAFF in 1986 to supplement this, particularly as a means of encouraging farmers to take land out of food production, either as part of or in addition to the set-aside scheme. In the longer term the environmental benefits of such a scheme could be positive, but the uptake of the Farm Woodland Scheme has been very low so far, since it entails farmers tying up their land in timber growing for unacceptably long periods of time.

The ALURE package also provided grants to aid diversification of the farm business, and feasibility studies for alternative business. The potential for diversification, however, falls into four broad areas, the first of which is tourism and recreation. Although these have been exploited to a degree, as was noted above, there is always the uncertainty here of not knowing when the market might become saturated in any particular area, and capturing income from recreation is always a problem.

Adding value to farm products represents a second process of diversification. This entails processing foods, marketing specific food qualities or developing 'pick your own' enterprises. These kinds of activity offer some potential since traditional rural imagery provides a potent marketing tool. Diversifying into alternative crops and livestock represents a third way of broadening the farm enterprise. These might include breeding goats or snails, or rearing sheep for milk, or deer for venison, or growing unusual crops such as borage or evening promise. Organic farming can be considered as part of this area of diversification, although in most respects it is a return to a mixed method of farming using crop rotation and involving the keeping of animals.

Finally, farms may develop ancillary resources on the farm. The potential of this is enhanced by the liberalisation of planning controls in that farm buildings might now be more readily used for homes, craft industries or tourist accommodation. But in addition areas such as woodlands or wetlands might be used for game, for timber or water-based recreation.

Generally, since the mid 1980s, then, the sovereignty of agriculture as a rural land use has been brought into question. Under a remit to promote more widely the economic and social interests of rural areas as a whole,

agriculture ministers under the Agriculture Act 1986 now have an explicit
duty for the conservation and enhancement of the natural beauty and
amenity of the countryside. Solutions to food over-production as well as
measures to diversify the rural economy will change the balance of
competing demands on the countryside, to make it more diverse in both its
environmental and economic base. These new measures, however, have not
been implemented long enough for their long-term effects on the
countryside economy and environment to be clear yet. However, as Table
5.1 shows, the sums of money to be devoted to diversification in 1987 were
minute compared to the £1800 million spent on supporting food production
in that year.

Table 5.1   Financial allocations to diversification

	£ million
ALURE package proposals made in April 1987	
Rural diversification under ALURE	25
Diversifying into woodland (Farm Woodlands Scheme)	13
	(for 33 000 ha)
On-farm diversification	3
Farm products marketing	2
Proposals made in November 1987 for additional funds	
Covering all ALURE schemes	1
Set-aside and extensification	
Proposals made in November 1987 for 1988/90 for extensification and set-aside	16
Conservation	
Annual allocations	
Original ESAs (6)	12
New ESAs (6)	7
NCC estimate of SSSI management agreement costs	15–20

Not surprisingly, these measures were subject to much debate and
criticism but first, and most notably, in a 1987 Countryside Commission
report, *New Opportunities for the Countryside*, produced by its Countryside
Policy Review Panel. In its proposals for wide-ranging changes to
agriculture, forestry, recreation and conservation, it took issue with
government over almost every aspect of its financial allocations to achieve
these objectives. The total cost of a package which would be effective,
according to the Countryside Commission's report, would be £320 million –
over seven times that of the ALURE proposal.

It would seem, then, that government had seriously underestimated the
cost and the extent of its new measures aimed at diversifying the rural
economy, at least as far as the Countryside Commission was concerned. It
is difficult to produce figures for later years that can be compared with
those in Table 5.1. The ALURE allocations often cover more than one year,
whilst since April 1987 other new initiatives have come on stream.
However, Table 5.2 does provide some idea of the levels of grant available
to farmers and landowners for diversification and conservation known to
be available at the time of writing (mid 1995). The message that comes
across is nonetheless reasonably clear – that the support for 'new
opportunities for the countryside' although much improved in certain
areas, such as ESAs, remains remarkably small.

Table 5.2   Environmental schemes available to farmers and total grants paid*

Farm Woodland Premium Scheme	£3.27m paid in 1991/92
Woodland Grant Scheme	£11.9m 1991/92
Nitrate Sensitive Areas	£0.9m 1991/92 (£6.7m planned for 1995/6)
ESAs	£11.6m 1991/92 (£343m planned for 1995/96)
SSSI Management Agreements	£5.8m in 1991/92 (England only)
Country Stewardship	£7.8m in 1992/93 (Countryside Commission, England only)
Hedgerow Incentive Scheme	£3.5m for first three years (from 1992)
Habitat Improvement Scheme	£3.0m (planned for 1995/96)
Wildlife Enhancement Scheme	£0.45m (planned 1992)

Note: *Above figures do not all relate to the same year(s).

Source: ATB-Landbase, 1994.

5.4   Summary

Since the mid-1980s a number of attempts to bring agriculture into a more harmonious relationship with conservation have been undertaken, at least in the most vulnerable of landscapes, especially through ESAs. But the key to the problem overall has been to find ways of cutting agricultural production, particularly of those commodities in surplus and which cannot be marketed. Schemes used to this end have included milk quotas and cereal production levies, extensification, set-aside, farm diversification and a range of alternative uses for rural land. However, the role of government in activating such a programme has been criticised by its chief advisory agency on the English countryside, the Countryside Commission. Thus, if the 'carrot' for diversification has remained small, so has the McSharry 'stick' provided by CAP reforms – so far!

6   The future

As well as the changes based on the agriculture industry that have come about since the mid-1980s, a number of other characteristics of rural areas have emerged, particularly in terms of changes to the rural community and its relationship with the land. This final section of the chapter briefly reviews two of these and speculates on how rural communities might develop beyond 2000.

6.1   Integrated rural development and beyond

So far this chapter has considered all of the competing demands for the countryside separately. This is a perfectly acceptable way of doing things, since, in the United Kingdom, each of the land-using sectors tends to be organised and administered separately. But the very fact that most rural activity is carried out in an unco-ordinated way is one of the principal causes of land-use conflict in the countryside.

Because of this, one type of initiative that attracted much interest in the 1980s was the development of a more holistic approach to the planning of the countryside, which became known as **integrated rural development (IRD)**. Such approaches had been tried in England in the 1960s, with the introduction of Rural Development Boards, but at that time government ministries jealously guarded their own sectoral interests and as a result only one was formed, for the North Pennines, and this was short-lived. However, by the 1980s, the European Commission helped support a series of experiments in IRD, the chief characteristics of which are summarised in Box 5.1. One of the best known IRD experiments was in the Peak National Park centred on the villages of Longnor and Monyash. Here the Peak National Park Joint Planning Board set up the experiment in an attempt to ease some of the problems that were outlined in Section 4 above.

Box 5.1   Integrated Rural Development – as perceived in the 1980s

Policies for the countryside have long been devised on a sectoral basis. In other words, they are separately designed to specifically serve the interests of agriculture, forestry, recreation, conservation and so on. Although they may be applied locally, the fact that they are originated by central government means that they are 'top down' in their approach. Integrated rural development seeks to stand much of this more traditional way of thinking on its head to positive effect.

Integrated rural development (IRD) as a concept can be summarised in three key words – *individuality, involvement* and *interdependence*.

'Individuality' means looking at rural areas as they are and recognising in their individuality a source of economic strength, social identity and environmental character. Different parts of the country are manifestly unalike. If public policies tend to treat rural areas as if they were the same, administrative convenience may be served, but there is a danger of masking or even destroying the individual identity of these different areas.

'Involvement' means trying to involve local communities in thinking about their own future and in working out and putting into practice their own ideas for improving that future. Such work can concentrate on solving problems, but people are encouraged to think positively – to improve their economic position, their social life or their environment.

'Interdependence' means looking at individual rural areas as a whole. Rural areas do not consist of a set of absolutely distinct interests. Society wants rural areas to provide food and a whole range of other resources to meet its needs. It also wants it to offer an attractive environment for recreation, opportunities for the conservation of wildlife and reasonable living conditions for a significant proportion of the population. The achievement of any one of these objectives can affect others for good or ill. IRD aims to eliminate any actions that may produce harmful side-effects. More positively, it tries to devise measures that encourage actions which create benefits for social, economic and environmental interests simultaneously.

In this experiment, arrangements were made to remove to a large degree the usual sectoral policies and funding arrangements for a small rural area, and encourage a number of agencies to work together and help to fund a common set of policies. Thus policies of MAFF, the (then) Development Commission, the Countryside Commission, the (then) Nature Conservancy Council and so on were co-ordinated, and so too were approaches to financing projects. This led to some interesting innovations in both community projects and environmental schemes well before the current over-production and diversification measures had come to fruition. Thus, rather than the normal MAFF payments to farmers to increase production in a marginal agricultural area, a consensus policy was devised that allowed, for example, environmental payments for the reconstruction and restoration of dry-stone walls and the growth of wild flowers in meadows. In community terms, too, small industrial developments were initiated and community facilities enhanced.

Although an experimental scheme, this 'tale of two villages', as it was called, generated some useful practices that were adopted Europe-wide, including a special agricultural package to assist poorer Mediterranean areas, known as the 'Integrated Mediterranean Programme'. However, the 1990s have seen new but related initiatives from the EU.

These initiatives, under what is known as the LEADER programme, can be seen as a response by the European Commission to the reform of the CAP, and are aimed at offering support to rural communities in what are termed 'fragile rural areas' or Objective 5b areas. In the UK these cover parts of west Devon, Cornwall, central and west Wales, north and west Scotland and much of Northern Ireland.

The schemes operate in much the same way as IRD, the aim being the provision of 'bottom up' solutions to rural development, enabling local groups and agencies to come together to diversify their local economies, create jobs, and, more often than not, improve the local environment. But as Box 5.2 indicates, there are other innovative aspects that go beyond IRD.

Box 5.2   Features of LEADER programmes – solutions for the 1990s

They are expected to:
- be managed by local action groups in bringing together significant players from the public, private and voluntary sectors, in a spirit of partnership;
- demonstrate a real inter-agency integration of objectives, resource deployment and activities at local level;
- involve, mobilise and respond to the needs of the local population and thereby embrace the concept of community development;
- be innovative in some clear respects, be it in the management of the programme locally or in the economic processes or products being supported;
- add value to local resources, interpreting that concept to include the human, environmental and cultural endowment;
- respect and foster local distinctiveness and diversity;
- come together in a Europe-wide LEADER network and thereby foster the exchange of experience and the cross-national transfer of innovation.

Source: Moseley, 1995.

Under the first LEADER programme which ran from 1992 to 1994, thirteen UK local projects were approved and funded, six in Scotland, four in Wales, two in England and one in Northern Ireland, out of a total of 271 across the Union. The financial outlay under the programme as a whole totalled about £900 million, although the second programme which runs from 1994 to 1999 is likely to be supported by over £2 billion. The probable outlay on UK-based schemes in this second round is difficult to assess at the time of writing, although with the extension of Objective 5b areas in south-west England in 1994, the proportion of the take of the total is likely to rise significantly. Nevertheless, the total sum of more than £2 billion to be spent over four to five years still has to be seen against the £35 billion spent in the UK alone in agricultural support in a single year, that of 1995.

6.2   The enterprise economy and the urban countryside

Remoter rural areas may therefore be nurtured into economic growth
through a system of integrating and co-ordinating competing demands on
the countryside of the United Kingdom. But the pressures and the
prospects for the future of communities and land uses in lowland area
closer to the larger urban centres may be very different. Here, although
some of the relaxation of planning controls under the 1987 Circular has
since been clawed back, much more pressure for development for both
housing and high-tech industry is becoming apparent and this is leading to
many areas taking on the appearance of an urban countryside.

The likely development in this direction has been termed 'Spread City',
shown in Figure 5.6. This means that essentially most of south-east England

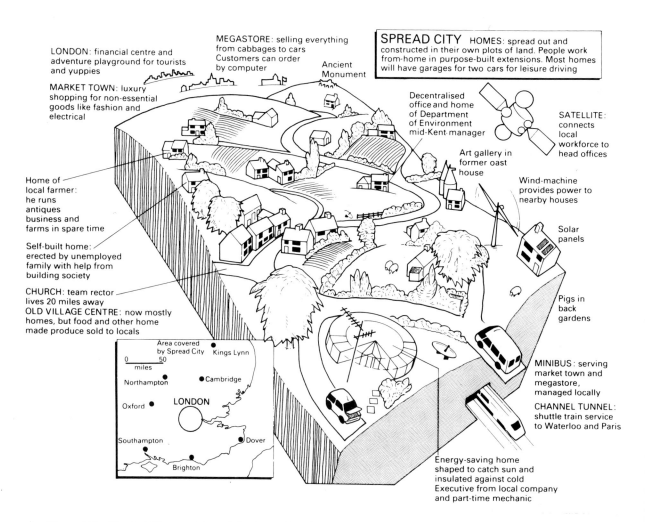

▲ Figure 5.6   Spread City – a vision of the future? By early in the twenty-first century will we have a semi-urbanised
countryside similar to that shown in this diagram, stretching from Cambridge to Bournemouth and from Oxford to
Dover? In a report published in June 1995, it is a prediction that the Council for the Protection of Rural England were
prepared to endorse.

> **Box 5.3   The Electronic Village: a vision for the future**
>
> Accessibility – whether to education or jobs – is increasingly becoming a problem for some country people. But one solution could be a growing network of village projects, in the United Kingdom and across the world, that are linked by the latest communications technology.
>
> The concept of the 'tele-cottage' was launched in 1985 in a small village in a remote area near Sweden's Norwegian border. Vemdalen, a winter ski resort and forestry area, had only limited employment opportunities and a falling population. But the tele-cottage, the concept of a Swedish businessman, has changed all this. The tele-cottage is equipped with fifteen personal computers and word-processors, telex, fax and teletext equipment and satellite television receivers, all financed by local and central government and locally raised funds.
>
> In it:
>
> • children and adults have opportunities to learn new technologies
> • formal education and self-study in computer uses relevant to business and private needs can take place without the need to travel to distant colleges
> • tradespeople can try out computer programs and be trained in skills such as book-keeping, and they can use telex, fax and electronic mail service facilities
> • access is possible to international databases, broadcasts and other users of electronic communications
> • people can find a social meeting place since the cottages are equipped with a lounge and kitchen
> • local people can work for employers many miles away via the computer technology without travel costs or wasted time.
>
> Apart from Sweden, tele-cottages now exist in Norway, Finland, Austria, France, Ireland and Germany. In the UK the first tele-cottage was opened in 1989 at Warslow funded by Staffordshire County Council and the Rural Development Commission. Since then a number of schemes have been supported by British Telecom, including a project with the Highlands and Islands Development Board to establish tele-cottages in Islay, Stornaway, Orkney, Shetland and Argyll. The European Commission now wants to use a wide range of advanced telematics to sustain and develop rural communities across the Union.

will have an urban mentality and lifestyle. Most shopping, whether you live in a town or not, will be done in out-of-town megastores. In spite of recent (1995) planning presumption against such developments, a very substantial number of these already have planning consent. As a result, village shops will sell only antiques and luxury items. Alternative technology will lead to smaller more sustainable 'homesteads' and many people will work from home or from 'tele-cottages' as they have been called. Here, computers, telephones, fax machines and eventually satellite 'interactive' television will obviate the need to meet face to face to do business. As Box 5.3 explains, it could also be the way forward for non-commuting rural communities away from large metropolitan areas.

6.3   Rural welfare

Given the strong possibility of this kind of 'Spread City' vision in less remote areas, it is now incumbent on government and its agencies to extract the benefits from such developments, and minimise the drawbacks. One of the prevailing features of the past fifteen years of 'market economics' policies has been the lack of any really significant policies to enhance rural welfare. 'Spread City' could be a welcome vision for some, but remains a vision of affluence. In parts of lowland Britain, many rural-dwellers have either drifted or been displaced to the town since the War and the cost of country property has led to an incoming, single-class rural society (as was noted in Section 2.4). If such a trend continues, there may ultimately be no need for policies of rural welfare, since all the less well-off will have left.

But for people now working, as opposed to only living, in the countryside, wages are still lower and job opportunities fewer than in the towns. In the remoter areas unemployment can be high or employment, at best, seasonal, and underemployment can be a particular problem for women. In some areas where house prices remain comparatively high, in spite of an overall fall in the early 1990s, significant rural deprivation and further out-migration can result. This situation is already leading to shortages of labour for rural work and an unbalanced economy. So much so, that in some parts of south-east England, villagers have to 'bus' people in from the nearest town to do house-cleaning and gardening jobs.

As the dependence on the countryside for work diminishes and the less affluent move out, the downward spiralling effect on services that was referred to earlier must continue. Village shops will close, and public transport diminish and so on, because the orientation of the affluent country-dweller is towards the town or its peripheral megastore for shopping and this is invariably reached by private car.

There is thus a policy challenge for the 1990s and beyond, not only to bring the land uses of the countryside into greater environmental harmony at a time of food surpluses, but also to ensure a balanced community in the countryside to avoid the creation of affluent ghettos and the danger that the countryside could end up as a place of private affluence but public deprivation.

References

BLUNDEN, J. and REDDISH, A. (eds) (1996) *Energy, Resources and Environment*, London, Hodder and Stoughton/The Open University (second edition) (Book Three of this series).

MOSELEY, M. J. (1995) 'Policy and practice: the environmental component of the LEADER programme, 1992–4', *Journal of Environmental Planning and Management*, Vol. 38, No. 2.

NEWBY, H. (1985) *Green and Pleasant Land? Social change in rural England*, London, Wildwood House (second edition).

SARRE, P. (1996) 'Environmental issues in Cumbria', Ch. 1 in Sarre, P. and Reddish, A. (eds) *Environment and Society*, London, Hodder and Stoughton/The Open University (second edition) (Book One of this series).

Further reading

BLUNDEN, J. R. and CURRY, N. (eds) (1985) *The Changing Countryside*, produced by The Open University in association with the Countryside Commission, London, Croom Helm.

BLUNDEN, J. R. and CURRY, N. (1989) *A Future for Our Countryside* Oxford, Blackwell (second edition).

BLUNDEN, J. and CURRY, N. (1996) 'Analysing amenity and scientific problems: Broadland, England', in Sloep, P. and Blowers, A. (eds) *Environmental Policy in an International Context, 2: Conflicts*, London, Edward Arnold.

MACEWEN, A. and MACEWEN, M. (1987) *Greenprints for the Countryside? The story of Britain's national parks*, London, Allen and Unwin.

NEWBY, H. (1985) *Green and Pleasant Land? Social change in rural England* London, Wildwood House (second edition).

Answers to Activities

Activity 3

Beneficiaries have been:

(a) the public who, before entry to the EEC, enjoyed 'cheap' food subsidised by Exchequer (deficiency) payments and stable levels of production;

(b) large cereal producers who became wealthy as a result of preferential price support for their output;

(c) upland farmers who were able to stay in business as result of Less Favoured Areas and Farm Modernisation Directives;

(d) land-owners, who have seen land values rise faster than inflation partly as a result of agricultural price support;

(e) farm machinery, fertiliser and pesticide manufacturers taking advantage of the replacement of labour by capital and the move to monoculture systems.

Losers have been:

(a) the public, who after the United Kingdom joined the EEC saw a transition from 'cheap' to 'expensive' food, with inflated prices now falling directly on the consumer, to the disadvantage of the least well-off;

(b) agricultural workers, many of whom have had to leave the land and, for those who stayed, low wages compared with industrial rates;

(c) upland stock farmers, compared with others in the agricultural sector, since their levels of support were generally less favourable;

(d) the wider rural community, since agricultural policy has paid little heed to the non-farming population, the increasing affluence of farmers often existing side by side with growing rural deprivation;

(e) the conservationist, with a reduction in the variety of flora and fauna, especially in lowland England.

Activity 4

28 000 hectares in extent, and lying astride the boundaries of Norfolk and Suffolk, the Broads is a delicately balanced system of shallow, reed-lined lakes (or broads), slow-moving wide rivers, marsh-grazing lands and fens. But in this area a number of conflicting demands – from recreation, agriculture, liquid waste disposal and the changing socio-economic character of the area in general – can be seen to have interacted with consequences for its wildlife and landscape.

For example, looking at the diagram, the discharge of sewage effluent containing phosphorus into the river system, combined with nitrogenous fertiliser run-off from the surrounding fields, has created a nutrient-rich medium. As a result the algal content of the waters has increased massively, leading to oxygen loss, damage to plant life and, in turn, a decline in the diversity of flora and fauna. Also, the greater use of powered pleasure-craft on the waterways (which feeds back into the need for more local sewage disposal) has churned up the mud, again with adverse effects on the reed-beds, and thus has added to the problems for wildlife. Without reeds to support banks of the broads and rivers, these have had to be artificially secured with piles, thus diminishing the traditional character of the area.

You can follow other interactive pathways through the diagram for yourself. But note that 'traditional economy' in the context of fen management involves the harvesting of reeds as material for thatching. Failure to do this eventually allows the growth of woody plant species which ultimately invade the open broads.

For a more detailed study of the Broads as an interactive system of conflicting interests, see Blunden and Curry (1996).

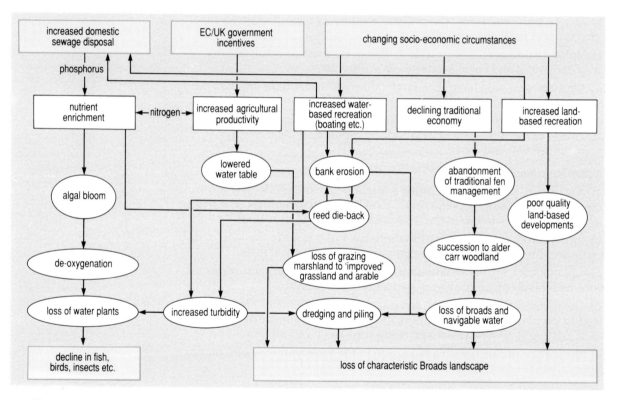

▲ *The Norfolk Broads: a web of interactions.*

1 Introduction

The previous chapters have provided an understanding of how populations, development and the environment interrelate. This chapter will look at how structures of the world economy, and especially trade, can alter these relationships.

World trade is about the exchange of goods and services between one country and another. Trade between countries has always taken place: some archaeologists point to evidence of trade in Neolithic times. However, the current patterns that are the focus of this chapter can be traced to the sixteenth century when European explorers sought new sources of wealth.

Through world trade the economic base of a country can be altered and possibly enlarged. Trade provides access to economic resources in other countries, so that people have access to what cannot be found, made or grown in their own locality – for example, supermarkets in Britain stock a wide variety of fruit and vegetables that cannot be grown there (such as oranges, bananas, mangoes and avocados) – but, at a more significant level, trade also provides access to valuable minerals such as oil and gas, diamonds and copper. For some countries trade is also a way of increasing the quantities of goods available, so that the people can use more than their own country produces: for example, Japan imports rice from South East Asia and Russia relies on wheat from the United States. Populations thus become less and less reliant on their immediate environment and what it can provide.

Economic development has depended on the ability to trade. This makes trade of central importance to understanding the relationship between economic development and the environment. Some environmentalists, in trying to identify how economic development alters the environment in ways that are damaging, have sought to isolate the effects of trade. However, trade is only one element of a complex and expanding economic system.

The question central to the trade and environment debate, and the question around which this chapter is structured, is how does world trade affect the environment? Are environmentalists correct when they claim that trade compounds environmental problems?

The first part of the chapter provides the background to this question. It looks briefly at the development of a world economic system in Section 2 and goes on to examine the different views in the trade and environment debate in Section 3. The second part of the chapter then examines the evidence of possible links between trade and the environment. Section 4 looks at the influence of trade on production methods, asking what the relationship is between world trade and environmentally damaging agricultural production processes. Section 5 examines the links between trade and consumption, asking how world trade affects patterns of consumption. Finally, Section 6 looks at the significance of trade as a way of distancing the consumer from the producer, and the environmental effects of this.

As you read this chapter, look out for answers to the following key questions:
- Does world trade compound environmental problems?
- What is its influence on environmentally damaging production processes?
- What is its effect on patterns of consumption?
- What are the environmental effects of distancing the consumer from the producer?

Although the trade and environment debate can be applied to the world trade in any commodity, the focus for our analysis will be agricultural products. Agriculture is an obvious area where human activity shapes the environment. It also allows us to apply our understanding of the evolution of agricultural production methods to the issue of world trade and to widen the analysis begun in previous chapters of the link between populations, agriculture and the environment.

2   The emergence of a global economy

2.1   Introduction

The modern world economy is a capitalist market economy where goods and services are distributed according to the mechanisms of demand and supply. Everything that is produced within this system is done so to be sold for profit. Countries are thus linked through their economic relations with one another. There are also a number of international institutions that influence how trade is structured on a world scale. After the end of the Second World War, three such institutions were established: the World Bank was initially responsible for the reconstruction of European economies (later being adapted to cover the development of the newly independent colonies); the International Monetary Fund (IMF) was set up to oversee and ensure there were stable exchange rates; and the General Agreement on Tariffs and Trade (GATT) provided the framework for international trade. Another significant trend has been the increase not only in the number of transnational companies, but in their size and scope. As a result of these changes, the trade in goods and services is no longer just between countries, but includes trade within transnational companies. Moreover, the production of goods is no longer limited by national boundaries: see Figure 6.1.

The origins of the modern **global economy** can be traced back to developments within Europe (and later North America) from the fourteenth century onwards. At this time the European economies were moving from agrarian/feudal societies to industrial/capitalist societies. There were three key processes involved: the enclosure of land; the industrial revolution; and the increased specialisation of production, in particular at the international level. All these processes depend on the ability to trade. This section will present a brief sketch of these developments to show the central importance of trade to the modern world economy. It focuses on events in Britain because it was the first country to experience this transition. Countries throughout Europe developed along similar paths, however, and through colonisation most countries in Asia, Africa and South America had been integrated into this pattern of development by the nineteenth century.

2.2   The roots of the global economy

For most people living in North America and many countries in Europe in the late twentieth century, the concept of common land means very little.

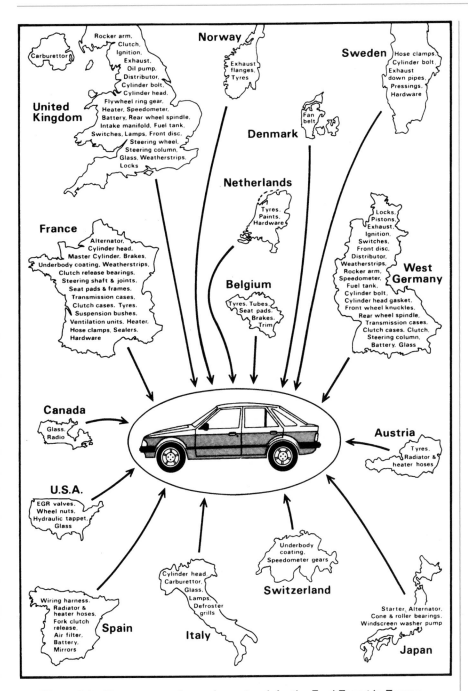

▲ Figure 6.1 The component sourcing network for the Ford Escort in Europe

Nearly all the land in Britain, for example, is either privately owned or is public space used for recreational purposes. In the past, however, 'the commons' referred to natural resources such as land, rivers and woods that everyone could use and for which everyone was responsible. These commons provided such amenities as grazing land, firewood and water for irrigation. Communities had complex rules and norms that established who could use them: these were often in the form of social norms and traditions or local bye-laws.

Economic development and the **industrialisation of agriculture** in Europe and North America between the fourteenth and the nineteenth centuries were underpinned by the conversion of the commons into privately and publicly owned property. In Britain, for example, until the end of the fifteenth century the majority of the population supported themselves by subsistence farming. Land was cultivated using an 'open-field' system, whereby individuals owned scattered strips of unfenced, arable land, and had access to common land of heathland, meadow, moorland or woodland that was used by everyone for pasture, peat or fuelwood (see Figure 6.2). The exact details may have varied between localities but the important point about this system was that farmers used both their own plots and the commons to supply produce for their own needs. Trade did not play a very important role and world trade was only of interest to the very wealthy.

Between the middle of the fourteenth century and the nineteenth century, European agriculture began to change in two important ways: new methods of production were developed that in turn required new systems of land ownership. Innovative agricultural methods – such as new ideas of crop rotation, the inclusion of new crops such as turnips and clover, and better methods of storing crops – offered the potential for farmers to grow crops surplus to their own needs that could be sold for profit. However, small strips of land were not always big enough for the newer methods of farming and some of the new technologies required considerable investment – of finance, labour or ideas – on the farmers' part. Those farmers who had their own capital to invest in these new methods began to press for a move away from strip farming to larger farms. The richer farmers began to fence off areas of common land for their own use. This disadvantaged the poor who relied on the commons to supplement their own plots of land. Many poor farmers became unable to earn a living from their own farms and were forced to sell up, becoming either landless agricultural labourers or drifting into the towns in search of employment there. The wealthy farmers benefited from these changes since they were able to buy the newly available strips of land to amalgamate with their own land in order to create larger units that could accommodate more specialised methods of farming. Increasingly, land was cultivated to produce surpluses that could be sold either at home or abroad: agricultural production became based less on principles of subsistence and more on trade.

These changes in agricultural production and the enclosure of common land took place alongside the industrial revolution. Cheap labour was needed for the factories and this could be supplied by the poorer farmers who had been excluded from the land and were forced to seek a living in the towns. At the same time food was required to feed these factory-workers. The larger farmer could specialise in the production of wheat to sell to the growing urban populations. The high level of demand for wheat from the growing urban population kept prices up. But by the mid nineteenth century there was growing pressure from industrialists to lower the price of basic foodstuffs, such as grain, so that wages could be kept low and thus the costs of production kept down. The solution lay in the availability of cheap wheat from other countries, in particular North America. Import restrictions on foreign grains were abolished in Britain in 1846, opening up the British market to imports of cheap grain from North America. In this way world trade played an integral role in Britain's industrialisation.

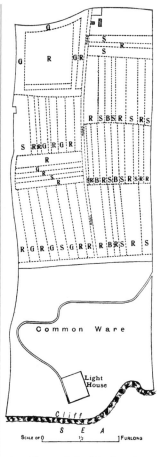

▲ *Figure 6.2 Medieval farming system: strip farming and common grazing in the South Common Field, Swanage (here unusually surviving to 1829): the initial G indicates the Parson's land; other initials denote the owners of the scattered strips. This is a simplified illustration: each person would have farmed many more and smaller strips.*

Here can also be seen the beginnings of an **international specialisation of production**, with North America concentrating on the production of wheat and Britain (followed by other European countries) specialising in manufactured products, made possible by the establishment of international trade.

This international specialisation can be seen most clearly in the development of economic relations with the colonial possessions in Africa and Asia. Industrialisation in Britain and other European countries needed both access to cheap raw materials and a large market for the final product. Political control of their colonies in Africa and Asia provided a convenient way for the industrialising countries to ensure both of these. In countries such as India, manufacturing was suppressed to ensure a monopoly supply of such goods from Britain. The resulting trade between the colonial countries and their colonies underpinned the international specialisation of production.

Colonisation also introduced methods of agricultural production that forced a European form of economic development on these countries through the enclosure of common land. Land was turned to plantations that could grow crops on a commercial scale for profit. These crops were often agricultural products such as cotton, rubber and jute used by manufacturing industries in Europe. Plantations were often developed on fertile land that would otherwise have been used by farmers to grow food crops for subsistence (a process which is occurring in places such as South and Central America even today).

These changes – the enclosure of common land, increased specialisation of production, the industrial revolution, and the international specialisation of production – were at the root of changes in agricultural production. Moreover, they could not have occurred on the scale they did without a corresponding growth in world trade.

Activity 1

How important was world trade to the industrialisation of agriculture? You will find an answer to this Activity at the end of the chapter.

2.3   Recent developments in the world trade in agricultural products

Since 1950 there has been a further centralisation of control of the land and industrialisation of agricultural production. As we have seen in earlier chapters, modern farming methods often rely on the use of inputs such as fertilisers, herbicides and pesticides as well as hybrid seeds and machinery. The initial investment needed for this form of farming is high and cannot easily be sustained by the smaller farmer. An inevitable consequence of this is that smaller farms go out of business and the land is merged with the larger concerns. This means fewer but larger farms and control of the land by a few powerful producers. Farmers have come to rely on trade both through their need to import chemicals, machinery and seeds and their dependence on the world market to sell on their surpluses.

The presence of transnational companies has also had an important effect. Nowadays, economic power is in the hands of just a few corporations. These companies supply foreign exchange, technology, skills and access to markets. In the agricultural sector farmers rely on transnational companies for technology and inputs (such as seeds, fertilisers, pesticides and machinery), as well as for purchase and credit. Although most of these companies are not directly involved in the growing of crops, they are powerful in determining what is grown and the methods used. According to the World Bank, transnational companies now account for 70% of all world trade, over 40% of which is the exchange of goods *within* transnational companies. A report by Christian Aid (Madden, 1992) found that: the biggest five transnational companies trading in cereals now control 77% of world trade in cereals; the top three companies trading in bananas control 80% of world trade; the top three cocoa companies control 83% of world trade in cocoa; and the top four tobacco companies control 87% of the world tobacco trade: see Figure 6.3 and Box 6.1.

Box 6.1   The power of the transnationals

• Seventy per cent of world trade is now controlled by just 500 corporations, which also control 80 per cent of foreign investment and 30 per cent of world GDP.

• Shell Oil's 1990 gross income ($132 billion) was more than the total GNP of Tanzania, Ethiopia, Nepal, Bangladesh, Zaire, Uganda, Nigeria, Kenya and Pakistan combined. Five hundred million people inhabit these countries, nearly a tenth of the world's population.

• Cargill, the Canadian grain giant, alone controls 60 per cent of the world trade in cereals. Its turnover in 1990 was the same as Pakistan's Gross National Product.

• Just 13 corporations supply 80 per cent of all automobiles: five of them (General Motors, Ford, Toyota, Nissan and Peugeot) sell half of all the vehicles manufactured each year.

• US corporations spend more than one billion dollars yearly on advertising. The average US citizen views 21,000 television commercials every year.

Transnational Corporations (TNCs) epitomise the logic of enclosure. Disembedded from any one culture and any one environment, they owe no loyalty to any community, any government of any people anywhere in the world. They are the most blatant example of what the anthropologist Roy Rappaport has called the 'special purpose institution'. Such institutions – from the military to government departments and international agencies – are driven by the desire to promote their own interests, to perpetuate themselves and to increase their power and influence. Decisions are not made because they are of benefit to the community or on environmental grounds but because they serve the institution's particular vested interest.

Employees are similarly disembedded from the real world. When acting for the organisation, company loyalty takes precedence over the moral and cultural restraints that mediate the rest of their lives. Dennis Levine, a Wall Street high-flyer who was imprisoned for insider trading, captures the detached world in which much corporate decision-making takes place: 'We had a phenomenal enterprise going on Wall Street, and it was easy to forget that the billions of dollars we threw around had any material impact upon the jobs and, thus, the daily lives of millions of Americans. All too often the Street seemed to be a giant Monopoly board, and this game-like attitude was clearly evident in our terminology. When a company was identified as an acquisition target, we declared that it was "in play". We designated the playing pieces and strategies in whimsical terms: white knight, target, shark repellent, the Pac-Man defence, poison pill, greenmail, the golden parachute. Keeping a scorecard was easy – the winner was the one who finalised the most deals and took home the most money.'

The power wielded by these organisations is greater than that of many, if not all, governments and makes a mockery of certain countries' claims to democracy. With the world as their gaming-table, TNCs are beholden neither to local communities nor to national electorates, but can dictate policy through their control of markets and the economic havoc they can cause by withdrawing support from a government. As such, they are the chief obstacle to the resolution of our environmental and social problems. If incalculably more money has been spent in the last 40 years on nuclear power rather than solar energy, for example, this is not because communities or electorates have favoured nuclear over solar; it is because TNCs, acting in alliance with state corporations, stand to benefit more from nuclear energy, whereas solar power has a potential to put control of energy back into the hands of the community.

Source: The Ecologist, 1993, pp. 79–80

Around 20 companies account for the bulk of trade in agricultural products

Cereals
Cargill
Continental
Ferruzzi
Louis Dreyfus
Bunge & Born

Bananas
United Brands
Castle and Cooke
Del Monte

Cocoa
Sucres et denrées
Ed. and F. Man
Cargill

Tea
Unilever
Twinings

Tobacco
Philip Morris
R.J.R. Reynolds
Gulf and Western
American Brand
B.A.T.

© Delpeuch – Solagral

▲ Figure 6.3

2.4   The theory of world trade

Trade is based on the twin pillars of international specialisation and comparative advantage.

International specialisation

In the modern world economy few, if any, countries are self-sufficient, and most rely on imports to meet some of their needs. The international specialisation of production allows for producers within any one country to specialise in the production of certain goods and to import others. According to classical economic theory, how production is divided and what producers specialise in will depend on local advantages such as climate, geology, workforce or technology. For example, to relate this idea to agricultural production, the farmers in the West Indies are better able to grow bananas than the farmers of Britain.

The economist, Adam Smith, writing in 1776, presented the idea that by specialising in just one production process producers can increase their output. Specialised production enables workers to concentrate on one task or one set of tasks, using the most appropriate technology. This results in more efficient production and increased productivity. Smith did not believe that specialisation should be limited to production within a country. If countries specialised in different areas of production, the economic benefits could be applied at an international level.

Adam Smith (1723–90)

Without trade, specialisation would not be possible. In order to invest in the production of one particular commodity, the producers must be confident that they can buy in other things that they need. At an international level it is of little use for farmers in the West Indies to specialise in growing bananas if they cannot purchase other basic foodstuffs. Trade also provides a larger market to absorb the surplus production. As more goods are produced, they require a larger market. Trade provides access to consumers in other countries willing to purchase the surplus production.

The rationale behind trade is that producers in one country should specialise in what they are best at and can produce more cheaply. These goods are then exchanged for other things that can be produced most cheaply in other countries. In the example of the West Indian and British farmers, the West Indian farmers have an advantage in the production of bananas, but Britain has an advantage over the West Indies in the production of agricultural machinery.

Comparative advantage

There are cases where one country can produce most things more cheaply than another. The incentive to trade in this case can be explained by the theory of **comparative advantage** put forward by another economist, David Ricardo. This theory states that if a country can produce everything it needs more cheaply than other countries, it should invest its labour and capital in the areas of production where it is most efficient and import the things it is less good at producing: see Box 6.2.

How world trade works in practice

According to these economic theories (of international specialisation and comparative advantage), trade is central to economic growth, increased world production, and an efficient distribution of resources. It will, therefore, benefit everyone. However, in practice these theories are compromised by four important factors:

- the ability of capital to move freely between countries
- the increased importance of technology

David Ricardo (1772–1823)

Box 6.2   Comparative advantage: an example

An easier way of understanding this is by thinking of trade between two individuals (Whitehead, 1982). Imagine there are two people: Ms Wig used to be an excellent typist before she took an OU degree and eventually became a practising barrister; she now earns £48 an hour. She uses a good secretary, Mr Keyin, who charges £12 an hour. If Ms Wig needs 10 pages of case notes typed up she could do it herself and it would take an hour. If Mr Keyin were to do it, it would take two hours. However, an hour of Ms Wig's time is the equivalent of £48. For this amount of money she could employ Mr Keyin for four hours. Even though he is half as fast, she would get twice as much typing done for the same amount of money as if she had done it herself.

The international trade example, used by David Ricardo when he first explained the theory of comparative advantage, is of trade between Portugal and England. In this semi-fictitious example, Portugal produces cloth and wine with less labour (and therefore less cost) than England, but it takes Portugal more labour to produce cloth than it does to produce wine. Ricardo argued that it is more profitable for Portugal to invest in the production of wine and import the cloth from England. Just as Ms Wig can both type and practise the law better than Mr Keyin, Portugal can produce cloth and wine more efficiently than England. In the same way as it proved to be a more cost-effective use of Ms Wig's time and skills to specialise in the practice of law and to pay Mr Keyin for any typing work she needs done, it is a better use of human and capital investment for Portugal to specialise in the production of wine and to import cloth from England.

- government policies to manipulate trade, and
- the policies of international organisations.

There are now few restrictions on moving money between countries: investors will choose the most profitable investment regardless of national boundaries. In economic terms one can say that investors are not motivated by comparative advantage but by absolute profit. To return to Ricardo's model of Britain and Portugal, capital would follow the most profitable investment, regardless of where it was. It would therefore be likely that British investors would put their money into wine production in Portugal rather than the production of cloth in Britain.

Moreover, what countries choose to invest in is no longer dependent on static factors of production such as natural resource endowments or the abundance of labour. Instead the use of technology has become much more important. Countries that have a technological advantage in the development of a product will have a trade advantage over other countries in that product. But these advantages are short-lived as technology is quickly diffused or copied in other places.

In addition, the role of governments in manipulating trade advantages has been ignored in the economic theories of trade. Governments erect **trade barriers** – *import tariffs* and *non-tariff barriers* (such as quotas, health and safety conditions or possibly even environmental conditions) – that exclude certain products from other countries. For example, the countries of Europe will only accept a limited amount of agricultural produce from outside the European Union. They may also implement policies that support and encourage domestic industry to invest in particular kinds of production by creating *tax incentives or subsidies*. The Common Agricultural Policy of the European Union is a classic example of this, but a system of agricultural subsidies also operates in the United States.

Finally, patterns of trade may be changed by international organisations. For example, under the World Bank's Structural Adjustment Programmes (see Section 4.3 below) developing countries can borrow money on the condition that they implement certain economic policies.

Central to these programmes is the requirement that countries receiving the loans should reduce their barriers to imports and increase their exports. World Bank staff often advise governments on what should be exported, that is, what products they should specialise in.

The theories of international specialisation and comparative advantage offer some useful concepts in understanding the way in which countries have developed their economies. However, it is important to look more widely than the picture that these theories present. There are some important political considerations and, in particular, questions of economic power.

2.5   Summary

This section looked at the role of world trade in the global economy. Trade has been a cornerstone in the core processes of economic development – the enclosure of common land, the industrialisation of agriculture, the industrial revolution, and the international specialisation of production. Without international trade, these processes would not have taken place. It has become clear, therefore, that trade plays a central role in economic development. It is because of this that trade is also central to the tensions between economic development and environmental protection. The next section will look at the trade and environment debate.

3   The trade and environment debate

3.1   Introduction

The importance of trade to economic growth was established in the last section. This understanding of the role of trade in economic development should have provided an appreciation of why trade has become central to the debate between environment and development.

The trade and environment debate is essentially about how the principles and practices of world trade can be reconciled with the need to protect the environment. The opposing interests made headline news in 1991 when environmentalists in the United States forced a ban on imports of Mexican tuna on the grounds that the Mexican fishermen were using methods that killed too many dolphins: see Figure 6.4. The conflicts appear

Global conservation threatened as Gatt declares war

A US law protecting dolphins has been overturned in the name of free trade. As John Vidal reports, the decision now threatens environmental legislation worldwide

The scarcely reported item was superficially unremarkable. An obscure General Agreement of Tariff and Trade (Gatt) dispute committee sitting behind closed doors in Geneva last week ruled that a US law banning imports of yellow-fish tuna from Mexico was contrary to international trade rules.

On the surface it was a blow for Flipper and the environmentalists because the ban had been originally imposed to prevent the death of dolphins who swim above the tuna and get caught up in the Mexican industry's voracious drift nets. Even so, in the great scheme of things it was very small potatoes to the layman and virtually ignored by the media.

But lurking below the Gatt's decision, taken after a three-week hearing of governmental evidence, and only leaked by accident, are a shoal of far-reaching implications. These throw in doubt a whole series of existing or proposed environmental laws and international agreements passed to protect individuals and species from health risks, over exploitation and inhumane deaths.

The decision—due to be rubber stamped by the full Gatt Council on October 8—was the first test of whether environmental concerns can be a factor in restricting a country's imports. By answering "no" the Gatt appears to be demanding that countries ditch many hard-won, publicly-supported environmental agreements.

Taken to its extreme, say Gatt critics, the Geneva precedent means that the Gatt could be used by any national government in the name of free trade, to overturn certain trade measures designed to protect endangered species or eco-systems. The ruling, in effect, says that environmental laws involving trade cannot extend beyond national borders.

Out, in theory, could go all national and international laws banning the imports of African elephant ivory or the trade in any other endangered species; it suggests that the Montreal Protocol which requires governments to impose trade sanctions against countries which refuse to phase out ozone-depleting chemicals is all but unworkable; it could mean that a whole series of environmental agreements on pollution between the US and Canada will have to be modified; it appears to say that no country may have a law to protect the environment or a species outside its own geographic territory; it throws into doubt the forthcoming Dutch ban on the import of certain sorts of tropical timber (which the EC was set to follow); and it adds confusion to the whole 1992 UN Earth Summit on environment and development, which has been building up to international agreements on energy conservation, biodiversity and toxic waste management.

Beyond that, the Gatt decision runs contrary to the western environment ministers' statements in Bergen last year on environmental reform, the influential 1987 Brundtland Report which laid down the tenets of sustainable development and even the thrust of much proposed EC and OECD environmental policy.

The consensus building this week in Britain and the US is that the Gatt is out of step with the growing international belief that extraordinary measures must be taken to protect the global environment. "It is a breathtaking attack on the progress made in the last 10 years," said a spokesman for Public Citizen, Ralph Nader's umbrella group which speaks for every blue chip environmental and consumer protection agency in the US.

"It's a major setback because it totally disregards national legislation designed to provide environmental protection for common resources," said WWF International policy analyst Charles Arden-Clarke.

But if the Gatt has deliberately stirred a hornet's nest of potential constitutional and legal problems, it is clearly with US connivance. The Bush Administration went through the motions of defending its tuna ban laws in Geneva but its superior and overriding commitment to free trade certainly coloured its defence. Free trade critics believe the US was quite happy to be ruled out of court, if only because the Administration has stated many times that it would like to end similar trade restrictions applied against its own producers. Hardline trade officials believe that environmental and consumer health regulations are little more than trade barriers.

The US and other countries will face severe opposition from the vociferous, well-organised and increasingly co-ordinated environment lobby which is already maintaining that this is the most serious attack on conservation and the sustainable use of natural resources in the last 30 years. As the implications sink in, they say, so any environmental credibility enjoyed by western governments will wither.

But if the Gatt has temporarily united the traditionally warring US and European environmental factions, it has also drawn the spotlight on itself at a critical time in its negotiations to conclude the Uruguay round of trade talks. All but stalled some months ago following European opposition to US attempts to reduce European farm subsidies, the talks were put on top of the western agenda by the G7 meeting in London in July and are continuing in Geneva with the express intention of concluding them by Christmas.

The outcome of the ambitious Uruguay round will dictate the terms of world trade for the next decade or more, and as that realisation seeps in, so opposition to the blanket free trade policies that the Gatt espouses is mounting.

World trade has only recently been on the agenda of groups such as WWF and Greenpeace, consumer groups and international charities. Having fought for legislation to protect natural resources for years, they are now uniting in fury against the Gatt which they say is the only world body with no remit for the environment.

The Gatt has traditionally kept a low profile, seeking little attention and largely shunning debate with the pressure groups which countercheck other world bodies and national governments. Its seemingly innocuous decision on tuna fishing is now liable to lift the lid on a can of worms and the Gatt may find itself subjected to the sort of virulent criticism that forces the World Bank to put the environment high on its agenda. If that happens, say its growing number of critics, perhaps Flipper may not have died in vain.

◀ *Figure 6.4*

Source: The Guardian,
6 September 1991, p. 9.

to polarise around two positions. On the one hand, environmentalists emphasise environmental protection while, on the other, the proponents of **trade liberalisation** support more trade and fewer import and export controls. Everyone involved would claim to be acting in the best interests of the environment, but they hold different values, beliefs and assumptions. These are reflected in their different understandings of the links between economic activity and environmental damage. As a result they propose different solutions to the problem of reconciling world trade with environmental protection.

There is, of course, a broad range of opinion on both sides. This section uses a political-economy approach to highlight the areas where there is a fundamental difference of opinion.

3.2   *The liberal economic approach*

Liberal economics dominates economic and political thought and policy. It has increased its dominance in the world since the collapse of the Soviet Union and the discrediting of alternative (Marxist) approaches. Many of the values, beliefs and assumptions of liberal economics are taken for granted in modern societies and rarely questioned. It forms the basis for government policy and the policies of international institutions such as the World Trade Organisation (and before that the General Agreement on Tariffs and Trade) and the World Bank. The arguments for greater trade liberalisation come from a liberal economic understanding of the world (Gilpin, 1987).

The liberal economic understanding is based on three assumptions: that the world comprises independent, rational individuals; that the market is a naturally occurring phenomenon and requires no regulation; and that economic growth is essential to the welfare of individuals.

The first assumption builds up a conception of the *individual* as the basis of society. The central focus for the activities of the individual is the market-place and, therefore, individuals are usually seen as economic actors such as consumers. Liberal economic individuals are rational beings, making decisions to satisfy certain values at the lowest cost to themselves. They have full access to all the necessary information but their decisions are always made under conditions of scarcity. That is to say, there is not enough for everyone to have everything. Individuals face choices between two (or more) things. As rational creatures, they base their decisions on a view of economic efficiency, using scarce resources to achieve the maximum benefit at the least cost.

The second assumption of liberal economics is the idea of *the market* and of *price mechanisms*. Markets are seen to emerge spontaneously as a consequence of the organisation of societies. As people organise to live together they need a market in which they can exchange goods to satisfy their needs. The pressures of demand and supply will regulate the price and quantity produced. The regulation of the market is thus a natural result of the market's own internal logic, often referred to as 'market forces'. The operation of self-regulating market forces provides the most efficient allocation of resources. It also provides the channels through which knowledge and technology are distributed: for example, the scientific and technical knowledge needed to clean up environmental pollution can be sold on the world market.

The third assumption of liberal economics is *economic growth*. Liberal economists believe that economic wealth is essential to ensure the welfare of all individuals. They see growth as gradual and continuous and believe that the long-term trend will always be increased growth. Economic growth is essential to support the provision of collective goods such as welfare benefits, health services, schools and also the protection of the environment. However, liberal economists also believe that the gains from economic growth will not be distributed equally to all individuals. Instead, people will be rewarded in terms of their relative contribution to the generation of that wealth. Put in economic terms, they talk about 'absolute gains' and 'relative gains'. An *absolute gain* would be overall world economic growth; *relative gains* would be the different amounts countries and individuals within countries benefit relative to one another: certain individuals or countries will gain relatively more (or less). At a global level, trade under free market conditions is seen as an important engine behind economic growth. The gains from trade will not be equally felt by all countries (there will be relative gains), but all countries will be better off (benefiting from the absolute gains).

3.3   Liberal economics in practice

In practice, however, the principles of liberal economics are honoured more in the breech than the observance by the rich countries of Europe, North America and Japan. It was intended that the three institutions set up after the Second World War to manage the international economy (GATT, the World Bank and the IMF) would support a new international economic order based on the principles of liberal economics. However, one common observation on these institutions that emerged was that they were 'liberal regimes, but with a lot of cheating taking place on the domestic side' (Ruggie, 1982, p. 214). In practice there was a tension between the liberal economic principles of comparative advantage and trade liberalisation and a desire to ensure that countries could protect against the possible harmful effects of international specialisation. The fear was that a commitment to the complete liberalisation of trade might result in cheap imports that domestic producers would be unable to compete with and would result in the contraction or loss of certain industries, rising unemployment and social unrest.

GATT was originally signed before most countries of Africa and Asia had won political independence and was, therefore, drawn up to apply mainly to the countries of North America and Europe. Although the text of GATT contains a commitment to the liberalisation of trade above all else, it also contains a number of exceptions that allow for discrimination against imports that may directly compete with domestic producers. Since it was first drafted, many more countries have signed it. GATT was replaced by the World Trade Organisation (WTO) in 1995 (although the full transfer between the two organisations was not completed until January 1996). However, the WTO contains the same apparently opposing aims of trade liberalisation and a lingering desire to protect national economic and political interests.

There are other factors that compromise the liberal economic ideal of trade liberalisation. Since the 1980s there has been a growing movement towards trading blocs: the European Union, the North American Free Trade Agreement and the Association of South East Asian Nations are examples of this. These are blocs that promote trade liberalisation for a region but tend to maintain a relatively high level of barriers on goods from outside that region.

Another factor is that economically powerful countries are able to engage in bilateral talks with weaker countries to promote their own interests. In the USA, Section 301 of the 1984 Trade Act – 'Super 301' – does exactly that. It permits the United States to take retaliatory action if a country denies access to US corporations 'unfairly' or fails to protect the intellectual property rights of US corporations. In effect, it overrides the USA's commitments under the WTO.

Moreover, the trade within transnational companies (moving goods from one subsidiary to another) now accounts for 40% of world trade (Lang and Hines, 1993, p. 34). These transactions lie outside the jurisdiction of the WTO (which is an agreement between countries), but the sheer volume and value of such movements make them very important to the economies of individual countries.

3.4   The liberal economic position in the trade and environment debate

The liberal economic perspective sees the environment as something external to human society that provides people with goods and services. These goods and services may take many different forms. We talk about **environmental goods** when we are talking about plants, animals, clear air or clean water, and **environmental services** when referring to the way in which forests absorb carbon dioxide or the sea disposes of sewage and oil pollution.

The problem for the liberal economic approach is that the environment exists outside the market: to a large extent the goods and services provided by nature cannot be bought and sold in a conventional sense and are, therefore seen as 'free'. For example, most people do not expect to have to 'pay' for clean air.

A good example of the liberal economic approach to the environment is the report from GATT on Trade and Environment: see Figure 6.5, which is a press release from GATT.

Activity 2

Read Figure 6.5 and note down how the assumptions outlined above are reflected in the GATT position on trade and environment.
You can compare your answer with the one given at the end of the chapter.

▼ *Figure 6.5*

Source: GATT Secretariat, Geneva, 3 February 1992.

Expanding trade can help solve environmental problems, says report

GATT rules are not an obstacle to environmental protection, but trade weapon could be counterproductive

Increased world trade leads to higher per capita incomes, and with that the freedom and incentive to devote a growing proportion of national expenditure to the environment. "The opportunity for countries to trade in world markets for goods and technologies facilitates the implementation of needed environment-improving processes at home." The available evidence suggests that this is indeed happening.

GATT rules do not prevent governments adopting efficient policies to safeguard their own domestic environment, nor are the rules likely to block regional or global policies which command broad support within the world community. At the same time, trade measures are seldom likely to be the best way to secure environmental objectives and, indeed, could be counter-productive.

These are among the key conclusions in a study on Trade and the Environment released today. The study, by the GATT Secretariat, is part of the International Trade report which will be published in full by GATT at the beginning of March.

There is, then, an acceptance that environmental damage is occurring but, in relating this to world trade, liberal economists and proponents of trade liberalisation argue that the exact links between world trade and damage to the environment are not known. Their response to heightened concern about the environment has three strands.

Firstly, they emphasise the ways in which trade benefits the environment through increased economic growth. According to this perspective, world trade and environmental protection are mutually supportive: there can be no trade without the sustainable use of natural resources and there can be no environmental protection without the wealth that is generated by trade. They also point to indications that a growing awareness of environmental problems, and the resulting pressure to devote resources to protecting the environment, will only be present when there is an increase in personal income levels.

Furthermore, liberal economists argue that trade is an essential channel for the distribution of the knowledge, skills and technology needed to protect the environment. An often cited example used to support this view is that of the environmental degradation in the former Soviet Union. (Opponents of the liberal economic approach would argue that whilst there has been an obvious lack of environmental protection in the former Soviet Union compared to the levels of protection that now exist in the countries of western Europe, this is not necessarily a direct result of the differences between market economies and a centrally planned economy. It may be a consequence of the different stage of economic development which the different countries have reached.)

The second strand of the liberal economic response to environmental concerns is to stress that any link between world trade and environmental protection is only indirect. According to the liberal economic view, world trade concerns the movement of goods across national boundaries; therefore discussions about trade policies can only relate to import barriers or export supports. Concerns about how those goods are produced and the possible effects on the environment of the production processes are a matter for the governments of the countries where they are produced. If the correct regulations are in place at a national level, then trade will be environmentally neutral. If no environmental policies are in place, then trade may magnify the effects of production and consumption patterns on the environment, but they are not the direct cause.

The final strand of the liberal economic response to environmental concerns is to warn of the possible damaging effects of introducing trade measures to protect the environment without multilateral agreement on these measures. The liberal economic assumption that economic growth is vital to the generation of social welfare, including environmental protection, and the belief that economic growth is only achieved through trade liberalisation, makes it inevitable that this approach will reject the need for trade measures to protect the environment. Where environmentalists call for an import ban on tropical timber or on ways in which tuna fish are caught that also kill dolphins, a liberal economist will argue that current trade structures do not need to be altered to accommodate environmental concerns. The biggest fear of those who support greater trade liberalisation is that protection of the environment will be used as a cover to introduce **trade protectionist measures** that are, in fact, aimed at nothing more than the sponsoring of domestic industry.

According to the liberal economic approach, the problem of reconciling environment and trade can be overcome by bringing the environment inside the free market. To do this, the different goods and services provided by the environment must be owned by an individual in order to be traded on the market. This then takes the liberal economists into discussions about property rights. At the moment property rights over environmental goods and services are either absent (for example, no-one owns the ozone layer) or unclear and under dispute (for example, rain forests are claimed by some to be owned by everyone in the world for their services in converting carbon dioxide into oxygen, while governments in countries where the forests are located claim they are owned exclusively by them).

> *Activity 3*
>
> Think about the attitude of environmentalists to economic development and write down how they might criticise the liberal economic approach. You could read (or re-read) *Sarre and Brown* (1996).

3.5   *The environmental perspective*

The environmental perspective is united in one basic criticism against the current trading system: that it does not take sufficient account of environmental issues. Beyond this, there are a number of different views. The chapter by Sarre and Brown uses Eckersley's (1992) classification that divides approaches to the environment into: resource conservation, human welfare ecology, preservationism, animal liberation and ecocentrism. This section follows this classification quite closely but puts the five categories that Eckersley uses into two broad groups: the *radical environmentalists* (that encompasses ecocentrism, preservationism and animal liberation) and the *reformists* (that broadly relates to Eckersley's resource conservation and human welfare ecology). No classification is perfect, but these two broad categories represent an important difference in how environmentalists approach the trade and environment debate.

The category of radical environmentalists is defined by a common view that there needs to be a complete reshaping of the existing global economic and political order. The label of reformist is applied to those who adopt a compromise position in the trade and environment debate: they try to work within existing global economic and political structures, suggesting various ways they might be reformed to better protect the interests of the environment.

A radical environmentalist approach

It is the rejection of the liberal economic perspective that unites the more radical environmental groups in their criticisms of the current trading system. They argue that the reason why the current trading system does not take sufficient account of environmental issues is because of the assumptions and methodologies of the dominant liberal economic model.

First, radical environmentalists reject the assumption that society is made up of individuals acting autonomously and rationally with no sense of social responsibility. They present a counter-view of people operating in ways that indicate a concern for social and environmental issues and that prioritise community well-being above economic growth. The Ecologist's *Whose Common Future?* argues that the promotion of the idea of the self-contained individual undermines other belief-systems that emphasise social responsibility and responsibility to the environment. Instead it wants to see a return to values of 'mutual support, responsibility and trust that sustain the commons' (The Ecologist, 1993, p. 197). Similarly, Lang and Hines articulate a radical call for a **'new protectionism'** that 'aims to protect the environment by reducing international trade and by reorienting and diversifying entire economies towards producing the most that they can locally or nationally, then looking to the region that surrounds them, and only as a last option to global international trade' (1993, p. 3).

Secondly, radical environmentalists continue to attack the assumptions behind the liberal economic conception of the market. They challenge the idea that the market is the most efficient mechanism for the distribution of resources. The central point here is that liberal economists view the market as something that is politically neutral, whereas radical environmentalists stress the political implications of the way in which the market operates. In the view of radical environmentalists, world trade is a process of unequal exchange whereby the richer and economically powerful countries are able to apply policies to their advantage. They argue that, in practice, trade has resulted in **overconsumption** in the advanced capitalist states and increased levels of poverty in poorer countries. Moreover, the recent developments in the global economy have led to an increase in the number and scope of transnational companies and to the emergence of large trading blocs. Decisions about how natural resources should be managed are now taken by a few powerful corporations and countries.

The third criticism the radical environmentalists make of a liberal economic perspective is that the promotion of a 'free market' approach undermines the accountability of institutions and actors. By conceptualising the market as politically neutral, the liberal economic perspective fails to tackle the question of accountability: who takes decisions affecting trade and to whom they should be responsible. They characterise the dilemma by comparing the role of individuals as consumers with the role of individuals as citizens. Citizens of a country are involved in the process of decision-making, even if only through a representative, whereas consumers can only choose whether to buy something or not, with no means of influencing what is on offer.

Finally, radical environmentalists challenge the priority given to economic growth. They reject the importance of trade as an engine for growth for three different reasons. First, they point to the fact that the economic benefits of trade are not equitably shared between countries or between people within a country (because of the reliance on market forces to distribute the wealth). Second, they argue that economic growth generated by trade takes no account of the ecological limits to that growth. Their view is that increases in wealth generated by trade will lead to increased levels of consumption of natural resources beyond the levels that are ecologically sustainable. Furthermore, trade alters consumption patterns away from local products to goods that are produced abroad, which, according to radical environmentalists, has negative environmental impacts. Finally, they dismiss the idea that economic growth is needed to finance environmental protection. Even where increased economic growth can be seen as a catalyst for greater environmental protection, radical environmentalists argue that it is often not possible to clean up the damage to the ecological systems that has already occurred. In any case, it always costs more to repair or 'clean up' environmental damage than to prevent it.

The solutions of radical environmentalists vary. Some end with calls for a fundamental restructuring with no specific policy recommendations (see Figure 6.6), while others present a more specific agenda (see Figure 6.7).

A CONCLUDING REMARK

It is customary to conclude a document such as this with policy recommendations. We are not going to do so. Our reasons are many but two of them have been expressed admirably (although in another context) by Philip Raikes in the introduction to his book *Modernising Hunger:*

'It becomes increasingly difficult to say what are practical suggestions, when one's research tends to show that what is politically feasible is usually too minor to make any difference, while changes significant enough to be worthwhile are often unthinkable in practical political terms. In any case, genuine practicality in making policy suggestions requires detailed knowledge of a particular country or area; its history, culture, vegetation, existing situation, and much more besides. Lists of general 'policy conclusions' make it all too easy for the rigid- minded to apply them as general recipes, without thought,

criticism or adjustment for circumstances.'

Like Raikes's book, our document is 'full of implicit conclusions' and explicit demands, but to formulate them as 'policy recommendations' would be to go against the case we have attempted to make. It would suggest that there is a single set of principles for change; and that today's policy-makers, whether in national governments or international institutions, are the best people to apply them. We reject that view.

A space for the commons cannot be created by economists, development planners, legislators, 'empowerment' specialists or other paternalistic outsiders. To place the future in the hands of such individuals would be to maintain the webs of power that are currently stifling commons regimes. One cannot legislate he commons into existence; nor can the commons be reclaimed simply by adopting 'green techniques' such as

organic agriculture, alternative energy strategies or better public transport – necessary and desirable though such strategies often are. Rather, commons regimes emerge through ordinary people's day-to-day resistance to enclosure, and through their efforts to regain the mutual support, responsibility and trust that sustain the   commons.

That is not to say that one can ignore policy-makers or policy-making. The depredations of transnational corporations, international bureaucracies and national governments cannot be allowed to go unchallenged. But the environmental movement has a responsibility to ensure that in seeking solutions, it does not remove the initiative from those who are defending their commons or attempting to regenerate commons regimes. It is a responsibility it should take seriously.

▲ *Figure 6.6* Source: *The Ecologist*, 1993, pp. 196-7

A new political agenda

The political challenge is to *make* the era of free trade close, yet most politicians are frankly at sea about trade issues. They may believe that free trade is a good thing, but when pressed, most have never read a word of GATT documents. The huge task we offer them is to begin to redirect the world's economies. In the short term, we want politicians to work towards the following long-term goals:

- A shift in politics away from promoting the free trade package ... to giving priority to equity and the protection of the environment.

- A new trade policy. The world needs less, not more international trade. Much trade is ecological madness with ships and wagons traversing the globe passing others carrying similar goods, or goods which could be produced more locally, going the other way.

- Promoting regionalism. Every region, by which we mean both localities within countries and regional groupings of neighbouring countries, should be economically diverse and production organized on a local level, rather than the current tendency to globalize production. The long-term goal should be to achieve as much regional self-reliance as possible,to minimize the

distance goods travel unnecessarily.

- A new competition policy. Global intervention to control TNCs is a priority. This is unlikely to come through the UN which in 1992 downgraded, *de facto* gutted, its own centre for studying TNCs, but it is more likely to come from local or regional groupings of nation states bent on protecting the economic health of their areas.

- New trade mechanisms to control and monitor trade should have as a goal the need to protect the environment adequately and to reverse the growing inequality of the world. We'd like GATT to become GAST, the General Agreement for Sustainable Trade.

- Research on developing transitional strategies to move trade into the new regional patterns.

The task may be formidable, but the movement for such a new vision for trade is already emerging. The GATT talks have generated opposition from an unparalleled global network of public interest groups – a coalition of environment groups, public health workers, citizens, farmers and others.

▲ *Figure 6.7* Source: Lang and Hines, 1993, p. 14

A reformist approach

The reformist approach covers a broad spectrum of views from environmental economists to environmental **non-governmental**

organisations (NGOs) such as the World Wide Fund for Nature (WWF). Some environmentalists that I have labelled as reformists share some of the radical environmentalists' criticisms of the liberal economic approach. What distinguishes them from the radical environmentalists is the solutions that they offer to the problem of reconciling environmental concerns with the demands of economic growth.

The reformist position, like that of the radical environmentalists, rejects the liberal economic view of the environment as something external to society. It sees the environment as an integral part of society. This forms part of the argument for a greater emphasis on sustainable development and the need to pass on an equal stock of environmental resources to future generations. However, most reformists do not place the same emphasis on the role of communities. The reformist position differs most from the radical environmentalists in its view of the market. There is an underlying acceptance of the market in the agendas of reformist environmentalists. They argue that the current trading system is having a damaging effect on the environment because the market fails to allocate an economic value to the natural resources that are used. With pressure from increasing populations, increased production and increased consumption and with no market to regulate the use of environmental goods and services, they are being used up at a faster rate than they can regenerate.

The possible solutions to this problem are diverse, reflecting some of the differences within this broad category. Most reformists advocate some form of valuation such as the internalisation of **environmental costs**. They argue that if the costs of production included the full environmental costs (translated into a monetary figure) the market would provide an efficient way of allocating resources. For example, the environment would benefit if logging companies in South East Asia included in the cost of logs not only the costs of leasing the land and of felling the trees, but also the environmental costs of replanting, cultivation and compensatory measures for environmental side-effects of a temporary loss of tree cover (such as soil erosion and flooding). The money generated could be spent on measures that would ensure logging was sustainable. Also, the price of the logs would reflect their scarcity: they would no longer be a cheap product and consumption would decline. As a spin-off, technology may be developed to provide alternatives to logs or alternate and less wasteful ways of using the logs. WWF(UK) claims to have worked with the British retailers B&Q to develop a demonstration forestry project that puts the idea of the internalisation of environmental cost into practice. As a result, B&Q invested in a small-scale, community-based forest management project from which they could buy timber. However, there are enormous political and practical problems in implementing the internalisation of environmental costs on a global scale.

Some reformists are more critical of the market. They advocate regulation of the market to implement their environmental standards. This may take the form of national standards that producers must adhere to, for example the implementation of penalties for the pollution of water or air. In terms of world trade, it may require counterbalancing trade measures. To return to the example of logs from South East Asia, if the Malaysian government were to implement standards relating to replanting and flood prevention measures, the cost of felling logs in Malaysia would increase. If other countries that exported tropical timber did not have similar standards, the cost of logging in Malaysia would be significantly higher than the cost of logging in the other countries. There would be an incentive for logging companies to relocate to countries without such strict standards in order to keep their production costs down. Some reformist environmentalists call for the World Trade Organisation to rule the absence of environmental standards to be 'illegal subsidies'.

Taking this one step further, some reformists would like to see the active use of the market to implement *environmental standards*. That is to say, they advocate the use of trade barriers to promote sustainable production processes. A famous example of this was the import ban in the United States against tuna from Mexico on the grounds that the fishing methods killed too many dolphins. In this case Mexico was able to appeal to GATT to have the ban lifted. Under GATT rules, countries cannot discriminate between products on the basis of production process. Reformist environmentalists like WWF, whilst holding on to their acceptance of the market, would like to see the rules of GATT's successor, the World Trade Organisation, altered to allow for trade policies that discriminated against products produced in ways that damage the environment.

Linked to this, some reformists would like to see an acceptance of trade measures that may be needed to support the implementation of multilateral environmental agreements. The two most famous examples are the Convention for International Trade in Endangered Species (CITES), which uses trade measures to halt the consumption of flora and fauna that are in danger of extinction, and the Montreal Protocol for the Protection of the Ozone Layer, which prohibits the trade in CFCs and products containing CFCs between countries who have signed the Protocol and countries who have not.

Finally, environmental NGOs that fall short of a radical agenda, such as WWF, nevertheless share the radical environmentalists concerns about accountability. Part of their agenda advocates a greater use of standards and regulations to supplement the use of market mechanisms in order to promote policies that protect the environment. In promoting these standards they also emphasise the need to make the institutions that would agree and implement the standards responsible, accountable and transparent. They criticised GATT for its decision-making structures that appear secretive, undemocratic and susceptible to domination by the powerful economic actors. Similar criticisms are now being raised against the new World Trade Organisation.

Activity 4

You might find it useful to summarise the three positions in the trade and environment debate – liberal economic, radical environmentalist and reformist – highlighting their differences and their similarities. Then compare your summary with that given at the end of the chapter.

The trade and environment debate presents two very different views. If presented with the question 'does world trade compound environmental problems?', they would come to opposing answers. The next stage is to look in more detail at three subsidiary questions behind the main question. The following section provides more detailed examination of the role of trade in shaping production processes.

4 The relationship between trade and environmentally damaging production processes

4.1 Introduction

The previous section looked at the different approaches to the trade and environment debate. This section will explore one area of disagreement in more detail: how far is world trade responsible for agricultural production processes that are environmentally damaging?

Activity 5

You may find it useful to refresh your memory about the environmental impacts of different methods of production. Look in particular at Section 4 in each of Chapter 2, Chapter 3 and Chapter 4 above.

Write a brief summary of how you think trade has contributed to the environmental impacts of wetland rice production and of temperate farming systems. Some points are listed at the end of the chapter.

Radical environmentalists believe that trade, by supporting a particular form of economic development, is, in part, responsible for the emergence of certain methods of production. To be specific, economic development has resulted in the industrialisation of farming and the increased specialisation of agricultural production. This has created large surpluses that are exported and involves production methods that damage the environment. The reformist position accepts that the industrialisation of farming cannot be prevented, but argues that the current trade rules prevent governments from introducing regulations that would promote methods of production that are less harmful to the environment. However, the link between trade, methods of agricultural production, and the damage to the environment are contested by those who support the current economic structures. Those holding a liberal economic view of world trade argue that trade has only an indirect effect on production methods and that inappropriate government policies or the lack of environmental regulation play a far more important role in determining what production methods are used.

This section explores these opposing claims in more detail. It will look first at the production of a basic foodstuff, grain (wheat, barley, corns, oats etc.). Has world trade encouraged the use of environmentally damaging farming methods to produce grain? One of the direct effects of trade in basic foods is an increased reliance on imports. But in order to pay for these imports a country must increase its exports to earn foreign revenue. This section goes on to examine the production of two crops grown almost exclusively for export – flowers and tobacco – and asks what the effects are on the environment of producing crops for export. Finally, it considers whether trade policies prevent the introduction of farming methods that would be less damaging to the environment. What happens when

government policies to promote production methods that are less harmful
to the environment clash with world trade policies?

4.2   *The effect of trade on the production of grain*

An important feature of trade is that it links up producers and consumers
in different countries through a world market. Through these links the
economic structures and decisions taken in one country can have an effect
on the options available to producers in other countries. So, for example,
grain producers in the USA are able to sell their grain at a low price on the
world market. This is often welcomed by governments looking for a cheap
solution to the problem of feeding expanding urban populations and by
consumers, but it may cause problems for the producers in those countries.
The existence of cheap grain on the world market will exert pressure on
producers to compete with the low prices.

So, why are some producers able to sell at a low price? The price a
producer can charge will depend on the costs of production, the availability
of subsidies, the level of demand and the competition from other
producers. In agricultural production the costs of production are
determined by a number of different things: local conditions such as soil
fertility and climate; the production methods used; the availability of other
resources needed for the production process such as energy; and
government policies.

Grain producers facing competition from cheap supplies on the world
market have three ways of responding: they may try to compete with these
cheap supplies by cutting the costs of production; they may decide to stop
producing grain and invest in another crop; or they may put pressure on
their government to introduce some form of price support mechanism or
import barrier.

Industrial farming methods are used to cut costs of production. In some
cases this can mean the adoption of intensive farming with a greater use of
new seed varieties, new machinery, fertilisers, irrigation and pesticides to

boost yields (see Chapter 4, Section 4.3). Put simply, by using fertilisers and pesticides and by cutting down on labour through the use of machines, farmers can produce more for less. Other forms of industrial farming use extensive systems. In areas where land is plentiful, farmers are able to adopt methods that have a low ratio of output per hectare but do not require large and costly inputs. The grain belt of the United States is one such example; another is the practice of cutting down rainforests in South America to provide land for cattle. Environmental damage is caused by converting areas rich in biodiversity to ranches or wheat fields. Extensive farming methods are also vulnerable to pressures to produce more from the land. This can lead to overgrazing and subsequent soil erosion (see Chapter 4, Section 4.4). Environmentalists argue that through the operation of the world market, trade can be seen to exert pressure on farmers to reduce costs by adopting modes of production that have a damaging effect on the environment.

The second response to low prices on the world market is to stop farming grain altogether and rely on imports. Some countries will find it cheaper to import grain than produce it locally. This has become more common for developing countries where governments are looking for an immediate solution to the demand for cheap food from a growing urban population. However, there are social, political and environmental impacts

"But if we don't sell our food, how can we earn the money we need to buy it back again?"

of abandoning the domestic production of basic foodstuffs. In order to generate income to pay for food imports countries must export more. Exactly what they export will depend on the natural resources of the country, the world market, and the policies of international economic organisations such as the World Trade Organisation and the World Bank. Some countries are able to specialise in mining, manufacturing or services; many of the poorer countries turn to producing crops for export. This may be the result of limited natural resources but the existing patterns of specialisation established during the period of colonial rule are also an important factor. These trade patterns are often reinforced by trade barriers in former colonial countries that restrict imports that compete with domestic products including many manufactured products. Environmentalists argue that trade, through current patterns of international specialisation, encourages large-scale agricultural production for export (*cash crops*) and that this has a damaging effect on the environment. The environmental impacts of growing crops for export are examined in more detail below (in Section 4.4).

There are, however, important economic and political implications of abandoning the production of food for domestic consumption. If a country is to transfer human and financial capital to the production of cash crops instead of basic foodstuffs, it must feel confident that it will always be able to import the other foods needed for the population. Developing countries have become increasingly dependent on food imports: imports of cereals by developing countries rose from 5 million tons in the 1950s to 100 million tons in the early 1980s (Watkins, 1992, p. 65). In the mid-1980s the Philippines was almost self-sufficient in rice; by 1990 it was importing near to 600 000 tons. The exporting country can easily exploit this situation to achieve political and economic ends (see Box 6.3). Moreover, it leaves the importing countries vulnerable to changes in the world market supply or price. For example, if the current surpluses of grain on the world market were to be reduced or eliminated, or if shortages or increased demand meant the world price rose substantially, the countries now dependent on the availability of cheap grain imports would be unable to meet their needs.

The third option open to farmers facing growing competition from cheap imports of grain is to put pressure on their government to introduce policies that would discourage imports and encourage domestic production. This may mean putting up trade barriers such as import duties on foreign grain or increasing subsidies to farmers or both. The United States and the European Union have, in fact, done both (see Box 6.4). However, developing countries are often severely constrained by other economic structures, as will be shown in the next section.

Box 6.3   *The control of food for political ends*

The United States, the world's largest exporter of grain, is able to dictate economic and political policy to many countries through their reliance on food imports from the USA. In 1960 the US Vice President declared that, 'If you are looking for a way to get people to lean on you and be dependent on you, in terms of their co-operation with you, it seems to me that food dependence would be terrific' (quoted by The Ecologist, 1993, p. 221).

Twenty-six years later things did not seem to have changed that much: John Block, the US Agricultural Secretary, in 1986 commented that, 'The idea that developing countries should feed themselves is an anachronism from a bygone era. They could better ensure their food security by relying on US agricultural products, which are available, in most cases at a lower cost' (Watkins, 1992, p. 70).

Box 6.4   Government subsidies and agricultural surpluses in Europe and America

In Europe and the USA prices for agricultural production have been kept below the actual costs of production through a system of government subsidies. In the early 1970s the United States operated a system of paying 'deficiency payments'. Through this scheme, grain companies bought grain at low prices, that were even lower than the costs of production. The difference between the market price for grain, and the price farmers claimed they needed to cover the cost of producing the grain, was made up by a direct payment from the government. The net result for the world market is that the grain companies could sell the grain at below the cost of production.

In Europe the Common Agricultural Policy (CAP) provided a system of price support to farmers that offered prices for agricultural products at between 35 and 40% above the price levels in countries that did not operate the same level of subsidies (for example, Australia, Argentina and New Zealand). In response to this, agricultural production increased dramatically and large surpluses began to accrue. This in turn led the then

European Community to subsidise the export of agricultural surpluses. In 1992 the cost to taxpayers of CAP was US$48, some 40% of which went on subsidising the export of cereals and other primary commodities (Gardner, 1993, p. 2).

One way of calculating the value of these subsidies is the Producer Subsidy Equivalent. The OECD calculated that in 1991 the PSE for the European Community (as it was then known) was 49% of the value of output; for Canada it was 41%: see Figure 6.8.

Since 1992 there have been some important reforms. Between 1989 and 1992, the US government reduced its level of subsidy for cereal production (Gardner, 1993, p. 15). The European Union also reformed the CAP, by switching subsidies from the market (price support) to direct support of farmers. In addition, the conclusion of the Uruguay Round of GATT negotiations in 1994 required that developed countries reduce the total amount spent on export subsidies by 36% and the amount of domestic support to agricultural producers by 18% (World Trade Organisation, 1995). Whilst these measures have been cautiously welcomed, the exact benefits to developing countries are not yet known.

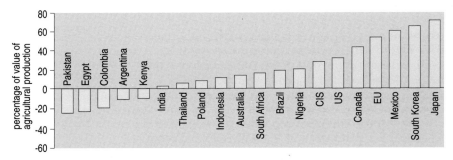

▲   Figure 6.8   Producer subsidy equivalents

Source: Farmers' Link, 1993, CAP Tales, p. 2, from OECD data

4.3   The trade–debt trap

Many developing countries are caught in a trap of high debt repayments, and declining **terms of trade**.

In the 1970s many countries in the developing world, particularly those that had no oil to export, were faced with **balance of payments** deficits. In response to this, these countries took out large loans with private banks (often actively encouraged by the banks themselves). In order to meet the repayments on these loans, the debtor countries had to raise foreign exchange. This meant increasing export earnings. For many of the poorer countries the main export was commodities – either mineral deposits or agricultural products.

However, from the end of the 1970s the price of commodities declined:

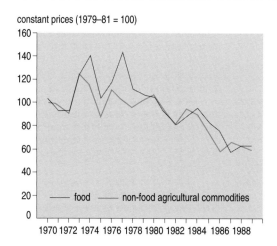

constant prices (1979–81 = 100)

▲ *Figure 6.9   Index of international farm commodity prices, 1970–89.*

Note: Food commodities include beverages (coffee, cocoa, tea), cereals (maize, rice, wheat, grain sorghum), fats and oils (palm oil, coconut oil, groundnut oil, soya beans, copra, groundnut meal, soya bean meal), and other food (sugar, beef, bananas, oranges). Non-food agricultural commodities include cotton, jute, rubber and tobacco.

see Figure 6.9. The graph shows that although the price of commodities has fluctuated, since the end of the 1970s the overall trend has been downwards.

In response to this problem countries tried to diversify their exports and since 1980 there has been a dramatic increase in the importance of manufacturing: see Figure 6.10. However, the graph shows that manufacturing still makes up about half of the value of exports from developing countries. Moreover, the share of manufacturing industry is not evenly spread among countries. The poorer countries, particularly those in Africa, remain dependent on the export of one or two products. Figure 6.11 shows that, in 1991, 26 African countries depended on the export of one product, 9 relied on just two products and 10 on three to five products.

There are three important reasons for the persistence of a dependence on commodities for foreign exchange. First, developed countries have maintained historical trading links established during the period of colonial rule through a system of trade barriers. In this way developed countries have limited the entry of goods that would directly compete with producers in their own countries. The Uruguay Round of GATT negotiations reduced these trade barriers, particularly those relating to agricultural products. However, it is still not clear to what extent this will benefit developing-country producers. Secondly, the persistent dependence on the trade in a few products is a result of a lack of resources and technology needed to develop new areas of production. Agreement under the Uruguay Round of GATT made this situation worse. By introducing rules governing the trade in intellectual property rights (TRIPS), the agreement makes it possible for countries to patent all technological developments such as new seeds, new methods of storage and processing. These patents can now be treated in the same ways as goods with companies selling the fruits of their research on a world market. For developing countries that cannot afford to 'buy' this technology, this amounts to an additional barrier to development. Thirdly, the inability of

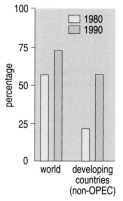

(a) Manufactured goods as % of exports (by value)

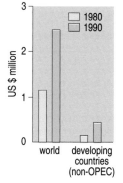

(b) Manufactured goods – value of exports

▲ *Figure 6.10 Manufactured goods as percentage of exports by value, 1980 and 1990.*

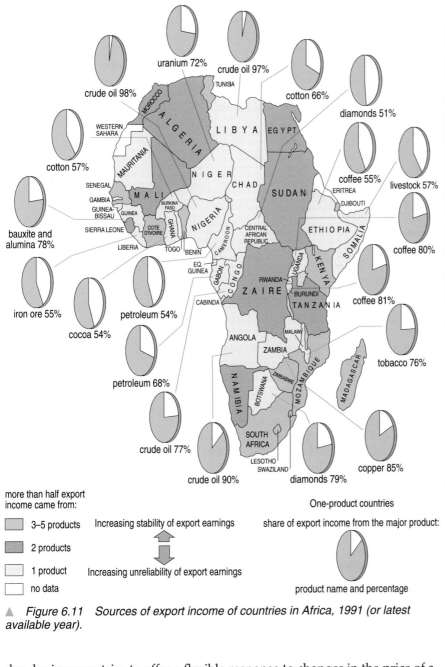

▲ Figure 6.11   Sources of export income of countries in Africa, 1991 (or latest available year).

developing countries to offer a flexible response to changes in the price of a commodity on the world market is a feature of agricultural production. If a country has invested heavily in one crop, it will take time to stop growing this and specialise in something else. The life-cycle of the plant, the machinery and the subsidiary skills must all be taken into account.

If developing countries are unable to switch production away from commodities into other forms of export earnings, their only option in the face of declining prices is to export more. However, according to the laws of supply and demand, any increase in supply is usually met by a further reduction in price. The situation has been exacerbated by competition from

substitutes. For example, cotton now competes with artificial fibres and sugar cane has lost some of its market recently to an artificial sweetener made from maize.

The combined effect of declining commodity prices and a seemingly unshakeable dependence on commodities for export earnings meant that countries in debt to private banks became increasingly unable to meet the interest repayments. At the beginning of the 1980s the situation was made worse still by a worldwide rise in interest rates. The World Bank, anxious to avert an international banking crisis, instigated a programme of lending – the Structural Adjustment Programme. Through this, the Bank offered cheap loans but with wide-ranging conditions: although different for different countries, the World Bank Structural Adjustment Lending required countries to reduce or abolish subsidies to domestic producers, reduce import barriers and increase exports. Countries were obliged to follow World Bank advice if they wanted these loans, but this was not always in their best interests: for example, the World Bank advised more than one country to specialise in the production of the same crop for export. For example, in sub-Saharan Africa, the fifteen countries that rely on cocoa for foreign exchange increased their output by 26% between 1985 and 1989, many of them acting on the advice of the World Bank. The inevitable result was a flooding of the market for that product and a subsequent fall in the market price.

4.4 The environmental impacts of producing crops for export

In order to pay for imports such as grain, a country must generate money from exports. Traditionally these have included tea, coffee, cocoa, rubber, tobacco, cotton and jute. More recently, developing countries have started to meet a growing demand for tropical fruit and vegetables and fresh flowers in developed countries. There are enormous social and environmental implications of switching investment into these crops.

By re-reading parts of the previous chapters in this book (for Activity 5), you should have reminded yourself of the environmental consequences of growing crops using industrial farming techniques. It can have a damaging effect on the fertility of the soil; the crops grown may require massive amounts of irrigation, and this eventually leads to salinisation of the soil; they may also be grown in areas that are not suitable resulting in irreversible soil erosion (Chapter 2, Section 4.2). The methods used may rely on high inputs of fertilisers and pesticides (Chapter 3, Section 4.2; Chapter 4, Section 3.2). These chemicals are then washed off with the rains and end up in the rivers and lakes, contaminating local water supplies.

Moreover, production for export is not as easy as production for local markets. A UK non-governmental organisation, Farmers' Link, reports that foreign buyers specify high standards for regular supply and require that the product should conform to standards of size and appearance. In order to meet these requirements, growers need greater inputs of fertilisers, pesticides, irrigation and seeds. These are usually imported, which in turn has implications for a country's balance of payments. What follows are two examples of the production of crops for export: flowers from Colombia and tobacco from Malawi.

Colombia: from wheat to carnations

In 1951 Colombia produced 78% of all the wheat it consumed. Twenty years later it was importing 89% of its domestic consumption of wheat. This meant an increase in imports from almost 40 000 tons in the early 1950s to nearly 400 000 in 1971: see Table 6.1.

This was partly a result of the United States trade and agriculture policies. During this period the USA pursued aggressive export policies. Under US Public Law 480 (PL480) the United States sold agricultural products at below cost price to 'friendly' countries and to those countries not competing with United States' exporters. The recipients of this food aid became the major importers of US wheat; Colombia was among these countries. This led to a counter-campaign amongst developing-country agricultural producers who sought to halt these agricultural export subsidies: see the advertisement taken from *The Wall Street Journal* addressed to the United States Government.

However, the changes in this period involved more than a switch from domestic production of wheat to a dependence on imports. Consumption patterns changed away from traditional crops that could be grown locally to wheat that was imported. The study by Dudley and Sandilands (1975) shows that wheat consumption in Colombia increased by 41%, whilst the

Table 6.1   Production, imports and consumption of wheat in Colombia, 1951–71, annual averages

	Production		Imports		Consumption
	Tons	Consumption (%)	Tons	Consumption (%)	(tons)
1951–54	139 750	78	38 900	22	178 700
1955–62	145 400	60	97 200	40	242 600
1963–70	99 000	33	205 500	67	304 500
1971	49 000	11	384 900	89	433 900

Source: Dudley, L. and Sandilands, R. (1975) 'The side-effects of foreign aid: the case of Public Law 480 wheat in Colombia', *Economic Development and Cultural Change*, Vol. 23, No. 2, Table 1, p. 331.

TO THE UNITED STATES GOVERNMENT

LET US PAY OUR FOREIGN DEBT

WE ARE   • eliminating all subsidies in the economic system.
   • increasing taxes.
   • seeding a very important area of grains and oilseeds.
   • reducing the fiscal deficit.

WE NEED • to improve our trade balance by increasing our exports.

WE CAN'T • face the unfair competition of your subsidized
   exports.
   • lose our markets fighting against G.S.M. 102 and
   the new E.E.P. Bonus.

WE WANT • to pay our debts but we need to earn foreign currency.

PLEASE   • **STOP YOUR SUBSIDIZED AGRIEXPORT SALES**

ARGENTINE GRAIN GROWERS ASSOCIATION
ADDRESS: Roque Saenz Peña 720   FAX/PHONE NUMBER (541)112287

Source: *The Wall Street Journal,* 5 October 1990.

production of local staples such as barley and potatoes declined. The Colombian government contributed to these changes by supporting a sharp fall in the price paid to domestic wheat producers. This fitted in with their need to provide cheap food for a growing urban population. Colombia was undergoing a process of urbanisation and the industrialisation of agriculture that was reinforced by cheap imports of American wheat. Other crops such as potatoes and barley went out of production and many peasant farmers became unemployed or marginally employed. This increased the drift to urban areas still further and added to the demand for cheap food.

Although an oil-exporting country, Colombia also relied on its cash crops, particularly coffee and bananas, to pay for the increased imports of food. More recently it has diversified into fresh flowers. It now exports carnations to the West (the US, Britain and Ireland). These carnations are grown in large greenhouses using fertilisers and pesticides that have been banned in the USA. The environmental effects of the use of pesticides on the workers have been widely reported – increased incidence of lung diseases, miscarriages and neurological diseases among the employees (see Figure 6.12).

Growing tobacco in Malawi

The growing of tobacco has serious implications for the environment: the cultivation of the plant depletes the soil nutrients and can involve intensive use of pesticides; the processing of tobacco can cause further environmental damage if it is cured using wood-fuelled barns; the final consumption of the product has a damaging effect on health. Moreover, it takes up land that could be used to greater benefit for the majority of the people, and it diverts agricultural labour away from the growing of basic foods such as maize.

The power of the flower

John Vidal

Mrs Susan Fairley went shopping in Chelsea last week. She wanted flowers for her lunch party, so she bought a dozen apricot roses. They were a fabulous price and, she says, in fabulous condition.

And so they should have been. These were pampered, jet-setting blooms, conceived in laboratories, their colour and shape designed by fashion experts and the marketing world. They had been fed almost entirely on chemicals, and lived in biologically dead soil.

Protected from nature by plastic and glass coverings, they had been doused in toxins and pesticides and had lived in precisely controlled heating systems. On the point of bloom they were refrigerated and flown halfway round the world to be sold in a Dutch auction house before ending up on Mrs Fairley's sideboard. Being an industrial product, they got tax breaks and cheap credits.

They came from Colombia, but could have been imported from Israel, Holland or one of the many developing countries which have leapt into flower-growing in the last decade. Colombia has cornered 10

per cent (up to 100,000 tonnes) of the market. The plains of Bogotá, where 80 per cent of the country's export flowers are grown on 4,000 hectares, earn private enterprise $267 million a year. It is the third export crop after coffee and bananas.

Floriculture employs 70,000 Colombians directly and 60,000 others depend on it. Workers earn about $120 a month with bonuses but it's widely accepted that it costs three times that to maintain a family. Where women work on the flower beds, men are the fumigators or supervisors.

If you talk to doctors and workers you hear stories of headaches, nausea, skin irritation, and rashes; dizziness, miscarriages, premature and stillbirths, malformations in babies, asthma, neurological problems, impaired eyesight and several sorts of cancer. No-one will say for sure that these are a result of working in the flower fields and no-one has conducted a comprehensive survey, so no-one can say that there's even a higher incidence of illness among the workers than elsewhere. But they can *tell*. When 40 women were checked, their list of complaints ran as follows:

headaches (26), skin problems (22), conjunctivitis (18), kidney ailments (17), vomiting (12), dizziness (10), numb limbs (7), stomach acidity, anxiety, breathlessness, miscarriages (4 each) etc. Not all 179 complaints in 40 people were directly work-related because the Bogotá plain – with its poverty, dust and contamination –is unhealthy anyway. But it is alarming.

Floriculture consumes more pesticides than any other agriculture sector. Because the produce is not eaten, residues are not considered important. Because a single bug in a consignment can lead to the US or Japan rejecting them all, producers wield the spray can liberally. Japanese and US standards dictate global pesticide regimes.

The pesticides are not handled with care, the fumigators tend not to receive training and notoriously dangerous substances, banned in many countries, are still used, if illegally. Organophosphates, another family of highly toxic chemicals, have not been outlawed. Many effects of the pesticides are long term, so their full damage may not have yet been seen.

Floriculture in Colombia has also lowered the water

table. People and flowers compete for water and the flowers tend to win. As a result many villages have water only for a few hours a week and the quality is unreliable because of the pesticide run-off.

Public protest has been widespread on the savannah. There have been 28 strikes and some municipalities want to limit further expansion of flower-growing. But in many villages, the mayor is a flower grower himself.

Several Western NGOs, including Christian Aid and the European Parliament, have taken up the cause of the flower workers but there is no sign of a boycott threat. Some companies, like Asocolflores and Rosex, are making efforts and have shown that the flower sector is capable of mending its ways if it wants to.

In the meantime, Mrs Fairley says she will buy some more Colombian roses in three days' time.

This article is based on The Game Of The Rose – The Third World in The Global Flower Trade by Niala Maharaj and Gaston Dorren (Jon Carpenter Publishing, Oxford).

Tobacco is, however, an important source of export earnings for certain countries, one of which is Malawi. Malawi was encouraged to grow tobacco by development assistance projects of the International Development Association (a part of the World Bank group). In 1992 it was the sixth most important exporter of tobacco. Tobacco is the single most important export for Malawi, accounting for over 40% of Malawi's export earnings; the next most important crop is tea, accounting for 10.5%.

However, the trade is controlled by the major transnational companies, in particular RJR Nabisco, British American Tobacco, Philip Morris and Gulf & Western (Lang and Hines, 1993). The crop is grown either on large estates (as is generally the case in Malawi) or under a system of what has

▲ *Figure 6.12*

Source: The Guardian, *24 May 1995, p. 5.*

Table 6.2   Depletion of soil nutrients by tobacco and other crops (loss in kg/ha)

Harvest of one ton per ha	Nitrogen	Phosphorus	Potassium
Tobacco	24.4	14.4	46.4
Coffee	15.0	2.5	19.5
Maize	9.8	1.9	6.7
Cassava	2.2	0.4	1.9

Source: The Ecologist, 1986, p. 125; taken from Goodland, R., Watson, C. and Ledec, G. (1984) *Environmental Management in Tropical Agriculture*, Boulder, CO, Westview Press.

come to be known as contract farming. Under this system, the main transnational companies offer individual farmers contracts to grow tobacco on smallholdings. They are given fertilisers, pesticides and seeds and guaranteed a price, providing these farmers a way in which to earn money.

The environmental effects of growing tobacco involve high costs, however, and these are not included in the calculations. For example, the cultivation of the tobacco depletes the soil nutrients to a larger extent than most other cash crops: see Table 6.2. In addition, tobacco is heavily reliant on pesticides: The Ecologist reports that some farmers are advised to apply pesticides sixteen times over a three-month period from five weeks before the crop is sown to six weeks after it has been put in the ground.

The other important effect of tobacco on the environment is deforestation. Many of the small farmers do not have access to modern drying systems and instead use wood-burning barns. Figures given by Madeley (1986) suggest that wood-burning barns need two to three hectares of forest to cure one tonne of tobacco. The wood is often taken from existing forest, woods, scrubland and savannah. This level of deforestation causes soil erosion and flooding. In response to concern about this, some governments (sometimes under pressure from the tobacco companies) have introduced replanting schemes. However, these are not always implemented. Moreover, small farmers do not have the land to grow the trees needed. If the average small farmer has four hectares of land to grow tobacco for cash and food for their family, it is not practical to set aside the 2.5 hectares needed to grow trees to cure a tonne of tobacco. Even where plantations are created to supply the wood, the effect may be that fertile land is cleared for monocrop plantations of eucalyptus trees.

As well as the direct environmental effects of growing tobacco, there is also the problem of people displaced from the land. They are often forced onto marginal lands such as hill-sides where cultivation leads to deforestation and soil erosion.

Nor does tobacco necessarily bring the financial rewards that are hoped for. The export of tobacco may look good for a country's balance of payments, but in reality the cultivation of tobacco requires the import of fertilisers and pesticides. Moreover, tobacco that is processed into cigarettes will require the import of packaging from one of the seven transnational companies that market cigarettes.

4.5   Problems of reconciling environmental measures with trade policies

A reformist approach accepts a link between production processes, trade and environmental damage. Reformists argue that trade policies clash with measures aimed at promoting less damaging methods of production. However, they believe that trade policies must be an integral part of the solution.

Some of the best examples of how trade can be seen to impede attempts to protect the environment are to be found in the production of timber and, in particular, tropical timber. The countries of South East Asia have relied on exports of tropical timber to support their economies. However, the price for raw logs does not include the environmental costs of extraction. Moreover more money is earned per tree cut if the timber is first processed before it is exported. That is to say, if the logs are turned into furniture or sawn into strips they will earn more foreign exchange than if they are exported as raw logs. However, import barriers and quotas in the importing countries prevent the export of the trees as anything except raw logs.

In 1985 the Indonesian government introduced measures that banned all exports of raw logs. Logs could only be sold for processing by companies operating within Indonesia. The Indonesian government justified the ban on the grounds that if the logs were processed within Indonesia and then exported, the Indonesian government would earn more foreign exchange and, therefore, have to cut down fewer trees. The importing countries protested and appealed to GATT on the grounds that the Indonesian export ban contravened the GATT rule that prohibited discrimination between domestic and foreign industries. Although it is by no means clear if Indonesia was motivated by environmental protection or by a desire to achieve greater economic growth through the promotion of domestic industries, environmentalists have used this case to argue that similar log export bans could be used to support forest management policies that would limit damage done to forests by logging concerns.

Since the controversial Indonesian log ban, environmentalists in developed countries have tried to lobby for trade measures that discriminate between imports of the same product that are produced in different ways. In 1993 environmentalists in Austria were successful in lobbying their government to introduce a scheme requiring all imports of tropical timber products to be labelled according to whether or not they had come from a 'sustainable source'. This was to run in conjunction with their own domestic campaigns informing consumers about the environmental damage caused by logging in an 'unsustainable' way. This also failed because it clashed with GATT rules. The Malaysian and Indonesian government protested that the labelling rule was unfair because it only applied to imported wood and not domestic timber. When they threatened to take their complaint to a GATT dispute panel, the Austrian government withdrew the regulations.

More successful has been a scheme set up by the UK branch of the World Wide Fund for Nature (WWF). WWF(UK) established a group of retailers that have pledged only to sell tropical timber from sustainable sources. This is in practice an import barrier that has been set up by the retailers. However, because it is a voluntary agreement, implemented between businesses, it avoids the jurisdiction of GATT (and its replacement, the WTO). WWF (UK) has heralded the ban as a great success. However, although WWF has its own system of monitoring, there are inevitably problems of establishing whether a producer is complying with certain environmental standards of 'sustainable' and who is defining 'sustainable'.

If trade policies are allowed to discriminate between goods produced in different ways on environmental grounds, they must overcome the ambiguity that surrounds the links between methods of production and trade. The claims of the liberal economic view that trade has only an indirect effect on methods of production is a matter of some contention.

However, it is still difficult to be certain that a particular trade policy would have a specific effect in terms of minimising environmental destruction. Moreover, the calls for trade barriers to impose environmental standards must also consider the structural reasons for the export of commodities that may be environmentally damaging to produce. Whilst analysts disagree about the importance of links between trade and the environment, a political solution to introduce trade measures to support environmental protection remains unlikely. In its absence, the dominance of the liberal economic approach embodied in the rules of international trade organisations ensures that trade is not used for environmental ends.

Activity 6

What sort of evidence would support the argument of liberal economists that trade does not have a direct affect on the environment and of radical environmentalists that it does? Do the reformist environmentalists have a viable alternative?
Compare your views with those given at the end of the chapter.

4.6   Summary

This section has looked at some of the ways in which trade can be linked to production methods that have a detrimental effect on the environment. Although trade does have specific and direct links to production (for example, the trade in tropical timber), the main contribution of trade to unsustainable methods of production is largely structural. That is to say, the importance of trade in determining what is produced and how it is produced lies in the role of trade in supporting the current economic system.

In the examples that this section has studied, trade provides a link between consumers and suppliers. We have seen that it is more than a passive link; the structures of trade can alter what is produced and where. The next section will look at whether trade can also alter what is consumed and where.

5   *The effect of trade on patterns of consumption*

5.1   *Introduction*

The previous section looked at how trade might encourage agricultural production methods that are harmful to the environment. This section will focus on the other side of trade and examine how trade has affected consumption patterns in ways that have harmful consequences for the environment.

Through international specialisation, the sum of things produced in the world is increased. More cereals, more cotton, more sugar are available now, in a global economic system where trade plays a central role, than if each country aimed for self-sufficiency. In addition, trade provides access to natural resources not found locally. It may be possible, through trade, to supplement local resources or to change what resources are consumed. The increased wealth that may result from this, and the increased availability of products, will have an effect on the levels of consumption.

This section looks at the implications for the environment of both these links between trade and consumption. The first half will examine the environmental consequences of high consumption levels in certain countries supported by trade. The second half goes on to focus on the role of trade in changing what is consumed and the effects that this has on the environment.

5.2   Trade, levels of consumption and damage to the environment

The most obvious connection between trade, consumption and the environment is the role of trade in facilitating an increase in consumption. Staying with the example of trade in agricultural products: trade provides access to agricultural products in other countries. Countries can import basic foodstuffs, non-essential products and industrial crops to supplement what is available locally. For instance, the amount of agricultural products imported into the Netherlands would require three times as much land to grow as the Netherlands has within its borders. Japan imports much of its agricultural consumption too: 70% of the corn, wheat and barley consumed, 95% of the soya beans and 50% of the wood (Durning, 1992, p. 56).

However, not all countries are able to import the equivalent amounts. Durning (from the American-based Worldwatch Institute) uses United Nations data to calculate that the average person in an industrialised country will consume six times as much meat, five times as much fertiliser, three times as much grain, ten times as much energy and ten times as much timber as someone in a developing country: see Table 6.3.

Table 6.3   Consumption of selected goods, industrial and developing countries, late 1980s

	Industrial countries' share of world consumption (%)	Consumption gap between industrial and developing countries (ratio of per capita consumption rates)
Aluminium	86	19
Chemicals	86	18
Paper	81	14
Iron and steel	80	13
Timber	76	10
Energy	75	10
Meat	61	6
Fertilisers	60	5
Cement	52	3
Fish	49	3
Grain	48	3
Fresh water	42	3

Source: Durning, A. (1992) *How Much is Enough? The consumer society and the future of the Earth*, London, Earthscan Publications, Table 4-1, p. 50.

Increased consumption is promoted by the liberal economic order in which we all live. If economic wealth is generated through the market, the assumption is that trade is needed to sustain the economy. Put in crude terms, if no-one buys, no-one sells, and if no-one sells then no-one will work. On a global scale developing countries have developed a dependence on trade. In the previous section we saw that Colombia relies on flowers, coffee and bananas to pay for its imports of wheat. Those who defend the benefits of trade argue that developing countries would be economically disadvantaged if developed countries stopped buying their exports (Durning, 1992, p. 106).

Most environmentalists (both reformists groups, such as WWF, and environmental economists as well as the more radical groups such as The Ecologist and those that support calls for a 'new protectionism' (see Section 3.5 above)) are concerned that the liberal economic approach ignores the ecological limits to economic growth. If environmentally damaging methods of agricultural production are used, eventually the land will no longer support the crops on which the economy depends. In some countries this has already happened. Thailand was once a major exporter of timber and timber products. It has now cleared most of its forests and imports timber for its own use.

The response to this has often been to develop technological solutions to resource degradation or depletion, for instance, new fertilisers, hybrids or genetically altered seeds and new farming techniques. The green revolution discussed in Chapter 3 is one such example of using technology to supplement the natural resource base. As Chapter 3 shows, these bring new problems for the environment. Radical environmentalists dismiss this approach as insufficient to prevent the more serious environmental problems that are occurring as a result of current patterns of economic development. In criticising the World Bank's *World Development Report 1992* for its emphasis on the ability of wealth to pay for science to 'clean up' environmental problems, The Ecologist argues that, 'The document inspires little confidence … that technical substitutes for water, air, genes and soil will be found quickly enough' (1993, p. 101).

5.3   *Trade, patterns of consumption and damage to the environment*

The environmental impacts of consumption depend on what is consumed as well as how much is consumed. The obvious and dramatic effect of the increase in trade has been the increase in consumption of imported goods. Trade has also given the wealthy countries and the wealthy in poorer countries access to luxury products that may not be available locally. Tea, coffee and cocoa fall into this category as does the trade in carnations and tobacco examined in the previous section. For the most part, these crops are grown on large plantations with a large chemical input, displacing subsistence farmers and disrupting local ecosystems. Out-of-season fruit and vegetables are another example. The recent developments in food packaging, storage and transport have made it easier to transport fruit and vegetables around the world in order that rich consumers can always have access to tropical and luxury fruits and vegetables, whatever the season.

Moreover, trade has supported the production of cheap foodstuffs that have been grown using methods that have a negative effect on the environment. The availability of cheap imports of basic foodstuffs has shifted consumption away from crops that have been produced with minimal environmental impact. For example, where once basic foods such as cereals came from farms using methods of production that did relatively little harm to the environment, nowadays the majority of grain consumed is grown on large farms where trees and hedgerows have been uprooted, and chemical fertilisers and pesticides used.

The shadow of consumption

The consumption of agricultural products that have been imported will have an environmental impact in a number of different countries. Take, for example a box of strawberries that is bought in a supermarket in Milton Keynes. The strawberries may have come from Mexico; they may have been grown using chemicals from Germany and a tractor made in Japan from components produced in another South East Asian country; the box will probably be made from plastic imported from the United States. The production of each one of these inputs will have a significant impact on the environment. So, the consumption of strawberries in Milton Keynes will have involved production processes that alter the environment in Mexico, Germany, Japan, South East Asia and the United States.

World trade has enabled wealthy consumers to dictate what is produced in the world regardless of the effect on the local ecology. One example of this is the import of frogs' legs: see Figure 6.13 (overleaf).

'Western-style' consumption is spreading to the urban elites in developing countries and from there to the rest of the population of those countries. Durning (1992) defines this style of consumption by its high demand for meat, packaged foods, soft drinks, consumer hardware (such as washing-machines, dishwashers, television and video recorders), private cars and throwaway materials. The growth in transnational companies and the competition to capture foreign markets has led to aggressive marketing strategies to promote processed foods and drinks. The essence of such marketing strategies is encapsulated in a quote from the president of Coca Cola: 'When I think of Indonesia – a country on the Equator with 180 million people, a median age of 18 and a Moslem ban on alcohol – I feel I know what heaven looks like' (quoted in Durning, 1992, p. 71). Government policies and policies of international institutions have also played a role. For instance, countries operating a World Bank Structural Adjustment Programme are required to reduce or to abolish import controls. This has left countries that take out Structural Adjustment loans vulnerable to marketing strategies of producers in richer countries.

5.4   The problem of illegal trade

Demand for certain products can mean that they will continue to be traded even when international agreements or national policies exist to limit or stop such trade. The trade in endangered species and products from endangered species, such as ivory or rhinoceros horn, is one such example. In South East Asia the export of logs is limited and regulated, but many trees are illegally felled and exported to Japan. No figures exist to quantify the extent of this problem but the appearance of such goods in the importing country indicates that it goes on.

The problem of illegal trade highlights both the need to change the behaviour of consumers and the limited use of trade rules in preventing levels and patterns of consumption that damage the environment.

Trade marks

John Vidal

Raymond Blanc, chef-patron of Britain's premier French restaurant, hasn't a clue where his cuisses de grenouille come from. Indonesia? Poland? Probably not France or Britain, where polluted watercourses all but dried supply to a trickle years ago.

'They are the only food which comes without labels,' says Blanc, the owner of Le Manoir aux Quat'-Saisons, outside Oxford.

Blanc cooks only the best wild frogs in foaming butter 'with a hint of garlic', and his customers lap them up. At £22 a plate (12 legs), that's twice what a Bangladeshi would earn for catching 100 frogs.

In 1977, Bangladesh, along with India and Indonesia, was the darling of the burgeoning frog trade, supplying restaurants in Europe and the US. Its vast paddy fields, farmed mainly by smallholding peasants, were one of the world's greatest sources of favoured hind limbs and there were thought to be one billion frogs in the country – not enough to provide the West's appetite for 6,500 tons of legs a year, but a resource to be exploited nevertheless.

In 11 years, Bangladeshi farmers scooped them out in bucketfuls while small boys butchered the frogs on crude blades. By 1988, more than 50 million Bangladeshi frogs a year were being exported, mostly to the US which favoured the abundant bull-frog. One year later, Bangladesh was thought to

have had only 400 million left.

The trade only stopped when Friends of the Earth Bangladesh helped persuade a sceptical government that it was economic and environmental suicide. Frog exports peaked at $10.5 million a year, but research showed the real price of the trade was phenomenal.

Taking frogs from the wild, it was pointed out, could have devastating consequences. Frogs are insectivorous and each one can eat more than its weight (about 200 grammes) in waterborne pests every day. Fewer than 50 frogs are needed to keep an acre of paddy field free of insects: they keep malaria and other illnesses at bay; they protect crops and are a natural biological control agent. Frog waste, too, is a fine organic fertiliser.

Remove the frogs, said the scientists, and the only way Bangladeshi farmers could protect their crops and their livelihoods was with pesticides. Indeed, from 1977–1989, Bangladesh imported more than $89 million of some of the world's worst quality chemicals. Much came from Bhopal, before the accident that killed 287 people in India in 1984.

The figures didn't add up. By 1989, Bangladesh was importing an extra 25 per cent of pesticides a year to cope with its frog loss. Government figures showed it was spending $30 million a year to earn $10 million. While large farmers could buy pesticides cheaply in

bulk, the traditional small-holders were paying about $5 in pesticides for every $2 they earned from catching 100 frogs. No one has estimated the cost of polluted waters or the health effects.

There was a further twist in the tale. Who should be exporting the frogs' legs to the West but, Friends of the Earth discovered, some of the very same companies that were importing the chemicals.

It was a Faustian contract that the government tore up in 1989. A temporary export ban was imposed, and extended further in 1992. It had immediate effects: within a year Bangladeshi pesticide imports had declined by 40 per cent and frog numbers had started to increase.

Five years later, the pesticide companies and frog traders are lobbying hard to resume trading. India, too, has banned frog exports, but there is much poaching and an illegal trade flourishing between the two countries, with companies labelling the frogs as 'frozen food'. Policing is difficult, if not impossible, and as long as there is a demand for frogs in the West there will be poaching and the temptation of quick 'profit'.

Frogs are one illustration of the issues involved in world trade. No international market – be it frogs, timber, copper, fish or beef – yet accounts for the environmental damage incurred in the production or transport of the goods. In economic jargon, these costs are 'externalised' – passed

on to society in general or to future generations to pay.

So Costa Rica, for example, felled its forests, which played a vital role in preventing floods, guaranteeing water supplies and protecting fisheries, turning them to poor quality grazing land to rear beef for the mighty US hamburger industry. Twenty years of economic 'advancement', on the back of massive loans for forestry exploitation, has left the country with terrible resource depletion that hinders its future prospects. Economists believe Costa Rica has lost $4 billion over 20 years through the sale of its natural assets for rock-bottom prices.

Conventional national economic indicators, such as gross domestic product, actually celebrate pollution, environmental damage and the sale of natural capital. In the case of those Bangladeshi frogs, short-term export-based indicators suggested that the country had become richer. The bankers were happy and economists talked of an export-led boom.

Earning a few dollars by collecting frogs had short-term attractions. But the trade was culturally devastating, undermining people's ability to survive on the land. It pushed many Bangladeshi farmers further to the economic margins, contributing to loss of land ownership and migration to the cities.

Raymond Blanc's customers might understand; many others don't.

▲ *Figure 6.13* Source: *The Guardian*, 17 June 1994, Environment Guardian, p. 16.

Environmental groups have to some extent taken up the challenge through their two-sided approach that seeks to raise awareness of the environmental consequences of certain patterns of behaviour whilst at the same time lobbying governments and international organisations for trade measures.

5.5   Summary

This section has examined how trade influences what we consume. Trade is important in supporting and promoting more consumption. Environmentalists fear that the support of trade will lead to levels of consumption that cannot be sustained by the Earth's resources. Trade also changes what we consume. We appear to have become more environmentally destructive consumers. Moreover, what is consumed in one country can have environmental effects throughout the world.

6   The separation of producer from consumer

6.1   Introduction: consumption at a distance

The previous sections have examined the extent to which world trade has affected methods of production and patterns of consumption in ways that damage the environment. The trade and environment debate has yet to reconcile the opposing views of environmentalists and supporters of trade liberalisation. The separation of producer from consumer provides an additional concern for two reasons. Firstly, world trade involves the transport of goods across national boundaries. Secondly, trade acts as an intermediary between the producer and the consumer. Specialisation has distanced the consumer from the source of the goods that are being consumed. As a result, we have reached the point where we neither consume what we produce nor produce what we consume.

6.2   The environmental costs of transport

World trade is defined as the movement of goods across national boundaries. The transport of agricultural products across the world has now become an accepted part of the global economy but it has serious implications for the environment. Several industries are involved. Firstly, containers, lorries, trains, ships, planes must be manufactured. This involves, among other things, the use of iron, steel, paint and energy. It produces pollutants and wastes that are discharged into the air, rivers, land and seas. The use of these carriers involves the construction of ports, roads, railways and airports. In addition further heavy equipment is needed to load and unload the goods on and off the carriers. In addition there is the energy required to power these vehicles, using up more resources and emitting more pollution.

    For food products there is also the packaging and processing necessary to prepare the goods for transport across long distances. This uses plastics and cardboard, preservative sprays and food additives. It means more pollution through the additional waste that extra packaging generates (think about how much you throw away when you buy a box of strawberries) and pollution from the use of chemicals, not to mention the direct effects on health of some food additives and preservatives.

6.3   The emotional and spatial distance between consumer and producer

If the consumer becomes separated from the point of production, they will find it difficult to understand the implications of their actions for the environment. People have little or no idea of how the products they consume are produced and even less idea of the environmental impacts of the production methods used. This is most clearly seen in the production and consumption of food products. Traditionally, the consumption of food has depended on what is available locally and what is available in the different growing seasons. Food has been surrounded by the rituals associated with production and consumption (for example, the Christian tradition of 'harvest festival'). Distanced from the agricultural production of food, consumers lose their conception of food as local produce bound up with a local community and it becomes a commodity in the same way as a car, a book or a coat. The damage that is done to the environment by the production of this food is invisible. Moreover the cost of damage to the environment, both in terms of the financial costs of cleaning up environmental damage and the social costs to human health and welfare, are not included in the price paid by the consumer. This reinforces the separation between consumer and producer and further isolates the consumer from the consequences of different production methods.

There are also implications of this separation for the trade and environment debate. The problem is one of establishing cause and effect. This brings us back to the original questions posed in this chapter: how does world trade affect the environment? The links between production and consumption are mediated by trade. There is no direct link between consumer and producer, but equally there is no direct link between trade and production or trade and consumption. In reality the three are interconnected. This makes it hard to establish either the extent to which trade can be held responsible for the environmental damage caused or the extent to which other mediating factors such as government policies and decisions by individual producers or consumers are more important.

Activity 7

Next time you are shopping, look at the country of origin of the products. Of the things that you buy that are not produced in this country, why have they been imported? What role in the global environment issue do you play as a consumer?

7   Conclusion

This chapter set out to look at how structures of the world economy, and in particular trade, shape economic development and the environment. Trade supports and motivates a particular form of economic development that has come to dominate the world. This kind of development, based on liberal economic principles, has an enormous impact on methods of production and patterns of consumption. Through the enclosure of land, industrialisation and the international specialisation of production, countries have become locked into certain patterns of production and consumption. Many people in the world neither consume what they produce nor produce what they consume. The evidence suggests that trade is an important factor in supporting these patterns of production and consumption. One conclusion is, therefore, that trade is an important mediating factor between populations, the environment and economic development.

The trade and environment debate is based upon a challenge to the dominant liberal economic ideology. This ideology stresses the important role of trade in supporting a form of economic development that benefits populations and ultimately the environment. This assumption has been challenged by radical environmentalists who argue that liberal economic trade systematically disadvantages many people, and is responsible for destroying the harmonious relationship between people and their environment. A third approach, offered by the reformists, calls for a compromise that would modify the current economic system. This modification represents a pragmatic solution to restore a balance between the demands for economic growth, on the one hand, and protection of the environment, on the other. However, the reformist solution fails to satisfy concerns about equity and accountability within the international economic system. All too often the trade and environment debate is concerned with the role of trade in isolation from other political, economic and social factors. In trying to look at the evidence behind some of the claims in the trade and environment debate, this chapter has shown the problems of this approach.

One important manifestation of economic development has been urbanisation: a drift from the countryside to towns and cities. The following chapter will look at this in more detail.

References

DUDLEY, L. and SANDILANDS, R. (1975) 'The side-effects of foreign aid: the case of Public Law 480 wheat in Colombia', *Economic Development and Cultural Change*, Vol. 23, No. 2.

DURNING, A. (1992) *How Much is Enough? The consumer society and the future of the Earth*, London, Earthscan Publications.

ECOLOGIST, THE (1993) *Whose Common Future?*, London, Earthscan Publications.

ECKERSLEY, R. (1992) *Environmentalism and Political Theory: toward an ecocentric approach*, London, UCL Press.

GARDNER, B. (1993) *The GATT Uruguay Round: implications for exports from the agricultural superpowers*, Briefing paper for the Catholic Institute for International Relations and the Sustainable Agriculture, Food and Environment Alliance (SAFE).

GILPIN, R. (1987) *The Political Economy of International Relations*, Princeton, NJ, Princeton University Press.

LANG, T. and HINES, C. (1993) *The New Protectionism: protecting the future against free trade,* London, Earthscan Publications.

MADDEN, P. (1992) *A Raw Deal,* London, Christian Aid.

MADELEY, J. (1986) 'Tobacco: a ruinous crop', *The Ecologist,* Vol. 16, No. 2/3, March/April.

RAIKES, P. (1988) *Modernising Hunger: famine, food surplus and farm policy in the EEC and Africa,* London, Catholic Institute for International Relations.

RUGGIE, J. (1982) 'International regimes, transactions, and change: embedded liberalism in the post-war economic order', *International Organisation,* Vol. 36, No. 2.

SARRE, P. and BROWN, S. (1996) 'Changing attitudes to nature', Chapter 2 in Sarre, P. and Reddish, A. (eds) *Environment and Society,* London, Hodder and Stoughton/The Open University (second edition) (Book One in this series).

WATKINS, K. (1992) *Fixing the Rules: North–South issues in international trade and the Gatt Uruguay round,* London, Catholic Institute for International Relations.

WHITEHEAD, G. (1982) *Economics Made Simple,* London, Heinemann.

WORLD TRADE ORGANISATION (1995) *Focus,* No. 1, January/February.

Further reading

ANDERSON, K. and BLACKHURST, R. (eds) (1992) *The Greening of World Trade Issues,* Hemel Hempstead, Harvester Wheatsheaf.

DALY, H. E. and COBB, J. B. (1990) *For the Common Good: redirecting the economy toward community, the environment and a sustainable future,* London, Green Print.

GEORGE, S. (1982) *Food for Beginners,* London, Writers' & Readers' Publishing Co-operative.

RAIKES, P. (1988) *Modernising Hunger: famine, food surplus and farm policy in the EEC and Africa,* London, Catholic Institute for International Relations.

SAURIN, J. (1994) 'Global environmental degradation, modernity and environmental knowledge', in Thomas, C. (ed.) *Rio: unravelling the consequences,* London, Frank Cass.

WILLIAMS, M. (1994) 'International trade and the environment: issues, perspectives and challenges', in Thomas, C. (ed.) *Rio: unravelling the consequences,* London, Frank Cass.

WORLD COMMISSION ON ENVIRONMENT AND DEVELOPMENT (1987) *Our Common Future* (The Brundtland Report), Oxford, Oxford University Press.

Answers to Activities

Activity 1

Trade supported the changes in agriculture and land ownership by enabling farmers to sell surplus production at a profit. As the industrial revolution started, farmers were encouraged to specialise in order to meet the growing demand for basic foodstuffs. As industrialisation developed and the international trade in food became more important, farmers were under pressure to use the emerging methods of mechanisation in order to be able to sell at lower prices. The net result was the emergence of larger farms managed for profit and not just to provide the farmers' immediate needs. The industrialisation of farming was part of a wider process involving the enclosure of land, the industrial revolution and international specialisation with each set of changes reinforcing the others.

Activity 2

The report is based on the liberal economic assumptions that trade is central for economic growth and that economic growth will benefit the environment. (Anderson and Blackhurst (eds, 1992) made an important contribution to the GATT report and their book is useful if you want to know more about the ideas expressed in the GATT report: see the list of further reading.)

Activity 4

The important differences are between the liberal economists, who emphasise the importance of economic growth and the use of market forces, and radical environmentalists who emphasise the importance of society and the environment. Whilst the liberal economic approach sees trade and environment as mutually supportive, the radical environmentalists call for fundamental reforms to the international economic system. They want to see a redefinition of human welfare in ways other than through economic growth, an eradication of the inequalities in the current market system and greater accountability in larger economic actors such as international institutions and transnational companies.

The third approach, the reformist position, offers a compromise. However, reformists retain the liberal economic assumptions about the importance of the market and of economic growth. Whilst the reformist position recognises the need for some intervention, this is only within the limits of a liberal economic order. Because of this, they can be seen to represent a modified form of liberal economics.

Activity 5

Trade presents a problem if producers using cheaper but more environmentally damaging production methods can sell their grain on the world market. This puts other producers under pressure to reduce their costs of production or switch to another crop. Liberal economists would point to the role of national governments in subsidising imports. Export crops are grown to pay for imports. Liberal economists would argue that the problem is not that they are grown but that they are permitted to be grown in ways that damage the environment. The solution is for governments to implement more effective measures to limit the damage.

Activity 6

The liberal economic approach distances trade from environmental problems and emphasises the importance of individual government policies. In the case of Colombia, the liberal economists would argue that it was the government's decision to take advantage of cheap wheat from the United States. They would also say that it is the responsibility of the Colombian government to ensure that its workers are properly protected when working in greenhouses.

The radical environmentalists take a wider view of the effect of trade on the environment. They would use the evidence of US subsidies to suggest that the Colombians were forced into growing export crops. Radical environmentalists also emphasise the power and influence of transnational companies. They would point to the pressures that TNCs involved in the trade in grain, flowers and tobacco place on governments to reduce regulations so that the costs of production can also be kept low.

The reformist position would use the evidence of environmental damage caused by the production of grain, flowers and tobacco to call for changes in trade regulations that would discriminate against products grown in ways that harm the environment. Reformists would argue that if the full environmental costs were included in the price of the product then governments could afford to implement environmental standards.

1   Introduction

As you read this chapter, look out for answers to the following key questions:
* How have cities evolved in the past?
* How do their present problems compare in the developed and developing world?
* How can cities be made 'sustainable' in the future?

Our first impression must surely be the degree to which the individual city appears to have been not so much planned for human purposes as simply beaten into some sort of shape by repeated strokes from gigantic hammers – the hammer of technology and applied power, the overwhelming drive of national self interest, the single-minded pursuit of economic gain.

(Barbara Ward, *The Home of Man*, 1976)

The quiet Sumatran town of Bengkulu has 100,000 inhabitants, a small Chinatown of old shophouses, a lovely beach and brightly-painted outrigger fishing boats which are drawn up on the sand now that the old port has silted up, and a few crumbling graves with Scottish-sounding names.

('Intriguing Sumatra', British Airways' *Highlife* in-flight magazine, September 1995)

These quotations show two very different ways of thinking about urban environments. They hint at some of the complexity of cities – the ways in which different people use them and are affected by them, and the ways in which settlements impact on the environment and are affected by it. This chapter looks at urban environments locally and globally. It analyses how cities developed in the past, the process of urban change today and the prospects for the future. It describes neighbourhoods and mega-cities, cities in the rich North and the poor South of the world, cities decreed by kings, cities ruled by market forces, and cities forged by communist apparatchiks.

There are two reasons for taking such a sweeping, all-inclusive view of urban environments. First, it emphasises the global dimension in urban environmental change, the multiple threads that connect North and South. If you look again at the description of Bengkulu, you can see that it is being sold as a holiday destination for today's Europeans, the Scottish skeletons reprocessed as a romantic legacy of earlier trans-global bonds. As telecommunications, air travel and world markets spin a web around all parts of the planet, are all urban settlements becoming more alike? The second reason for the wide scope is because it allows us to explore the development of technologies and political devices which have made cities manageable, pushing back their 'limits to growth'.

The twentieth century has been the century of the city. At the end of the nineteenth century, only one person in ten lived in cities; by the year 2000 a city will be home for about half the people on the planet. Chapter 1 showed that the global population has increased rapidly over this same period; this means that the absolute number of people living in urban environments has been transformed.

Activity 1

Look at the data in Table 7.1. Which group of countries are the most urbanised? Which group are urbanising most quickly? Consider these questions before reading on.

Table 7.1   *Global change in urban population by national income, 1965–90*

Countries by income band	Urban population as a percentage of total population	
	1965	1990
Low-income economies	18	38
Middle-income economies	42	60
High-income economies	72	77
World	36	50

Source: drawn from Table 31 of World Bank (1992) *World Development Report, 1992: Development and the Environment*, Oxford, Oxford University Press.

The most urbanised countries are those with high-income economies, like the European Union countries, the USA and Japan. The least urbanised countries are mainly the poorer countries of the South, like Mozambique or Malawi. This group also includes some poorer northern hemisphere countries such as China, India and Pakistan. Within the 'low-income economies' group, 1990 urbanisation levels ranged between 5% (in Bhutan) and 56% (in China); so there are considerable differences even amongst the poor nations. (Such differences demonstrate the problems of using terms like 'third world' which imply broad similarity – see Box 1.1 of Chapter 1.) The middle group includes much of Latin America, eastern Europe and the Middle East. The low-income economies are urbanising most rapidly.

As cities are becoming so pervasive, and so dynamic, their significance to local and global environments becomes a key concern for the present and for the future. Does the staggering spread of urban environments demonstrate successful new technologies allowing cities to thrive and multiply in a way that was not possible in the past? Or are cities plundering and polluting the Earth to a degree that cannot last? This chapter explores what constitutes a 'sustainable city' in environmental terms and whether it is possible to plan and run cities so that they use resources in renewable rather than damaging ways.

We begin by asking what is an 'urban environment', and by looking at who reaps the benefits of city living – and who suffers the environmental costs. The subsequent sections build on these debates, providing more empirical discussions of urban environments in the past, present and future.

2   What is an 'urban environment'?

The aim of this section is to introduce concepts which will allow you to explore how urban settlements relate to the wider environmental context. Firstly, it asks what makes an environment 'urban', and looks at some of the different forms and functions of urban places. It then introduces a way of looking at settlements as ecosystems, and the questions this poses for the sustainability of such environments. Last, but not least, Section 2 suggests that there are social and political dimensions to urban environments, and that these need to be part of the the debate about sustainability.

After reading this section you should be able to:

(a)   give examples of different types of urban environment;
(b)   identify some of the indicators that might be used to measure the sustainability of urban environments; and
(c)   indicate in broad terms the way in which their environmental costs and benefits are shared amongst different groups of people.

2.1   Key characteristics of urban environments

What makes an environment 'urban'? There has to be a concentration of people and activities, and an intensity in the use of land which contrasts with more extensive uses in rural areas. To create an urban environment there must be investment of wealth and energy to provide shelter and infrastructure. Continued investment is needed to maintain these services. Ways must be found to supply safe food and water, and to remove the waste created by the concentration of people and activities. These tasks can be accomplished by individuals, but in urban environments there is a social or collective dimension. They are places where people live physically close together, sharing the same space and water sources, breathing the same air. Such proximity facilitates, and even requires, social organisation.

These basic features – size, intensity, infrastructure, investment and collective responsibility – determine the quality of urban environments. Since urban environments represent, but also require, large investments of resources, poor urban conditions are typically associated with poverty and disinvestment. Similarly, at a global scale, conditions in third world cities are symptomatic of an unequal world. The quality of an urban environment also depends on how people – living in close proximity – behave towards one another. To safeguard or improve conditions, constraints may be imposed collectively on individual freedoms – speed-limits and noise regulations are everyday examples. Crucially then, the quality of an urban environment depends not just on the individuals using it, but on how it is managed by them or for them. Power and relations between people are critical.

Constraints on individual freedoms are more than outweighed by the opportunities which urban environments offer. Concentration of people creates economic possibilities and benefits – access to labour, customers, employment, information and new ideas. The market square, one of the earliest and most enduring components of an urban environment, was a much more efficient and convenient way to organise food distribution than for individuals each to grow their own food, or for farmers to try to sell produce at the point of production. The market square also had a social function. It was a meeting place, a place to make friends, exchange gossip, pick up new ideas. In European history, just as in developing nations today, the city has been a focus for political change, making it possible for people to break free from the bonds of a feudal, rural existence. By organising together in urban environments, people have been able to provide a range of entertainment and cultural opportunities that would have been impossible amongst a scattered rural population.

Cities also have some environmental advantages. Concentration of population reduces the unit cost of providing basic services such as electricity, piped water, sanitation, collection of household waste, schools and health care. It can even be argued that the concentration of industry in cities facilitates effective environmental regulation and enforcement of pollution control. Recycling schemes, too, are easier to organise where there are many people living close to one another.

Activity 2

Consider these questions for each of the following photos: the bazaar in Lahore andKurfürstendamm, Berlin opposite; Nowa Huta in Section 3.10; and the street scene in Section 4.3.

- What would you visit each of these urban environments for?
- Who else might be sharing that same environment?
- Would their needs complement or conflict with yours?
- What sources of energy and raw materials are consumed to create and sustain these four environments, and where are the supplies from?
- What kind of waste materials will they generate and where might that waste be disposed of?
- What noises would you hear in them? What would each one smell like?
- What will happen to rainwater passing through each of these environments?
- How comfortable would you feel in each?

As the richer countries are the most industrialised and the most urbanised, it is easy to equate cities with industrial development. But not all cities, still less all urban environments, are industrial. As the photographs show, cities are also centres of administration, trade, commerce and information exchange, and cities had these functions long before the advent of modern industry. Transport of goods and people is a key feature of a city. Today, many cities also have a tourism role. To some extent, then, the environmental accountancy of the city in the modern world is that of the modern industrial society. However, to equate the city directly and exclusively with industry is to understate the city's environmental accounts.

Of course, not everyone lives in urban environments, still less in large cities. Within England, in particular, **anti-urbanism** is strong, an attitude which shuns the city, blaming it for a range of social ills. In contrast the countryside is seen as natural, harmonious, unspoilt, a place of peace. There is a long tradition of this type of thinking, which gathered force with the rise of industrialism, but endures to the present. Analogies with natural environments are used to depict urban environments in negative terms. We talk of the 'concrete jungle' (in the dense undergrowth of streets and alleys beneath a canopy of high buildings lurk danger and competition) or the 'urban desert' (where the endless brick, stone and concrete crowd out green and water, and social life is sterile). Such extreme natural environments threaten human survival, and so, by implication, does the city.

It is easy to romanticise rural life, whereas the reality is that living conditions have typically been worse in the country than in the city. For

▲   A bazaar in Lahore, Pakistan

▲   Kurfürstendamm, Berlin

Box 7.1   Defining an urban environment

Defining the urban environment is a complex challenge. Cities are dynamic entities and their composite environments and the quality afforded by them are determined not only by the fulfilment of the material economic needs of their citizens, but also by the social and environmental conditions which prevail. These social qualities include aspects such as the healthiness, attractiveness, and safety of urban areas. Cities, therefore, have many attributes that contribute in different ways to creating these desired qualities. For example:

• the level of pollution: air and water quality, noise levels, waste disposal;
• the type of land use: the pattern of built and open land, the mix of uses, the extent of land vacancy and dereliction;
• the building stock and infrastructure: its fitness for purpose, condition, the processes of renewal and conservation; and

• the type of townscape: the design of buildings and public spaces, landscaping, traffic management, litter and vandalism.

The definition of what constitutes the urban environment can be and is interpreted in a number of ways. The narrowest interpretation is one that is concerned primarily with the appearance of urban places and embraces building design, conservation, townscape, planning. A wider definition extends this to traffic safety, the condition of buildings and infrastructure. A yet wider definition is concerned with the sustainability of urban environments and embraces resource consumption ... Sustainable urban environments, therefore, are those which develop and grow in harmony with the changing productive potential of local, national and global ecosystems.

Source: Organisation for Economic Co-operation and Development, 1990, p. 20

---

example, Robert Burns (1759–96), the eldest son of an unsuccessful tenant farmer in Ayrshire, began to write his poetry to find 'some kind of counterpoise' to the life of demanding physical work, poverty and acute awareness of social disadvantage in which he grew up. He described country life as 'the cheerless gloom of a hermit with the unceasing toil of a galley slave'. The people behind the statistics in Table 7.1 – those who are moving from rural poverty to the cities of the low-income economies – know only too well what Burns meant.

Box 7.1 gives a definition of the urban environment from an OECD proposed policy on the environment of cities and introduces the notion of the sustainability of urban environments. The critical question, which runs through the rest of this chapter, is a highly practical one: how to plan, organise and manage urban environments, globally and locally, in the spirit of the Brundtland Report (World Commission on Environment and Development, 1987).

2.2   Urban environments and ecological sustainability

In order to approach this idea of sustainaibility in the context of cities, it is possible to analyse the city as an **ecosystem** (*Silvertown*, 1996). Like a natural ecosystem (such as a pond or forest), a settlement takes energy (e.g. fossil fuels), applies it to materials (e.g. wood, metals, even information), and creates products and by-products, which in turn can either be consumed or exported. Sustainable natural systems recycle the by-products, and use renewable sources of energy or materials. What happens in cities? To answer that question we have to understand the dynamics of the system. This means charting the flows of energy, materials and people, identifying the places where energy is stored and where it becomes used and dissipated. We need to look at the wastes generated in urban living and what happens to them. In this way we can begin to see how the process of creating and using cities impacts on the environment.

The concept of the **ecological footprint** is an important one when we view settlements in this way. Instead of defining a city as a built-up area, with a municipal boundary line around it, its ecological footprint takes in the total area of land required to sustain a settlement or urban region, for its fuel and food needs and so on. This **ecological space** must inevitably extend beyond the physical edge of the settlement. A related concept is the notion of **carrying capacity**. Again the idea is well established in respect of natural ecology where it means 'the population of a given species that can be supported indefinitely in a given region without permanently damaging the ecosystem on which it depends' (Rees, 1992, p. 125). For example, overgrazing would reduce that area's carrying capacity for the number of grazing animals it could adequately feed. In this sense, all urban regions appropriate the carrying capacity of places elsewhere.

The implication of these concepts is that they start to underline the importance of striving for sustainable urban environments, following the notion of **sustainable development**, defined in the Brundtland Report as 'development that meets the needs of the present without compromising the ability of future generations to meet their own needs' (World Commission on Environment and Development, 1987, p. 49). (Section 5.1 of Sarre and Brown (1996) discusses this idea.) However, it is not easy to establish just what constitutes sustainability in settlements. At the most basic level, no city can grow the food to feed its populace. As Chapter 6 showed, at every scale, from individual to nation, we no longer produce what we consume, nor consume what we produce. The city is a physical expression of this concept. The energy to sustain a city may have been mined from coal-pits beneath the city streets, but for most cities it is brought in from elsewhere. The raw materials to build the city – the wood, stone and bricks – at first may come from local sources, but as these are exhausted the contribution has to come from environments elsewhere, a process further extending the urban growth dynamic to new locations.

There is no standard, agreed checklist of what makes a sustainable city, although various attempts have been made. In the UK the Local Government Management Board (1994) has identified 101 possible **sustainability indicators** defining different aspects of a sustainable community. On the other side of the globe there is a set of urban environmental indicators for the Inner Metropolitan Region of Melbourne. These include: measures of the condition of environmental resources within the urban area (air quality, water quality, soil, flora and fauna); measures of the impact of urban activities on regional and global resources (resource use, waste/recycling/litter, and transport); and 'quality of life' indicators – open space, urban form, physical health, amenity, noise and security.

On the west coast of the USA the Sustainable Seattle Indicators Project puts indicators into ten groups: resource consumption, economy, natural environment, social environment, education, transport, public safety, health, culture and recreation, and community participation and involvement. For each of these they have developed what they call primary, secondary and provocative indicators. The *primary indicators* are seen as key measures of progress towards sustainability. Examples are non-renewable energy consumed per capita, solid waste generated per capita, or overall air quality. The *secondary indicators* include the data needed to build up the primary indicators and other important data about particular topics. Population density is one example, and another is the weight of waste generated compared to the weight of waste recycled. *Provocative indicators* are chosen to suggest social patterns or trends which might stimulate concern, such as the average distance food travels from the source to the consumer, or the number of salmon returning to spawn (Local Government Management Board, 1994, pp. 33, 39).

Indicators in part depend on what data are available, but they also reflect the ideologies of those inventing and using them, and there are likely to be differences between what governments see as key indicators and what 'deep green' activists perceive to be an adequate listing. Is it acceptable for a city to live beyond its own carrying capacity, leaving its ecological footprint on distant places? Chapter 6 discussed trade as an intervening factor between economic development and the environment. Rees (1992) argues that patterns of trade allow people to exceed the carrying capacity of their settlement unknowingly, and therefore to be oblivious to the consequences of further resource depletion. In this view, the city can be seen as 'a node of pure consumption existing parasitically on an extensive external resource base' (Rees, 1992, p. 128). But it can be argued that it is unrealistic to imagine that cities can service themselves and that some relationship with distant sites is both inevitable and not necessarily problematic, since cities 'give' as well as 'take'. The alternative would seem to be a settlement pattern consisting of small, self-sufficient villages.

2.3   Urban environments and social sustainability

While all lists of indicators of sustainable urban environments will include physical criteria, such as air or water quality, you will have noticed that several include social factors or suggest that citizen involvement in decision-making is a key requirement for creating sustainable living environments. Roseland (1994), for example, argues that 'Social equity is not only desirable but essential: inequities undermine sustainable development' and that 'Public participation is itself a sustainable

▲   *Independencia shanty town in Lima, Peru, an area of gradually improving housing (note the drains being laid), but obviously housing for those scarcely in the housing market.*

development strategy' (p. 77). **Social equity** is a key principle of the Agenda 21 for sustainable development agreed between more than 150 nations at the 'Earth Summit' in Rio de Janeiro in 1992. Such thoughts open up important questions about **social sustainability**, the idea that oppression, domination or exploitation of groups of people do not provide a morally or politically acceptable basis for development, and therefore cannot be considered to be sustainable.

The ecological space of settlements is such that they are part of a set of international relationships. As the quotation at the start of this chapter showed, 'the quiet Sumatran town of Bengkulu' draws people from across the globe. They are transported there by planes using a fossil fuel, oil. The environment within Bengkulu is designed to meet the needs of the foreign visitors, with hotels, tourist shops and so on. There are hundreds of Bengkulus, places where the command of energy and the spending power of people from the rich North – expressed through world markets – fashions the urban environment. Similarly, the day-to-day environments that are 'home' to Europeans and North Americans are dependent on cheap and readily available petrol. The need to protect assured oil supplies is of vital importance, as demonstrated by the 1991 Gulf War; the nature of urban environments thus has geo-political consequences.

The domination of global resources by rich countries and the ties that link cities in the North and the South of the world have their origins in **colonialism** and the development of world trade. At its simplest, colonialism involved the production and/or extraction of raw materials in the colonies, and then the processing of them into manufactured products in the imperialist countries. Some of these products were then exported back to the colonies. The requirements of this system gave a particular pattern to the urbanisation process in the colonies. Some of the ports developed into substantial cities, based on the export of raw materials and cash crops and the import of finished goods. Elsewhere cities were mainly

▲ *In contrast, a house in an élite area of Bolivia. Wealthy people in 'third world' countries have a standard of living equal to the rich of the first world.*

vehicles for administration and the exercise of military power. They were
under-industrialised because they were effectively linked into an urban
system of colonialism in which the manufacturing stage was concentrated
in the smokestack cities of the European colonial powers. King (1990) calls
this **dependent urbanisation**, as the role and development of the colonial
city was dependent on linked urban–industrial growth in the cities of the
colonialists' homelands.

At the local level, as well, urban environments can exacerbate social
inequity. Divisions of *class* are etched into the fabric of settlements. Rich
and poor live in very different environments within the same city, and
these differences are often reflected in health and life expectancy. The
places where we live now have developed around a set of assumptions
about the environmental and social acceptability of mass reliance on
extensive car use, and the dangers and restrictions which that imposes on
other people's travel.

Activity 3

Look at the queue at a local bus stop. Are the people in it a random
cross-section of your area in terms of age and gender? If they are not,
what does it imply for socially sustainable approaches to urban
transport?

Where did you play as a child and how did you travel to school?
Do children today have more or less freedom than you had in their
use of the urban environment?

Gender is an important social divide. Most decisions about urban
development are taken by men, yet men and women are likely to
experience urban environments in different ways. For example, men are
more likely to have the use of a car to get to work than are women, and
therefore the extent to which an urban environment is designed to meet
needs of car-drivers rather than pedestrians, for example, is indicative of
wider patterns of male domination within the society.

Another major social division is *ethnicity*, which means the culture and
practices that distinguish a community of people who see themselves as
culturally distinct from other groupings. Such differences might be
reflected in religion, language, history, even styles of dress and patterns of
leisure and recreation. Ethnic differences are often associated with
differences in wealth and power, and can be the focus for antagonisms in
which territory can be a focus for conflict. Within a city, ethnic
neighbourhoods may be formed on a voluntary basis by the ethnic group,
appreciating the solidarity and access to ethnic shops and social and
religious facilities that can be gained from living in proximity to one
another. In other situations ethnic residential segregation may be forced,
with discrimination restricting people to a ghetto.

2.4   *Summary*

•   We live in an increasingly urbanised world, with the most rapid urban
growth now occurring in poor countries.

- Urban environments take many different forms, and have visual, social and ecological dimensions.
- Size, density, infrastructure, investment and collective responsibility are definitive features of urban environments and determine their quality.
- A city can be analysed as an ecosystem.
- The ecological footprint of a settlement extends far beyond the boundaries of its built-up area, as settlements appropriate the carrying capacity of places elsewhere.
- There is much interest internationally in developing sustainability indicators for urban environments, but no agreed checklist.
- In aspiring to achieve sustainable urban environments we have to understand the social and political relationships that shape the city and urban processes.
- Urban environments embody the domination of some social groups by others, yet typically also offer better living conditions and more opportunities than in the countryside.
- Class, gender and ethnicity are important social divisions, and influence the experience of urban environments.
- Urban growth occurs within a global urban system, which has its origins in colonialism.

3   Urban environments in the past

Cities are ... natural ecosystems ... They are not disposable. Whenever and wherever societies have flourished and prospered rather than stagnated and decayed, creative and workable cities have been at the core of the phenomenon ... Decaying cities, declining economies and mounting social troubles travel together. The combination is not coincidental.

(Jacobs, 1994, pp. 42–3)

Although the growth of modern cities has been inextricably linked to the spread of the industrial system, cities have a much longer history. The aim of this section is to explore the urban environments of the ancient world and those created by industrialisation, colonialism and communism, and to highlight the key thresholds in urban development and their environmental pre-conditions and consequences. This entails not just a focus on the technologies, though these are important, but also appreciation of the relationships between people which made possible different types of cities and different urban environments, a theme introduced in Section 2. At the end of this section you should be able to identify the key features of urban environments in the ancient city, the industrial city, the colonial city and the cities developed under communism. Hopefully you will also be thinking about the links between them, and what guides they provide to achieving sustainability in urban settlements.

3.1   The first urban environments

The world's first cities formed about 5500 BP in the valleys of the great
rivers of the ancient world – the Nile, the Tigris–Euphrates (Iraq today) and
the Indus (Pakistan). Chinese urbanisation began around 3000 BP. These
earliest cities developed in similar environments – river sites in hot plains.
Management of water would have brought people together, and provided
an agricultural surplus to support an urban population. There were villages
before there were cities, but a city was more than an enlarged village: it
developed when some people could be freed from the need to work the
land, and so could take on new roles and assume new powers. Under such
conditions the settlement took on new functions and achieved a new size.

> The feat of city building itself was made possible by the fertility and
> productivity of the great valleys, by the reproductive capacities of the
> small village, well nourished and oriented wholly to life, by the traffic
> of water systems, and by the availability of ample material means and
> energy for supporting whole classes exempt from both the ancient
> village tutelage and harassing manual toil.   (Mumford, 1961, p. 76)

Ancient cities were surrounded by walls or ditches, and were very small by
today's standards: for example, ancient Babylon never had more than
15–20 000 inhabitants, and covered only 3.2 square miles. Availability of
water and food would have restricted the city's carrying capacity. Mumford
suggests that there were social and communication constraints, too: 'the
number that could respond promptly to a summons from headquarters …
Early cities did not grow beyond walking distance or hearing distance'
(1961, pp. 79–80). Therefore the early cities were concentric, with a single
centre from which routes spread.
      These earliest urban environments typically had a spiritual significance,
expressed in the lay-out of areas and buildings. The city was capped by
monuments to gods and to the idea and realisation of the city itself.
Moholy-Nagy (1968, p. 44) comments: 'Man [sic] in the centre of the
universe was not a geographical fact but a truth. Once a city had been
founded and dedicated to a god representing its ideal and no other, it could
not be moved to a different location.' Such cities could not be moved, but
they could be obliterated. Taxila flourished in the Indus valley from 500 BC
(2500 BP) until 455 AD when the White Huns arrived. Before the Huns
destroyed the urban environment entirely, Persians, Alexander the Great,
early Buddhists, Sanskrit scholars had all passed through, and would have
carried away ideas about urban development learnt from Taxila.

3.2   The polis

As Europeans we traditionally turn to the *polis*, the ancient Greek city, as
the urban archetype. The *polis* was dynamic but stable, in balance with
nature. The topographical context of the Aegean cities was very different to
the plains that had cradled the first cities. Islands and mountains created
isolated communities, and influenced the form of Greek cities. Again there
were important economic motors. Iron, the alphabet and the invention of
coinage made possible the extension of urban civilisation.
      City development in this part of the world began in Crete, where
Minoan cities had drains and sewers and piped drinking-water. The main
phase of Aegean urbanism began between the eighth and sixth centuries
BC. The settlements were small, relatively self-contained, constructed from

local building materials and largely dependent on the surrounding countryside for food. Thebes, for example, controlled and relied on a territory of roughly 1000 square kilometres, Corinth a mere 880 (Benevolo, 1980, p. 57). When population growth exceeded thresholds of sustainability new settlements were founded; the existing polis did not sprawl to contain it:

> Even the boldest conquerors had to acknowledge the natural limits of the city. When Alexander's chief architect offered to build him the largest city of all time, that leader, who understood logistics as well as strategy, peremptorily dismissed the idea: impossible to provision such a city!   (Mumford, 1961, pp. 155–6)

The Greeks contributed some seminal technical ideas that facilitated urban development and planning. The Hippocratic treatise entitled 'Airs, Waters, Places' provided guidelines about orientation of buildings and streets, avoiding development in marshy sites, and the need for pure sources of water (though the disposal of excreta was not discussed). These principles were then carried forward by the Roman architect and planner, Vitruvius, in the first century AD, who, in turn, was a strong influence on neo-classical town planning in Europe 1600 years later.

But the idea of the polis was much more than a set of plans for roads and drains. The city became highly valued, a place of balance, harmony and democratic involvement, an image to be worshipped, not least through a new civic architecture. The city, and its controlled development, was the physical expression of the politics of the independent city-state and the ethics of a society where art and science were expected to serve human growth in balance with nature. The optimal size of the polis was debated. As numbers increased it was not only provisioning that posed problems; direct democracy also became impracticable, prompting Aristotle to argue for a limit to the population of a city. Plato imagined the ideal city; reason, art and geometry could make it reality.

Benevolo (1980, p. 60) concludes that the polis 'succeeded in achieving a precise and lasting realisation of the theory of human co-existence.' So, was sustainable urban development invented by the Greeks all those years ago? It depends on the importance you attach to social equity and what criteria you set for public participation. The so-called democratic Greek city-state was sustained by imperialism, serviced by a slave economy, and women had no vote and little power.

3.3   The Roman city

Mumford has described the Roman Empire as 'a vast city-building enterprise' (1961, p. 239). The Romans used a regular lay-out on a grid basis, and perfected drainage and sewage disposal systems. Such control of the environment was fundamental to the colonising methods of the Roman Empire. Bridges, roads and aqueducts were constructed, massive walls and lines of fortification were strung across the countryside, land was divided and farmed, and new cities were founded. Through the domination which it exerted over distant lands, Rome's own population had reached about half a million by the time of Christ's birth, and perhaps a million three centuries later. This was an urban environment on a scale unprecedented in the West, though it did not have a monumental layout. Until the second century AD there was no defensive wall to cramp the extension of the city, and the city covered some 2000 hectares (Benevolo, 1980, p. 149).

Servicing and sustaining an urban environment for a million people required not just control of colonies, but technology and organisation. The sewers were begun in the sixth century and were continually extended and enlarged. They carried rainwater, and waste products from public buildings and ground-level houses; buildings that were distant from drains disposed of waste into cesspits or rubbish heaps. Though most private houses lacked sanitation, aqueducts brought a water supply from the mountains to sustain an extensive system of sanitation in the public domain – fountains, latrines and baths. There were edicts making residents responsible for street cleaning, and traffic congestion was addressed by banning the use of carts in narrow streets between sunrise and sunset. Food supply was also an organised process that gave rise to the development of the port of Ostia.

Rome was a classic early example of the appropriation of carrying capacity (see Section 1.3). Rome's domination of and dependence on distant environments is shown in Figure 7.1. When the Empire began to crumble, the city of a million could no longer be sustained. Food supplies became disrupted and a large part of the population moved from the city into the countryside. Aqueducts collapsed, rendering the core of the city uninhabitable:

▼ *Figure 7.1   Rome reaches out*

Source: Adapted from Girardet, 1992, pp. 42–3.

> Rome was the zenith of Ancient European civilisation, but it achieved this status at a price. It had to reach further and further to supply itself

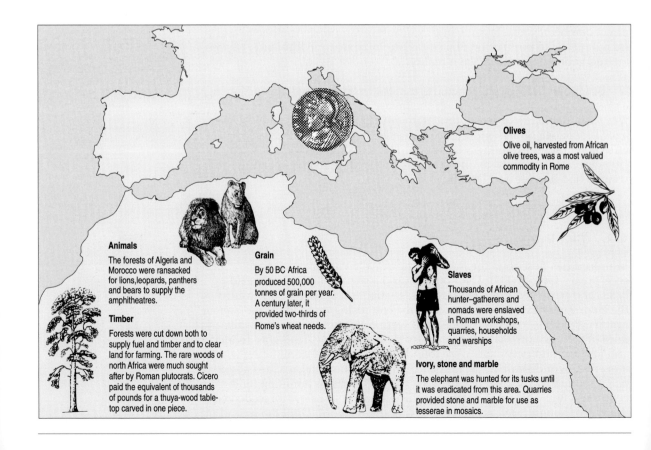

Olives

Olive oil, harvested from African olive trees, was a most valued commodity in Rome

Animals

The forests of Algeria and Morocco were ransacked for lions, leopards, panthers and bears to supply the amphitheatres.

Timber

Forests were cut down both to supply fuel and timber and to clear land for farming. The rare woods of north Africa were much sought after by Roman plutocrats. Cicero paid the equivalent of thousands of pounds for a thuya-wood table-top carved in one piece.

Grain

By 50 BC Africa produced 500,000 tonnes of grain per year. A century later, it provided two-thirds of Rome's wheat needs.

Slaves

Thousands of African hunter–gatherers and nomads were enslaved in Roman workshops, quarries, households and warships

Ivory, stone and marble

The elephant was hunted for its tusks until it was eradicated from this area. Quarries provided stone and marble for use as tesserae in mosaics.

with grain. Eventually the supply line became overstretched and
Rome collapsed, in part because of environmental decline: loss of soil
fertility and deforestation. The fertility of north Africa was shovelled
through Roman digestive systems, and then flushed into the
Mediterranean, never to be returned to the land. By AD600 many
cities of the empire had collapsed and today most of the 600 Roman
cities of north Africa, such as Leptis Magna, are piles of rubble in a
denuded landscape.   (Girardet, 1992, pp. 42–3)

3.4   Ancient cities: summary and the example of Chang'an

Giddens (1993) summarises the basic pattern of the cities of the ancient
world:

Cities were usually walled; the walls, primarily for military defence,
emphasised the separation of the urban community from the
countryside. The central area, often including a large public space,
was sometimes enclosed within a second, inner wall. Although it
usually contained a market, the centre was quite different from the

---

Box 7.2   Ancient cities: the example of Chang'an

Although every city is unique, the Chinese city of
Chang'an can be studied to illustrate some of the
key features of ancient cities. The Weihe River
Valley was one of the major sites for urban
development in the ancient world. The Zhou
dynasty (1100–771BC) built twin capital cities either
side of the River Fen (a tributary of the Weihe). The
Han dynasty (206 BC–220 AD) built a major city
nearby, on the south bank of the River Weihe. It
was called Chang'an, which means 'everlasting
peace'. As the plan shows, it was a walled city, with
four walls with three gates in each wall. Message
and meanings were imparted in passing through the
gates, with, for example, the crane, the bird of
destiny, symbolically depicted above a gate
(Moholy-Nagy, 1968 p. 164). The road layout
created a grid pattern within the walls, though, as
you can see from the plan, it combined different
sizes and shapes of blocks into a sophisticated
plan, as well as water courses and lakes. The
government was based in a walled central
enclosure, the 'forbidden city'. Other walled squares
were occupied by various trades and professions
and their housing.
    The arrangement of Chang'an reflected the
beliefs of the people. The four main quadrants
represented the quadrants of the heavens, the four

cardinal points and the four seasons. The royal
palace represented the polar star, which dominates
the universe as the palace dominated the city.
South was the most favoured direction, and so the
palace faced south. As Williams (1983, p. 414) puts
it, 'the city was a cosmo-magical symbol,
representing in highly formalised morphology the
interrelationships of man [sic] and nature.' The city
was planned by and for an imperial elite, but its form
was intended to express a sustainable relationship
with the natural environment.

Plan of Chang'an

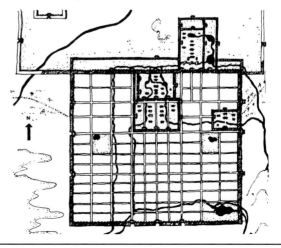

business districts found at the core of modern cities. The main
buildings were nearly always religious or political, such as temples
and palaces and courts … The dwellings of the ruling class or elite
tended to be concentrated in or near the centre, while the less
privileged lived towards the edges of the city, with some living
outside the walls but able quickly to get within them if they came
under attack.   (Giddens, 1993, p. 565)

Activity 4

Look back through this section on ancient cities and try to pick out
what features Chang'an had in common with cities from other
cultures and times. Use these to summarise for yourself the basic
features of cities in the Ancient World.

3.5   *The industrial city*

If the first creation of cities constituted a quantum leap for *Homo sapiens*, the
next change of comparable significance in urban environments did not
occur until the advent of the industrial city. In 1800 Britain was still a rural
society, with a total population of 10.6 million of whom only 20% lived in
towns of over 10 000. Half of these were in London, whose population of a
million, much the same as late Imperial Rome and the Tang Chang'an,
made it the largest city in the world. The first industrialisation was largely
in rural locations, as mills harnessed the renewable power of fast-flowing
streams, with cottages built within walking-distance of the mill to house the
labour force. The early entrepreneurs often lived close to their factories,
though in grander residences than those of their operatives.

As steam replaced water power, cities began to coalesce, and a new
word, '**conurbation**', was invented to describe this new form of settlement.
The great employers now lived in mansions at the edge of the town or in
the countryside, while their workers were in accommodation which a
contemporary journalist in Oldham described as '… filthy and
smouldering. Airless little back streets and close nasty courts are common;
pieces of dismal waste ground – all covered with wreaths of mud and piles
of blackened brick – separate the mills' (quoted in Foster, 1977, p. 84). By
1901 Britain's population had reached 25 million, of whom 77% were urban,
and London had grown to 6.5 million, the largest city the world had
known; in Germany, the population of the Ruhr increased from 237 000 in
1843 to 1.5 million in 1895 (Girardet, 1992, p. 50).

Urban birth rates were high, but so were death rates, especially during
the first half of the century. In 1841 the Registrar General calculated the
average expectation of life at birth as 41 years for England and Wales, 45
years in rural Surrey, but only 26 in Liverpool (Ashworth, 1954, p. 59).
Nevertheless people were moving from the countryside into a new type of
urban environment, the industrial city. Briggs (1968, p. 70) quotes from a
traveller in 1822, observing that the manufacturing system in Great Britain,
and 'the inconceivably immense towns under it' were 'without parallel in
the history of the world'. Manchester in the 1840s was 'the shock city of the
age' (Briggs, 1968, p. 96), with life expectancy at birth for those born there a
mere 24 years (Ashworth, 1954, p. 59). You may remember Box 1.3 in
Chapter 1 which describes Glasgow as an example of such urban industrial
growth.

How was this new scale of urban environment possible, and what were its social relationships? Above all there was a dramatic increase in the use of energy. Long-distance trade, coal and steam power provided the resource base for the transformation to an urban society. Fossil fuels and other resources were exploited to a degree that was unprecedented in geographical extent and in relation to the time taken to build up the resources. Production of the cities and in the cities also resulted in an unprecedented quantity and range of waste products which were discharged into the environment. Socially these environments were particularly deadly to babies and children – in 1875 infant mortality for England and Wales was 158 per 1000 births, but over 200 in places like Liverpool and Leeds (Ashworth, 1954, p. 59).

3.6   Air quality in the industrial city

Air pollution was a problem in big cities even before industrial growth. John Evelyn wrote the first text on air pollution in 1661, bemoaning conditions in London:

> That this Glorious and Antient City ... which commands the Proud Ocean to the *Indies*, and reaches the farthest *Antipodes*, should wrap her stately head in Clowds of Smoake and Sulphur, so full of Stink and Darknesse, I deplore with just Indignation.   (Evelyn, 1661)

The great increase in burning of coal with industrialisation made conditions even worse. British coal production was probably about 10 million tons in 1800, but by 1913 had increased to 287 million tons; though some was exported, home consumption over the same period escalated from 10 to 189 million tons. About a fifth of this was burnt in domestic grates (Clapp, 1994, pp. 15–16). When coal burns, it gives off waste gases, typically oxides of carbon and of sulphur; it leaves ash, soot and carbon particles; the fine ones are carried up into the air by the heat, while the heavy ones remain behind. Volcanic eruptions or strong winds had injected dust and foreign particles into the air, but the industrial revolution created air pollution in a new, long-term and systematic manner.

> The AIRE below is doubly dyed and damned;
> The AIR above, with lurid smoke is crammed:
> The ONE flows steaming foul as Charon's Styx,
> Its poisonous vapours in the other mix.
> These sable twins the murky town invest –
> By them the skin's begrimed, the lungs oppressed.
> How dear the penalty thus paid for wealth;
> Obtained through wasted life and broken health.
>
> (from William Osburn, 'A Poem read before Members of the Leeds Philosophical and Literary Society' (1857), quoted in Briggs, 1968, p. 139)

The earliest records date from 1881–85; during these years central London in December and January enjoyed less than one-sixth of the bright sunlight experienced in the small, university towns of Oxford and Cambridge (Clapp, 1994, p. 14). Table 7.2 gives some indication of the disparities in air pollution between urban and rural areas.

Table 7.2   Mean monthly deposits of air pollutants, 1914–16 (tons per square mile)

	Soot	Total solids
Sheffield	9.6	55
London (8 stations)	5.9	38
Manchester	4.3	32
Malvern	0.4	5

Source: Shaw, N. and Owens, J. S. (1925) 'The smoke problem of the great cities', pp. 64–6, 91, quoted in Clapp, B. W. (1994) *An Environmental History of Britain since the Industrial Revolution*, London, Longman, p. 14.

▲   *Sheffield in the nineteenth century: factories and terraced housing were built close together and all burnt coal, resulting in unhealthy, and potentially dangerous, living conditions.*

The contrast between the steel city of Sheffield and the rural spa town of Malvern is spectacular. The cities were dirty, dusty, foggy places where inhabitants breathed air containing considerable levels of impurities, and suffered a range of bronchial ailments as a result. Table 7.2 simply shows the concentration of pollutants, ignoring the issue of dispersal. The dramatic increase in carbon dioxide, as a by-product of coal burning, was largely unnoticed, with only a couple of scientists recognising the dangers of the greenhouse effect. (Clapp (1994, p. 23) cites Fourier in the 1820s and Arrhenius in 1896.) Acid rain damaged masonry and ironwork. As well as these generalised problems (which themselves varied slightly with the different local types of coal), there were the specific emissions of different types of manufacturing industry which varied from place to place.

Nevertheless some improvements were achieved. As transport technologies began to change and the middle class began to move to lower-density housing in the suburbs, so the concentration of smoke from domestic fires decreased. Though suburbanisation involved the loss of productive agricultural land and extended the ecological footprint of the conurbations, it led to more healthy domestic living conditions.

3.7   *Water supply and effluents in the industrial city*

Water power drove the first textile mills. Water was fundamental to the transport of industrial materials and products (whether by canal, sea, river or later by steam railway). It was an essential raw material in many production processes and it was a means of disposing of waste products (both solids in suspension and other substances in solution) and it was, of course, vital to sustaining the large concentrations of population.

Pollution of rivers by sewage and industrial discharges was widely noted in the industrial districts. There are numerous accounts of rivers which ceased to yield fish to anglers in the generation after the onset of industrial urban development in the 1820s. Let one quotation in respect of the Thames suffice. It was:

> . . . a repository for refuse from gasworks, unwholesome factories, slaughter-houses, cow-sheds, stables, and breweries and for the drainage of graveyards, yet which, unfiltered, was the only drinking water available for part of Rotherhithe.   (Godwin, 1859, pp. 52–3, quoted in Ashworth, 1954, pp. 68–9)

In the heatwave of 1858, the Houses of Parliament closed down for a week because of the intolerable odour of the Thames; such closures had also occurred in pre-industrial times.

▲   *'The "silent highway"-man' from* Punch, *10 July 1858.*

The polluted water supplies were a vital conduit for the transmission of water-borne diseases, most seriously cholera. The first epidemic in 1831 followed hard on the heels of the first surge of urban growth. The next was in 1848 and then 1853–4 (customers of the Southwark and Vauxhall Water company, served from the Thames at Battersea, proved particularly susceptible to the disease). Cholera typically first appeared in the slums, but would spread even into areas where the better-off lived.

Even more than with air pollution, significant steps towards sustainable living conditions were achieved with respect to water. The cholera epidemic of 1866 was the last in Britain as knowledge, technology and management improved the quality of urban water supplies, making it possible for dense populations to exist without falling prey to water-borne communicable diseases.

3.8   *The sanitary idea*

Any assessment of the environment of the industrial city needs to recognise two important facts. First, though conditions were bad in the cities they were often worse in the countryside, not least in respect of housing conditions and economic opportunities. The population of Ireland, predominantly a rural country, reached 8 million in 1841, a peak never since regained, as famine and emigration took their toll. Irish ethnic communities, exploited as cheap labour by employers, poverty-stricken and socially stigmatised, lived in desperate conditions in urban environments like those in Glasgow, but they did not go back to the land. Similarly, as Massey and McDowell (1994) show, the mills of Lancashire, worked by large numbers of female employees in deafening and at times choking conditions, provided social and economic opportunities for women that they were denied in rural East Anglia. Sustainability should not be equated with tradition and preservation of rural living.

Secondly, increased knowledge, better technology and enlightened political action resulted in major improvements to urban environments. The Health of Towns Commission was set up by government in 1843, and the Health of Towns Association was formed in 1844 to share information about urban living conditions and to work for changes in the law. Ashton sums up **the sanitary idea** as follows: 'overcrowding, inadequate sanitation and the abuse of safe water and food created the conditions under which epidemics could thrive. The response was seen to lie in housing standards and hygiene regulations, paved streets and publicly funded water and sewerage systems' (1992, p. 2).

A key part of the advance of public health was the recognition that state action was needed to regulate urban development and to ensure compliance with minimum standards. Slowly but surely the idea of laissez-faire, where any entrepreneur was free to impose environmental costs on others, gave way to environmental legislation and municipal power, embodied in vaulted town halls where the city fathers presided. Women had no vote, urban management was paternalistic, professional and top-down, the cities continued to stoke up acid rain and to dispatch a substantial proportion of infants to early graves, but the urban crisis that seemed so imminent in the 1830s was soothed. Cities flourished as never before, providing work, public libraries, elementary schools, and a range of entertainment and social contacts that could not be matched in the village.

Another part of the response to the unsustainable city of freely competitive industrial capitalism was the development of ideas and

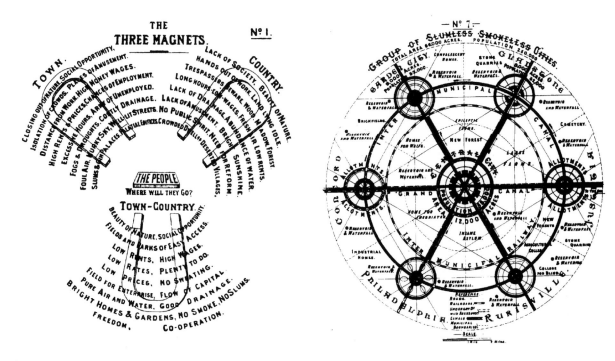

▲   Figure 7.2

(a)   The three magnets: 'the chief advantages of the Town and of the Country are set forth with their corresponding drawbacks, while the advantages of the Town-Country are seen to be free from the disadvantages of either' (Howard, 1970, p. 47).

(b)   The Social City: a cluster of Garden Cities, with a central city, all interlinked to form a 'Social City', maintaining rural as well as urban amenities.

experiments in town planning to create liveable cities. Such thinking was by no means restricted to Britain, but Ebenezer Howard's proposals for 'Garden Cities' published in 1898, as a response to the problems of London's growth and the depression of surrounding agricultural districts, was hugely influential internationally.

Howard saw that the problems of sustainable living in the big city and in the countryside were interrelated, and that social equity was integral to achieve 'a peaceful path to real reform', as his book is subtitled. Howard proposed a new type of settlement, the Garden City, with a population of 30 000 (plus 2000 in the 'rural estate'), but linked together around a central city of 58 000 to form a polynuclear system, the Social City. He argued that this new type of city could have beneficial effects and also be self-sustaining, first, by undermining the high land values and high rents of the big city and, secondly, by recycling the surplus gained from rents in the Garden City into social investment within that city for the good of the whole community.

Activity 5

How do Howard's ideas compare with today's concerns about the sustainable city? Compare your answer with that given at the end of the chapter.

By the turn of the century, when Howard was writing, mortality rates were falling: for example, the crude mortality rate for Edinburgh was 30.4 per 1000 in 1869, but 16.1 per 1000 in 1894 (Hague, 1984, p. 152). The reproductive cycle was perceptibly shifting from short (high birth-rates and high death-rates) to long (low birth-rates and low death-rates), in part aided by the developing division of labour under which women looked after the household. Despite these improvements, mortality, and especially infant mortality, was still highly skewed by social class. This was due to a spate of suburbanisation facilitated by the electrification of tramways in the 1890s, combined with land availability, resulting in the middle classes and the better-off artisans moving to healthier, lower-density conditions. Where philanthropists had built model new village communities, with low housing densities and large gardens, for example Port Sunlight (Lever Brothers) or Bourneville (Cadbury), the health of their residents was better than that of other workers in the nearby big cities.

Living conditions in the huge working-class neighbourhoods of the cities, however, were still sufficiently poor as to threaten the stability of the British Empire. The Boer War began in 1899, but a large percentage of the recruits from the cities had to be rejected as unfit for service in the armed forces: out of 11 000 young men from Manchester, 8000 were rejected and only 1000 were deemed fit for regular service (Hall, 1988, p. 33). A Committee was set up by government on 'Physical Deterioration'. It reported in 1904, blaming urban conditions for the general decline of the British 'race', and for Britain's decline as an imperial power. In Germany, only 42% of Berliners were passed fit for army service in 1913, compared to 66% of those from rural areas (Hall, 1988, p. 33). Urban environments and geo-politics were thus interlinked.

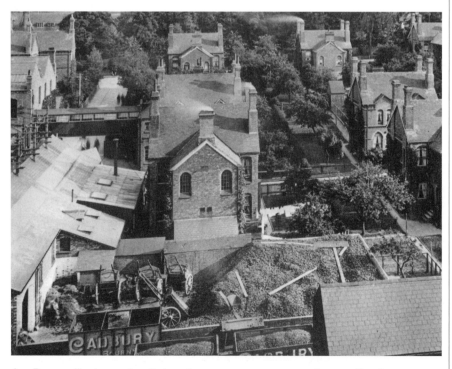

▲ *Bourneville: lower-density housing, more open space and separation from industry improved the living conditions of the industrial workers.*

3.9 Colonial cities

Gilbert (1981, p. 13) observed that: 'The grandeur of European, and indeed North American, cities at the turn of this century can only be explained in terms of the international division of labour and the evolution of a world urban system.' Though colonial rule was imposed on much of the world, societies that the Europeans colonised had widely differing urban traditions. There were whole regions in the Pacific and in Africa where urbanisation was non-existent, but there were great Moghul cities in the north of India, and China had more large cities than Europe. The *conquistador,* Cortés, reputedly was overwhelmed by the beauty and greatness of Montezuma's capital, Tenochtitlan, in what is now Mexico. It had aqueducts, excreta collection and composting for re-use as fertiliser, and zoological and botanical gardens. Like the cities of the ancient world, these were **pre-industrial cities**, built on religion. For example, Cuzco, the Inca city of Peru, was mystically connected to Inti, the sun god, from whom power ultimately derived.

Where major traditional cities existed, they helped to mould the features of colonial urbanisation, as did the background of the colonists, for Spanish, Portuguese, British, German, French, Dutch, Italian and Belgian expansion took different forms, and some colonies experienced more than one coloniser. However, Gilbert argues that variations became less marked in the nineteenth and twentieth centuries:

> ... the world came to be dominated by Europe and the United States. From this time on, the world began to look more similar. The cities

▲ *Simla: a hill-station in the Himalayas for summer retreat. Note the Anglican church at top centre.*

associated with these developments, which began to expand in the late nineteenth century, resembled one another. The tram and the railway, the suburban house, the occasional dash of town planning created passable European cities. The so-called modern cities contained all the advantages and extravagances of European urbanisation together with the additional disadvantages of general poverty. The new cities were frequently insanitary, badly located and tasteless.   (Gilbert, 1981, p. 22)

Colonialism brought disease. In Latin America, diseases from the Old World ravaged indigenous peoples who had no immunity, with the result that large areas became depopulated and their infrastructure collapsed. Elsewhere it was the Europeans who suffered from an array of tropical diseases. One response was a new type of urban settlement, the hill-station, a re-creation of a Home Counties village to which the Europeans could retreat from the scorching summer heat of the plains.

This vulnerability to disease, together with the needs of the garrison, resulted in the design of distinctive types of urban environments for colonial cities. In India in particular, a typical urban structure saw a tripartite city. The *cantonment* housed the soldiers, while the colonial civil servants, teachers and administrators lived in the *civil lines* nearby. These European areas were separated from the *native city* by physical barriers such as railway-lines, canals or natural topography. In the late colonial period (after 1920) the administrative ranks were 'Indianised' and this produced a new indigenous middle class, who occupied new neighbourhoods, intermediate in density and in location between the native city and the civil lines. The architecture of public colonial buildings could be bizarre in its disregard of local culture and tradition – the railway station, in Lahore, is a pastiche English medieval castle!

By the early twentieth century, development in these European areas was very much influenced by the Garden City ideas: indeed, in some cases the colonial areas anticipated Garden City styles. Low-density houses were built in spacious grounds, and the houses were served by wide, paved roads (typically wider than in the Garden City lay-outs back home), water and drains. The result was a fundamentally different urban environment compared with the traditional city of those countries.

Colonisation led to new urban growth. Indigenous cities acquired new roles, eventually including manufacturing industry. The displacement of traditional agriculture by cash crops resulted in increased agricultural production, falling mortality rates, population increase and, therefore, large-scale migration from the countryside to the cities. The population influx led to a deterioration of conditions in the cities, especially where the traditional city was still enclosed by a wall.

As in the imperial heartlands, water was a major problem in sustaining the new urban growth. The latrines that were traditional in the indigenous city were inadequate for the increased population, who used waste ground. The same source of water was used both for consumption and for disposal of waste. Untreated waste would infiltrate groundwater drawn from wells. During wet seasons unsurfaced roads and alleyways became quagmires that could not be used by carts. Drainage was often a problem where a city had been built on low-lying, marshy ground, and stagnant water was a breeding-ground for disease-causing organisms.

The new sanitary technology was needed with the transition to larger cities. Installing the pipes and sewers in traditional neighbourhoods was a challenging task. The traditional street patterns were typically a web of narrow, irregular lanes, not the straight, wide streets favoured by sanitation engineers. (See Plate 13.) There was a cultural dimension also, of colonists imposing their norms on an indigenous ethnic population, invading the privacy of residential areas, and riding roughshod over customs and the symbolism of traditional urban environments. The British exported to their colonial cities the measures and institutions developed to manage the industrial cities back home.

Lowder (1986) stresses problems with bureaucratic decision-making, and with the inappropriate nature of British town-planning solutions in the Indian context: 'solutions to city problems were perceived in engineering terms and based on principles of individualism and land-use alien to Indian societies' (p. 66). For example, in the 1860s British local authorities had set up Improvements Trusts to carry through urban redevelopment of the worst slum areas by a process of compulsory purchase, clearance and road widening, and re-building of better (but higher rent) houses. Such measures were often opposed by the slum-dwellers who were displaced by the process. The same model, now in national legislation (the Housing of the Working Classes Act 1890), was imposed on Indian cities. The Indian Town Planning Act of 1915 was based on the British Housing, Town Planning Etc. Act 1909. None of the thinking behind these measures was sensitive to the social and economic structures of the indigenous society, which was:

> . . . based on spatial communities whose cohesion was maintained by kin and caste ties lubricated by reciprocal services and charitable institutions. They included families of very different wealth: a rich household would be surrounded by the huts of the poor servicing its needs. Families clustered around symbols legitimising their status, such as temples, tombs and rich bazaars; more recent arrivals were relegated to the unconsolidated parts.   (Lowder, 1986, pp. 66–7)

To secure the ecological space of the colony to support the carrying capacity of the cities of the imperial heartlands, it was necessary to create a sustainable urban environment in the colonial city. Though the technology was there and administrative mechanisms had been tested and proven in urban Britain, the historical legacy of pre-industrial cities together with divisions of ethnicity confounded implementation. The inequities of colonialism, so vividly expressed in the form and structure of colonial cities, severely undermined the sustainability of such urban environments.

Activity 6

Today the colonial era is often criticised as being the origin of the burgeoning mega-cities of the third world. Make a balance-sheet of the ways in which the colonial city was moving towards or away from sustainability. Then compare your ideas with those at the end of the chapter.

3.10   The experience of the communist urban environment

If the cities of industrial capitalism and colonialism have left the poor and ethnic groups in the most hostile urban environments, are communist cities fundamentally different, an expression of the liberation of the poor?

Communist cities have developed in a range of situations, so generalisation
has to be treated with caution. In places like the former Czechoslovakia the
society was already highly urbanised before the communists came to
power. Russia and China, and much of eastern Europe, were rural-based
peasant societies. While the communist hierarchies of Europe collapsed in
1989, China remains ruled by the Communist Party (though its urban
development is now much more market-driven). Nevertheless, it is possible
to identify basic premises of communist urban development, and
distinctive environments resulting from their implementation.

Bater (1980) sees the application of Marxist–Leninist principles as
differentiating the Soviet socialist city from its western counterpart in the
following ways:

1   Nationalisation of all resources (including land);
2   planned rather than market determined land use;
3   substitution of collectivism for privatism, most apparent in terms of the
    absence of residential segregation, the dominant role of public
    transport, and the conscious limitation and dispersal of retail functions;
4   planned industrialisation as the major factor in city growth;
5   the perceived role of the city as the agent for directed social and
    economic change in backward and frontier regions alike;
6   cradle to the grave security in return for some restrictions on personal
    choice of place of residence and freedom to migrate;
7   directed urbanisation and the planned development of cities according
    to principles of equality and hygiene rather than ability to pay.
                                                                  (Bater, 1980, p. 5)

Soviet planners believed that an optimum size of city could be
scientifically calculated and then achieved through town planning. At first
the optimum size was put at 50 000 or 60 000 (remember Howard's figure
for a single Garden City had been 32 000). The idea was that cities of this
size would be able to provide all the essential goods and services, yet still
be small enough for a sense of community and to help create a socialist
ethos. The ceiling figure was then upped to between 150 000 and 250 000 by
the mid-1950s, and to 300 000 by the 1960s, after which the notion fell into
disrepute. Thus in 1959 there were only three cities of over 1 million
inhabitants in the USSR – Moscow, Leningrad and Kiev, seven more
crossed that threshold in the 1960s and eight more in the 1970s, by which
time the biggest cities were over 3 million in population. Industrial growth,
decreed by the economic plans, produced urban growth, despite long
waiting-lists for housing and restrictions on residence in the large cities.
Thus, under-urbanisation and long-distance commuting characterised
communist urban planning.

The planning that Bater refers to was very much top-down, and driven
by the push for industrial growth. Under the communists, local
government was a mere cipher for central government and the Communist
Party. Democratic civic culture atrophied. The Soviet model of the socialist
city was dispersed to satellite states. New cities were built around new
industrial activities – steel works, chemicals, uranium mining – often in
regions with no industrial tradition. There was scant regulation of the
industries. Nowa Huta in southern Poland is a classic example. Cracow was
an old, bourgeois university and trading city in an impoverished
agricultural region. In 1949 a new city was planned five miles away. Nowa
Huta would be based on the Lenin Steel Works, to create a large, new and
strong socialist, urban working-class society. The discharges from the plant,
together with the valley location which is naturally prone to fogs, resulted

▲ *Nowa Huta, Poland, showing the dominance of industry over residential areas, and the severe air pollution.*

in severe air pollution, and consequent damage to the historic buildings in Cracow.

A Prague journalist quipped, 'The air pollution, more than the existence of the Iron Curtain, brought about the revolution in Czechoslovakia' (quoted in Hague, 1990, p. 20).When the Velvet Revolution came in Czechoslovakia in 1989, 23 tons per square kilometre of sulphur dioxide were being deposited annually (largely from the burning of brown coal), 70% of the country's rivers were polluted, 40% of sewage was untreated, 75% of toxic waste was dangerously stored, and half the forests were dying or damaged.

What kind of urban environments made up these cities? High-rise office-blocks built by speculative developers in post-war western cities did not dominate the skyline of communist city centres. Unless it was destroyed in the war, the historic core remained – indeed the Old Town of Warsaw, which was destroyed by the Nazis, was faithfully reconstructed in its pre-war form. The streets and squares in the centre of the communist cities were blazoned with banners, red flags and slogans, not advertising hoardings. Statues of approved saints of the communist movement abounded. The nineteenth-century tenemental ring of mixed land uses received very little investment under the communists. This is where the poorest citizens lived, often in properties badly in need of modernisation. On the edge of the city, massive new high-rise housing estates were built from the 1960s onwards. Petrezjalka, the extension of Bratislava across the Danube, was planned to house 180 000 people, for example. Further out, and on sloping sites unsuitable for the high blocks, were one- and two-storey single-family houses, privately owned and generally self-built by the owners. Thus the socialist city did not achieve a totally egalitarian living environment, though the extremes were less than in capitalist cities, with facilities in new residential areas provided to a fixed set of standards. Cheap and frequent public transport, together with planned provision of facilites such as shops and community centres to set standards, were means of creating equal access to facilities.

Activity 7

Try to complete the boxes in the matrix to help you to synthesise the analysis of past urban environments.

	Ancient cities	Industrial cities	Colonial cities	Communist cities
Population of largest cities				
New ideas /urban technology				
Typical types of areas and layout of the city				
Political base – who had power and who was dominated or exploited				
Key ecological and social limits to sustainability				

3.11 Summary

• In reviewing cities of the past we have seen two major transitions: the first began some 5500 years BP and saw the growth of the first cities; the second came with the technologies of industrialism and produced a new scale and type of urban environment.

• Urban development depended on the production of a surplus from agriculture, and the domination by the city of land outside its walls.

• Cities of the ancient world occasionally grew to 1 million people; they were typically centres of religion and royal power; they were often consciously designed, usually on a grid basis, and the design expressed religious symbolism.

• Industrialism and colonialism brought about new types of cities, much larger than any previously seen, and much more demanding in terms of resources needed to sustain them.

• Water, in particular, was vital to these new urban environments.

• Gradually, technologies and institutional structures were developed to allow for some control of the disposal of urban wastes and therefore better urban environments were achieved.

• Communist cities have also been fundamentally about the process of industrialisation; they included good quality public transport, but showed a disregard for the environmental impacts of waste disposal and energy use.

4   Urban environments in the present

Cities are the single most complex products of the human mind. Their labyrinthine street plans are the most intricate human designs to appear on the face of the Earth, becoming the mind maps of their inhabitants. Cities are more than static structures of stone and concrete. They are also vast processors of food, fuels, and the many raw materials that feed a civilization. With their complex metabolisms they are huge organisms without precedent in nature; their connections spread across the globe.

(Girardet, 1992, p. 19)

… the city exists for one particular kind of citizen: the adult, male, white-collar, out-of-town car-user.

(Colin Ward, 1977, *Child in the City*)

4.1   Restructuring and the city

New types of urban environment are now emerging, and more urgent questions are being asked about the world's cities than at any time since the invention of the industrial city almost 200 years ago. The world is changing, and so its cities are changing too. New communications technology and the deregulation of international finance create a **new international division of labour**. Labour-intensive smokestack industries leave the old industrial countries and set up in cheap labour locations with relaxed pollution-control regimes. High technology industry and corporate services are concentrated in the richest countries, but also growing rapidly in the 'Newly Industrialising Countries', notably in the Far East. The powerful state that provided housing and public transport and enforced minimum standards for all was built to resolve the problems of the industrial city. Now the 'enabling state' sets the context for private-sector provision. There has also been a cultural change. New technologies, such as the video and the personal computer, privatise and individualise leisure and recreation, undermining the historic role of the city as a space for social contact. Last, but not least, the economic changes which began in the 1970s were completed by political changes, the fall of the Soviet Union and the effective end of communism as a global ideology.

In terms of European urban environments, if one event exemplifies the onset of a new era of qualitative change, it is perhaps the publication in 1990 of the European Commission's Green Paper on the Urban Environment. This argued for high-density, mixed-use cities, which could be energy-efficient by reducing travel distances and maximising provision for public transport. Suburban spread, in contrast, was seen as requiring high energy consumption and also providing a poorer quality of life. European governments are beginning to implement planning control policies specifically designed to locate major trip-generating land uses in such a way as to reduce travel. The Dutch are leading the way on this, but significantly the British government, which in the 1980s de-regulated public transport and actively encouraged out-of-town retail developments through its planning policies, was by the mid 1990s trying to protect town centres by advising planning authorities to refuse applications for new out-of-town shopping centres. It may be too late: more than 25% of shopping floorspace is now in out-of-town centres.

The aim of this section is to identify and analyse trends in urban environments today, and to highlight questions these pose as we look towards the twenty-first century. Evidence mingles with opinion. As you read through, you are invited to think about what needs to be done to make urban environments liveable and healthy for all, a means to conserving not destroying the global environment. Such thinking prepares the ground for the final section of the chapter, on future urban environments.

4.2   *The dispersal of low-density urban environments in the developed world*

Life is immensely more comfortable in rich cities today than in the past. Diseases that ravaged the early industrial city have largely been eradicated. We have pure water on tap; the air is not polluted by domestic smoke and industrial emissions. Housing standards have been transformed. Yet, in the most affluent countries, better-off people are moving out of the cities, and there are signs that economic activity and leisure activities are following them. The cities are spreading, but at lower densities than in the past: for example, population in the New York metropolitan region increased by only 5% over 25 years, but the developed area increased by 61%, consuming nearly a quarter of the metropolitan region's open space, forests and farmland (Lowe, 1992, p. 121). The fastest-growing settlements are the small towns within commuting distance of large cities. New types of urban environment have been created in the last generation – the business parks and out-of-town shopping malls – while derelict factories are 'regenerated' as heritage experiences.

Changes in the cities of the rich Northern hemisphere are not just of local or national significance, however; they matter globally. These are the places which account for a high proportion of the world's consumption of resources, not least energy, and they make a major contribution to global greenhouse gas emissions. They draw resources from across the globe, impacting on distant ecosystems, and their waste products reach the oceans, the atmosphere and the ozone layer. Though such impacts are due to overall consumption patterns in wealthy countries, the move towards low-density suburban settlement patterns is an important contributor. Roseland (1992, p. 25) claims that this western pattern of urban development 'is not only ecologically unconscionable, but economically inefficient and socially inequitable.' Areas decay or are abandoned, homeless people sleep in shop doorways, unemployed and elderly people, members of minority ethnic groups and others trapped in poverty face hostile environments which offer declining opportunities. Such problems are not caused by cities, nor are they unique to cities, but they are exacerbated by the spread of the city, and it is in the cities that they are most significant.

Not all commentators agree that city spread is so bad. Suburban environments provide open space and trees, perhaps a chance to grow your own fruit and vegetables. The development of new centres for shopping and employment on the edge of the city will bring such facilities closer to suburban residents, and so may reduce the length of energy-consuming journeys. New communication technologies – fax and electronic mail, for

▲   *Las Vegas: an example of suburban sprawl.*

example – will tend to reduce the need to travel, and will facilitate working from home (see Chapter 5, Section 6.2). They bring entertainment into the home too. Such trends undermine the rationale of the big city as the centre of work and culture, and stimulate the growth of small settlements, developed on sustainable principles, where people could participate more directly in decison-making than is possible in the conurbations. On this theme, a group calling themselves 'Global Tomorrow Coalition' have argued that:

> Large cities by definition are centralised, manmade environments that depend mainly on food, water, energy, and other goods from outside. Smaller cities by contrast can be the heart of community-based development and provide services to the surrounding countryside.

> Given the importance of cities, special efforts and safeguards are needed to ensure that the resources they demand are produced sustainably and that urban dwellers participate in decisions affecting their lives. Residential areas are likely to be more habitable if they are governed as individual neighbourhoods with direct local participation. To the extent that energy and other needs can be met on a local basis, both the city and surrounding areas will be better off. (Global Tomorrow Coalition, 1986, quoted in World Commission on Environment and Development, 1987, p. 243)

Activity 8

Where do you live now? What is the attraction of that place to you? What technological, economic or social changes might make you move to a different type of urban environment?

4.3   *Cities for cars*

Transport and communications have always been of fundamental
importance to the form and function of urban environment, as Section 3
showed. In the industrial cities that grew in the nineteenth century,
factories and warehouses jostled for space close to the railway and the
canal. Houses grew around them so that people could walk to work. The
major shops were concentrated in the most accessible part of the whole city,
where all the routes converged, the city centre. Smaller shops straddled
along these radial roads through the areas where people lived. The internal
combustion engine has changed every one of these locational imperatives.
Mass production for mass consumption made cars affordable for a
substantial proportion of households in the richer countries, and cheap oil
encouraged the use of these vehicles. Road transport is now the dominant
mode of shifting goods around and of servicing the city, and has added
new elements to urban environments – the filling-station, the multi-storey
car-park, the split-level interchange. How does the city for cars measure up
to the requirements for sustainability?

We need to consider the input of resources and the output of waste
products. Car manufacture and use particularly consumes steel, rubber, oil
and land. Of these, oil is probably the most critical as it is not a renewable
or recyclable resource and supplies are finite. The global market for oil is
fundamental to the form of today's urban environments. The waste-
products of the car-based city – emissions of carbon monoxide, for example
– contribute to urban air pollution which causes respiratory disease. The
urban and industrial haze from the cities of Europe and North America has
been detected as far away as the Arctic Circle (Girardet, 1992, p. 108). The
impact of road traffic is especially felt in the urban areas. For example, half
of all road traffic movements in the UK occur in urban areas, and urban
traffic is continuing to increase by about 1% each year (OECD, 1990, p. 27).
The peak traffic period is now from 8.15 until 9am, with cars delivering

▲   *The dominance of traffic in urban environments makes them unpleasant places
for pedestrians.*

children to school being a significant contributor. Those children who do walk to school are thus exposed to traffic hazards and the pollution caused by the cars carrying their colleagues, and so the incentive for parents to take them by car is increased.

Activity 9

Note down your own movements over the last twenty-four hours. Where have you travelled and how? Did you make some of the trips by car, and, if so, would you have been able to make the same journeys by other modes of transport? If you did not use a car, might you have done different journeys if you had used a car?

Compare your own pattern of movement and the means of travel with those of a partner, relation or friend. Think about the extent to which the place where you live has developed around the presumption of car use, and about the environment that is created. Does the comment from Colin Ward about the city existing for one type of person apply to your place of residence?

Although the car has conquered all urban cultures, there still remains a considerable difference between the relatively compact urban form of European countries and the dispersal of urban areas in North America and Australasia, where dependence on the car has become absolute (Kenworthy and Newman, 1990). The most **transport-efficient land-use pattern** combines a high-density, mixed-use city centre, and high-density residential corridors focusing in the city centre, but with their own activity centres at public transport interchanges. Though some Canadian cities like Toronto and Vancouver have sought to achieve this pattern, as a generalisation it is most likely to be found in Europe, where state intervention in the land development process has been stronger. The conscious planning to achieve such a pattern in the Stockholm city region is a particularly good example: see Figure 7.3 and the photograph of Farsta.

Car-based environments threaten the safety of others using that space. In cities in the past the street was the place where activity was most intense. It was a public environment, freely available to male and female, young and old; there were no owners to exclude the poor. The energy converted into speed by a car changes this; the environment becomes unsafe; play and movement adapt accordingly. Again planning intervention can make a difference, and achieve *traffic calming*. Environments can be designed around the needs of pedestrians. The Dutch *Woonerf*, for example, is a street where cars can enter, but are forced to be subordinate to pedestrians. There are ramps, 'pinch points' (where the street is narrowed, for example by tree planting), and the paving (there is no distinction between pavement and carriageway) shows that this is a pedestrian environment. Such schemes empower people to control the environment where they live.

Activity 10

Find out about planning and transport policy in your home area by asking your local council. Are there specific policies to help improve conditions for pedestrians and cyclists? Are there any schemes to calm traffic in residential areas? How adequate is the public transport system? Look at the neighbourhood where you live and think about ways in which traffic hazards could be reduced and more pedestrian-friendly areas could be created.

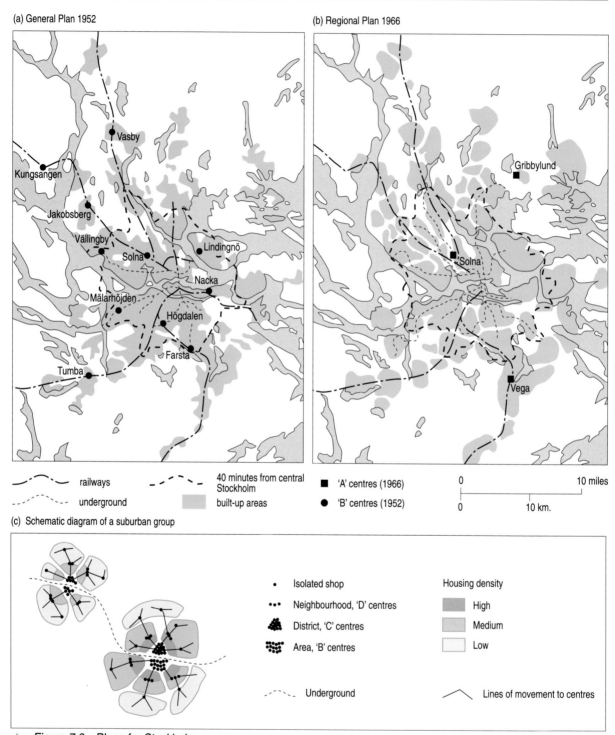

(a) General Plan 1952

(b) Regional Plan 1966

railways
underground
40 minutes from central Stockholm
built-up areas
■ 'A' centres (1966)
● 'B' centres (1952)

0   10 miles
0   10 km.

(c) Schematic diagram of a suburban group

•   Isolated shop
•••   Neighbourhood, 'D' centres
         District, 'C' centres
         Area, 'B' centres

Housing density
         High
         Medium
         Low

------   Underground   Lines of movement to centres

▲   *Figure 7.3   Plans for Stockholm*

The 1952 plan established the idea of planned suburban satellites, with a hierarchy of shopping centres, linked by the new underground railway system. This plan was largely implemented by the late 1960s, when a wider-ranging regional plan extended the principle through new developments along mainline railways and motorways radiating from the city.

Figure 7.3c shows the principle of the hierarchy of suburban shopping centres in the 1952 plan. The bigger 'B' centres are surrounded by high-density residential areas from which the inhabitants can walk to the shops or to the underground station. Densities fall away from the centre towards the edge.

▲   Aerial view of Farsta, Sweden, the centre of one of the planned Stockholm suburbs, designed in the early 1960s. It shows the high-density residential area with easy accessibility, by foot or car, to the shops (centre) and railway station (on the right).

4.4   Water in cities in the developed world

In some senses the problem of urban water supply, so critical a feature of the early industrial city, has been solved in the cities of the developed world. We turn on the tap and out pours clean water, while fish are returning to some rivers which have been barren for over one hundred years. It is easy to overlook the enormous demands on the water environment that the large, rich city makes. We rely on long-distance distribution of water to make the urban form viable. While smaller cities or those in very wet climates could rely on local supplies, for big cities and those constructed in hostile environments urban water is typically supplied from many miles away. The extreme examples are where there has been rapid urban growth in very dry climates such as Israel or in the south-west of the USA. There are, for example, 30 million people in California – more than the population of many nation-states. Las Vegas is perhaps the most extreme example: it is the fastest-growing North American city, all watered gardens, swimming-pools and green golf courses, yet it is in the middle of a desert. Levels of rivers have fallen and it will be impossible to provide for the predicted increases in water consumption over the next two decades.

Water supply is even an issue in the highly urbanised north-eastern states of the USA, where, despite the wetter climate, there is rationing and curtailment of non-essential water use in dry periods (OECD, 1990, p. 22). Even within the UK, the most urbanised areas with the highest water demand are in the driest parts of the country in the south-east of England, and the hot summer of 1995 saw rationing in some regions, including areas of relatively high rainfall such as north-west England and Yorkshire.

Water pollution is not a problem on the scale it was in the past, though there are no grounds for complacency, and there are concerns about the adequacy of water treatment. It is interesting to note that the European Union, aware of transnational flows of pollution, has taken an important role in pressing for improved standards. Improvements in the oxygen content and in microbiological quality have occurred in the surface waters of many OECD countries, though there are increased levels of nitrates, heavy metals and organic pollutants which may pose problems for aquatic environments and for people (OECD, 1990, p. 22). Action at European Union level forced the UK to address sewage treatment seriously, instead of disposing of untreated sewage into the sea; however, some experts now feel that abandonment of sewage sea disposal was a mistake because of the environmental costs of the Urban Waste Water Directive in terms of the carbon dioxide burden on the air and disposal to the land. In this, as in other aspects of urban environmental management, the setting and attainment of acceptable standards have to be balanced against the costs of achieving those standards, and affordability to consumers.

Improvement in water through municipal action was one of the great advances in managing the environment of the industrial city. Provision of a water supply had been undertaken by local companies (remember the Southwark and Vauxhall Water Company, drawing water from the Thames at Battersea, discussed in Section 3.7); with 'the sanitary idea', local councils increasingly took on the role. The creation of Regional Water Authorities in the 1970s recognised the extended ecological space involved in urban water provision, and looked towards a national grid. Underpinning this change (which has parallels in gas and electricity) was a presumption that government would facilitate a standardised provision, with cities as part of a national urban system, and equalised access for all people and all locations. To all intents and purposes, provision to domestic users was not based on a market relationship, while commercial users everywhere could be sure of much the same standards of water supply and waste disposal.

The national grid was never achieved. There had been a long period of under-investment, living off the capital invested in the nineteenth-century infrastructure systems, and such a grid was deemed to be too expensive to complete. The state restructured, as outlined in Section 4.1. Water in England and Wales was privatised, and reorganised in Scotland in a way that could lead to privatisation. This has led to a regionalised structure based on competitive but regulated markets in which investors seek a return on their capital. The new companies have attracted investment from overseas, resulting in foreign ownership of some of the water companies and a trend to merger, either with other water companies or other regionally based utilities. These processes, with variation in costs of water, are intensifying spatial and social inequalities; there has been criticism of the readiness of companies to cut off water supplies to those households who cannot pay their bills, a problem exacerbated by the introduction of water metering, which can increase costs to those least able to pay, such as families with small children or with disabled members.

The role of national and local government in the management of today's urban environments is therefore changing, as supra-national bodies and the private sector exert more control. In the case of water, this entailed the establishment for the first time of an independent regulator, the National Rivers Authority. With 7500 employees in 1995, and about to become part of the new Environment Agency with 9000 employees, it is the largest environmental regulator ever known in Europe.

4.5   Air in cities in the developed world

The air of our cities has been improved through legislation. In Britain the Clean Air Act, which became law in 1956, was dramatically effective, with industrial smoke emissions reduced by 74% within a decade (Clapp, 1994, p. 51). Implementation of the smokeless zones was greatly helped by the technology that facilitated production of town gas and availability of cheap oil supplies from across the globe. Coal was dirty and inconvenient inside the house and so environmental improvements were achieved by a mixture of legislation, available technology and consumer choice. On the debit side, town gas production sites were themselves seriously contaminated. Relocation of manufacturing industry from the old industrial countries to the newly industrialising countries has also reduced air pollution in the former. New, high technology industry, though often making massive demands on high-quality water supplies and electricity, does not produce the same kind of noxious emissions into the air that characterised the old smokestack industries. In these respects, then, the cities of the present appear to have overcome the air problems of past urban environments, and to provide lessons in how to cope with today's environmental problems.

Despite these advances, in the richest country of the world, the USA, between 40 million and 75 million people reside in areas failing to meet air quality standards for ozone, carbon monoxide and particulates. In other rich countries national air quality standards and World Health Organisation (WHO) recommended concentration limits are still exceeded in big cities and industrial conurbations, especially for carbon monoxide and nitrous oxides. Upward trends in nitrogen dioxide levels are reported, for example, in London, Frankfurt, Amsterdam, Stockholm and New York, and though national carbon monoxide levels are falling, Toronto, Chicago, New York, Los Angeles and Paris all showed measures exceeding the WHO guideline limit in the period 1980–84 (OECD, 1990, pp. 22–3). Road traffic is at the root of the problem.

4.6   Urban regeneration

One way to make cities more compact, and to absorb growth without consuming further land resources, is to re-use vacant and derelict land. Deindustrialisation has left substantial tracts of redundant land, generally close to the city centre. The results are degraded and depressing environments, which prompt further out-movement by those residents who can choose whether to stay or to move. Thus today's cities frequently show a pattern of **social polarisation**, with the affluent living on or beyond the edge of the city, while the poor, the elderly and the minority ethnic communities live in run-down **inner-city areas** where the quality of services deteriorates because of the limited spending power of the local population. This pattern is most pronounced in the cities of the USA, though the European Union has also identified 'social exclusion' as an issue to be addressed in the cities. In some western European cities the poor also live on the edge of the city, in the large social housing schemes built in the 1960s and 1970s.

The crisis of living conditions in the inner-city environments has been compounded by a weakening of the finances and powers of their local governments. This is not an even process; it is particularly marked in the USA and Britain. In the 1970s and early 1980s several cities in the USA had suffered such a loss in their tax base that they reached the edge of bankruptcy and had to make serious cuts in their spending on services. As the better-off left the cities, and industries relocated to suburbs or closed, the cities lost income, at the same time as in-migration of poor ethnic minorities and rising unemployment and poverty increased the demands on their coffers. New York had debts of over 1.2 billion dollars in the mid-1970s and banks refused to extend loans. To avoid bankruptcy, the city had to cut its payroll by some 20%, thus adding to the social and economic problems of its population.

In Britain the abolition of the Greater London Council and the metropolitan county councils in 1986, and of the Scottish regional councils in 1996, together with drastic reforms of local taxation structures (including the short-lived 'poll tax'), reduced the power of urban local government to intervene in the process of social and economic change. Though planning controls have prevented migration to suburbs, and to free-standing settlements beyond the suburbs, on the scale that has occurred in the United States, the process of social sifting has left behind environments of physical degradation and social despair. The Church of England (1985) called them 'urban priority areas'. One indicator of the restructuring that has taken place (see Section 4.1) is that the approach to such areas has been re-thought. During the 1980s in Britain the emphasis was very much on market-led solutions. In England and Wales the task was taken out of the hands of local government and given to government-appointed Urban Development Corporations. Property development was central to their approach, and they tended to operate on a basis of commercial confidentiality rather than community involvement. The needs of low income groups were only addressed indirectly through the notion that the benefits of new development would eventually 'trickle down' to them.

There was an important shift in emphasis in the 1990s towards the idea of partnership between central government, local government, the private sector and the local community. The depressed conditions of the property market also made it essential to look beyond property development for solutions. The emphasis turned to **urban regeneration**, a term which implies that as well as making physical improvements to the environment, viable social and economic conditions are also needed. Local Agenda 21 – the local action to put into effect the ideas from the Earth Summit in Rio in 1992 – introduced a new concern for sustainability into urban regeneration.

There has also been a change in thinking about what makes a desirable urban environment. New houses are built in traditional styles; the street has been rediscovered as a vital part of an urban environment (moving away from modernist notions of the city summed up in the architect Le Corbusier's dictum – 'Kill the street'). Changed technology and better pollution control make mixed-use areas possible again, whereas in industrial cities town planners had sought to improve environments by segregating housing from shops and industry. Architectural styles have become more eclectic and consciously idiosyncratic – 'post-modernist' – combining components and motifs from earlier architectural traditions.

Such changes create new local urban environments, many of which

▲   *Grocer's Warehouse in the Castlefield area of Manchester, one of the places where the industrial city (see Section 3.7) was born, now renovated for new purposes.*

pose as old ones! Examples are the gentrified neighbourhood, where the middle classes and young professionals seek the charm and identity of old working-class living and working areas. Conversions of warehouses into flats for single people matches demographic, economic, social and cultural changes. There are the **urban heritage areas** based on museums celebrating the former manufacturing industries of the city, and using the abandoned canals for recreational purposes – the Castlefield area of Manchester is one such example. By the 1970s it was a run-down area with derelict canals and buildings, waste ground and unattractive storage and industrial land uses. The city council saw its potential as an urban heritage park and this vision was subsequently backed by funds and marketing from the Central Manchester Development Corporation, one of the Urban Development Corporations set up by government in the 1980s.

Activity 11

Visit such an area if you can. How do you react to this type of urban environment? Do you find that it creates an important sense of history and place, retaining valued urban fabric of a human scale? Or is it a fake, an idealisation of the past, a kind of urban theme park? Try to think through the reasoning behind your feelings, and consider how well such an area fits the criteria of sustainability.

4.7   Global cities

Gentrified dockland neighbourhoods, the desolate inner city, car-based commuter suburbs, out-of-town shopping centres, homeless persons in city-centre streets begging for 'spare change', asthma caused by vehicle emissions, clean water on tap from a privatised water company (for those who can pay): the pieces of the local urban mosaic form a wider, global picture. Another part of that picture is a new role for some cities in the affluent world, a role as **global cities**, the nodes that integrate the global financial structures through which world trade is conducted, the command points in the world economy and in chains of urbanism which link the environments of cities around the planet.

There is disagreement about which cities qualify as global cities, though the claims of London, New York and Tokyo (the three main world stock exchanges which between them, because of their time-zones, allow twenty-four-hour trading) are unchallenged (see Hamnett, 1995). Sassen (1991) suggests that the more globalized economic life becomes, the more concentrated becomes its management in a few key centres. One of the influences in the restructuring of the world's financial system and its urban geography was the third-world debt crisis of 1982, which undermined the leading role of traditional transnational banks. The key factor, though, has been deregulation of national financial markets. The result, argues Sassen, has been to shift the centre of gravity of the financial industry away from large transnational banks and towards 'major *centres* of finance' (p. 19, her emphasis). King (1990) argues that:

> . . . the world city is increasingly 'unhooked' from the state where it exists, its fortunes decided by forces over which it has little control.

> Increasingly the city becomes an arena for capital, the site for specialised operations of a global market. Forced to compete with its major international rivals, obstacles to that competition are, independent of state policies, progressively removed. It is here where the interests of local populations are directly in conflict with, and are sacrificed for, the interests of international capital ... (which) require public sector spending on infrastructural facilities while simultaneously, they look for tax concessions to persuade them to stay.   (King, 1990, pp. 145–8)

The concentration of financial institutions in the centres of the global cities creates a demand for space: for example, there are 350 foreign banks with offices in New York, and 2500 other foreign financial companies. This demand is further fuelled by the need for offices for those who provide the services needed by banks and investment trusts – the lawyers, accountants, and others in what is known as the **producer services sector**. The result is a skyline in which the towers of these financial giants vie for supremacy. The high wages paid to those working in these temples of capital lubricate the gentrification process. Manufacturing simply cannot compete in the market for land and so traditional sources of blue-collar employment are squeezed out, to be replaced by the restaurants, hotels and boutiques required to sustain the global city. These provide substantial employment opportunities, but the jobs are typically part-time, low paid and involve anti-social hours. Thus **social polarisation** is a key feature of the global city. To the extent that ethnic minority migrant populations are likely to seek the kind of casualised low-paid jobs on offer, it is sometimes argued that the global city encompasses within it a 'third world' city.

At best we can try to manage such cities. The notion of planning them, in the sense that an ancient city like Chang'an was planned, no longer has real meaning. Is the global city a sustainable urban form? The crisis of the finance markets in 1987 certainly damaged the property industry in such cities, but the inertia and slow recovery means the cities' role has developed, not disappeared. (See Plate 12.) Would a major collapse of the world monetary system signal the end of such cities? Is the real threat a social one, caused by the juxtaposition of wealth and poverty? The adaptive capacity of the global capitalist system should not be underestimated, as recovery from the oil crises of 1973 and 1978 shows.

4.8   Third world mega-cities

Despite the dramatic changes that the cities of the affluent world are going through, the global urban agenda is now dominated by the urban environments of the third world. United Nations projections suggest that 'third world' urban population will grow by more than 700 million persons between 1990 and 2000, and that 80% of the world's population growth in the decade will be in urban areas (United Nations, 1991).

Part, but by no means all of this growth, is in cities of a staggering size. In 1960, eight of the world's ten largest urban areas in terms of population were in the first world, the other two being Shanghai and Buenos Aires. By the year 2000, eight of the ten biggest cities will be in Asia or South America, and they will be bigger than any cities previously known, with Mexico City an urban area of 30 million people. (See Plate 14.) In 1981 India's urban population was 156 million, but by the end of the century it will be between 350 and 400 million, there will be 20 cities of over 1 million people (bigger than any in the UK except London and Birmingham), and Calcutta, Delhi, Bombay and Madras will all number over 10 million (Girardet, 1992, pp. 74–5). However, some experts do suggest that urbanisation will now slow down, and that the mega-cities will not grow to the scale predicted. Their argument is that the attractions of urbanisation are waning, and government policies could reduce the attractiveness of cities still further (World Commission on Environment and Development, 1987, p. 237).

Though they contribute a large share of the third world's emission of greenhouse gases, relative to their share of the world's urban population, the contribution of cities in poor countries to global environmental problems is relatively small. In contrast, they do more serious damage to residents' health and to local resources and ecosystems than today's rich cities. Nor is rapid urban growth necessarily an unmanageable problem –

Table 7.3   Examples of rapid population growth in third world cities (millions)

City	1950	Latest	UN Projection for 2000
Mexico City	3.05	16.0 (1982)	26.3
São Paulo	2.7	12.6 (1980)	24.0
Bombay	3.0	8.2 (1981)	16.0
Jakarta	1.45	6.2 (1977)	12.8
Cairo	2.5	8.5 (1979)	13.2
Delhi	1.4	5.8 (1981)	13.3
Manila	1.78	5.5 (1980)	11.1
Lagos	0.27	4.0 (1980)	8.3
Bogotá	0.61	3.9 (1985)	9.6
Nairobi	0.14	0.83 (1979)	5.3

Source: World Commission on Environment and Development (1987) *Our Common Future*, Oxford, Oxford University Press, p. 237.

for example, many rich countries successfully constructed substantial new towns to good environmental standards by adapting Howard's idea for Garden Cities.

The issue is not just one of scale, daunting though that is, but also of rates of changes and the causes of change. Rural-to-urban migration is still a major factor in the rapid growth of these third world cities. Rural poverty is the main trigger of such movement, with the hope of better opportunities in the city. As in Europe of the 1830s, though conditions in the cities are bad, conditions in the countryside are worse. Broadly speaking, it is the poorest regions of the world that have the fastest urban growth rates. This means that the environmental issues of the third world city today are inextricably linked with **poverty** and overall population growth. The scale and pace of urbanisation would be difficult to manage even if financial resources were no object, and supplies of trained planners and engineers were unlimited. A dramatic increase in our capacity to produce and manage infrastructure and shelter is needed simply to maintain existing conditions. This is the challenge facing many of the countries already saddled with debt. Something like 1.2 billion urban-dwellers now have access to safe water: in the next generation we need to provide the same facility to another 3.7 billion.

Because it is effectively impossible for some of the world's poorest people to buy or pay rent for an adequate house, the form that much of the urban growth takes is homelessness or unauthorised **'shanty town' settlements** at the edge of the city, a base from which to search for employment. Lowe (1992, p. 130) suggests that between 70 and 95% of new housing in most third world cities is unauthorised. In terms of understanding the dynamics of urban development under these conditions, it is important to recognise that the growth of such areas is not so anarchic as it appears. There certainly have been invasions of land by poor people in some cities, notably in South America, and these tend to be dramatised in the literature, and often involved politicians providing protection in exchange for votes. More generally, a plot in such areas is purchased, though the transaction may be illegal, and consequently the purchaser may lack long-term security. In addition plot sizes are likely to be small, there will probably be no proper provision of water, sewers, electricity or schools (though these may come as the settlement becomes more established), and the lanes will be narrower than those in approved areas.

In short, as with so much of the 'third world' (which is not one world at all), it is difficult to generalise, and the form, development and quality of unauthorised settlements can vary widely. Certainly it would be wrong to suggest that all self-help housing built on authorised sites is inadequate, but equally it is misleading to depict such schemes as a cheap and cheery community-based solution to bad urban and rural environments. Building a house is never a cheap or an easy process.

4.9   *Water in third world cities*

Though water in the home is the best correlate of health status, it is difficult to obtain reliable data on the proportion of urban households able to enjoy a supply of water; official statistics often ignore unauthorised settlements. In Lahore, it is estimated that 60% of houses are connected to the mains, while for Bogotá the figure is 74% (Lowder, 1986, p. 123). In poor areas the figures are much lower: across the Buenos Aires metropolitan area 57% have access to running water, but in low-income communities the figure is

◀ Rimac shanty-town in Lima, Peru: fetching water from standpipes can be arduous and time-consuming.

5% (Hardoy, Mitlin and Satterthwaite, 1992, p. 103). The poor rely on public standpipes, private wells or surface sources, or buy from tanker trucks which tour some low-income areas (and typically this costs more in both money and time – usually women's time – than water from a supply piped to the house). Those denied easy access to water use less. Average consumption of water in litres per capita per day (lcd) is fairly constant between some third world cities (Calcutta 95 lcd, Lagos 85, Cuenca 82), but there is massive variation within the cities. The rich in Lagos and Lahore consume 447 lcd and 336 lcd, respectively, compared with 151 and 40 for citizens with only a single tap, but those reliant on a public standpipe use only 20 lcd, and World Bank projects for the very poor are designed to provide only 15–25 lcd (Lowder, 1986, p. 123).

If rapid urban growth puts pressure on existing supplies of water, whose supply will be sacrificed? Lahore's water, for example, comes from aquifers; Lagos depends on six bore holes. There is the risk that increases in withdrawal rates will allow seepages from nearby contaminated sources and, as aquifers do not have the self-cleansing properties of rivers, once polluted they are difficult and expensive to clean.

Water purification technology is available, but not really an economic proposition in situations where water is (and has to be) supplied at relatively low cost. All this has predictable impacts on health. Poor families without piped water are most likely to drink polluted supplies and hence to fall victim to dysentery and other forms of water-borne diseases. Women and young children are the main victims of these social inequities. Diseases that can be eradicated when there is an adequate water supply and sewage system today cause huge human suffering in third world cities. In Pakistan water-borne diseases cause up to 40% of illnesses and 60% of infants' illness. Diarrhoea is the biggest killer of infants in third world cities.

Out of India's 3119 towns and cities, only 209 had partial and only 8 had full sewage and sewerage treatment facilities. On the River Ganges, 114 cities each with 50 000 or more inhabitants dump untreated sewage into the river every day. DDT factories, tanneries, paper and pulp mills, petrochemical and fertiliser complexes, rubber factories and a host of others use the river to get rid of their wastes. The Hoogly estuary (near Calcutta) is choked with untreated industrial wastes from more than 150 major factories around Calcutta. Sixty per cent of Calcutta's population suffer from pneumonia, bronchitis, and other respiratory diseases related to air pollution. (Centre for Science and Environment, 1983)

If lack of a supply of water is a typical problem in a third world urban environment, lack of sanitation is still more common. This is particularly problematic where cities occupy low-lying, swampy sites in climates that combine high temperatures with dangers of floods. Bangkok, Lagos, Calcutta and Lahore are good examples, though by no means the only ones. The result is that sewage rises to the surface of the street when drains get blocked (a situation made likely by poor maintenance) or when water-levels in the rivers rise above the outfalls of the sewers. Poor areas of such cities typically have open drains or no drains at all; pit latrines and waste ground are used. Disease-bearing human wastes are the most widespread contaminants of water, and the United Nations Global Environmental Monitoring System reports poor and deteriorating surface water quality in many countries (World Bank, 1992, p. 45).

Thus, as cities grow at very rapid rates without adequate sanitation, they face a situation where surface-water supplies are likely to become increasingly unsafe unless treated. Part of the problem is that sewer networks are expensive to install, and that problems are compounded where buildings are dense and lay-outs irregular, as they typically are in unauthorised settlements, for example. Over and above that, there are problems of management and maintenance, compounded by remote bureaucracies and even corruption. Very often provision of water for agricultural use is subsidised. In low-income countries domestic use accounts for only 4% of water withdrawals, industry 5% and agriculture (mainly for irrigation) 91%, yet users, especially farmers, are usually paying less than 10% of the costs of provision (World Bank, 1992, p. 100). As ever, environmental solutions are intricately tied to questions of social equity and state policy.

While domestic effluent is a major issue we should note problems with **industrial waste** too (taken up in *Blunden and Reddish*, eds, 1996). Just one or two factories dumping waste into a river can have devastating impacts on downstream users of that water. Unauthorised settlements frequently develop on land close to factories, and so are avoided by those able to procure a more salubrious environment. Thus the risks are magnified for such residents.

4.10   Air in third world cities

As car ownership is less in developing than developed countries, we might expect fewer problems of air pollution than in the West. At present the situation is patchy. There are serious problems in some third world cities, notably Mexico City, where pollution levels are reported to be six times higher than acceptable thresholds, and in Manila, where the figure is three to four times. (See Plate 15.) Bangkok, Beijing, Calcutta, New Delhi and Tehran are also cited as problems, though less industrialised and less motorised cities have better conditions. The data in Figure 7.4 indicate both the magnitudes and range of urban air pollution for different types of country, as well as the trends. Crudely, they show that things are getting

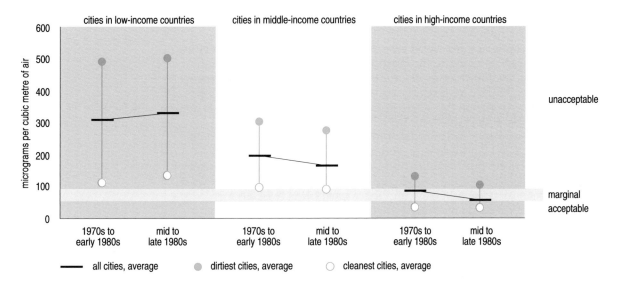

▲ *Figure 7.4   Urban air pollution levels and trends: concentrations of suspended particulate matter across country income groups (micrograms per cubic metre of air)*

Note: Data are for 20 urban sites in low-income countries, 15 urban sites in middle-income countries and 30 urban sites in high-income countries. 'Cleanest cities' and 'dirtiest cities' are the first and last quartiles when ranked by air quality. WHO guidelines for air quality are used as criteria for acceptability.

Source: World Bank (1992) *World Development Report, 1992: Development and the Environment*, Oxford, Oxford University Press, Figure 2.3, p. 51.

better in the rich countries and worse in the poor countries. In interpreting such data it is again necessary to stress the underlying dynamics of third world rapid urbanisation. If air quality continues to deteriorate in cities with growing populations, there will be a double effect – greater pollution levels affecting ever more people.

The consequences of air pollution are illness, particularly respiratory diseases, and premature mortality, with poverty and poor health care exacerbating the risks. One aspect of the problem is that air pollution, and car ownership, which contributes to it, tends to increase with economic growth. Thus if third world cities can achieve development, which might reduce poverty and hence the vulnerability to problems of contaminated water, one of the results could be a deterioration in air quality. This trend is already apparent, for example in cities such as São Paulo in Brazil and Mexico City. More optimistically, newer cars are less polluting than older ones, and an important factor in air pollution in less developed countries at present is poor maintenance and the age of engines in vehicles. For example, the average Pakistani vehicle emits twenty times the hydrocarbons, 25 times the carbon monoxide, and four times the nitrous oxides per kilometre of a US vehicle. If, as seems likely, air pollution will get worse in such countries before it gets better (as happened in the richer countries), the aim should be to minimise the amount of that increase, through proactive transfer of better technologies from the richer countries, for example.

4.11   *Transport in third world cities*

Transport problems in third world cities are likely to get worse as the cities grow, and local elites aspire to the lifestyle of the rich countries. In this respect, at least, the big cities in all countries seem to display a convergence, as all try to cope with extreme traffic congestion. Employment and shops tend to be concentrated in the centre of the city, while housing spreads out for vast distances. Public transport undertakings tend to be weak and inefficient. For example, in Lahore, a fleet of Volvo buses is usually out of commission due to lengthy repairs, and public transport of all types contributes only one-third of all motorised trips, and nearly all of these are by small, inefficient minibuses which can be a serious hazard to other road-users. Private enterprise provision of public transport is typically unregulated and often dangerous; it offers a variety of means of travel: minibuses, taxis, rickshaw-style vehicles driven by small power/high pollution engines, horses and carts.

Although the main means of travel are by foot or by bicycle, the public agencies aspire to invest in improved facilities for motorists, with engineers drafting ambitious plans for motorways, for example. The problem is that road transport is breaking down in many such cities. Not only are the roads literally crumbling under the weight of vehicles using them, but jams are a frequent experience, in part because of the lack of segregation of different types of traffic, as motorised vehicles jostle with animal-drawn carts and hand-drawn carts for the limited road space. Lack of parking controls and the equation between poor maintenance and vehicle breakdown exacerbate the problems. As well as frustration, accidents and delays to the emergency services are part of the costs. It seems clear that the current transport

▲   *A congested street in Ahmadabad, Gujarat, India, with a mixture of pedestrians, scooters, cars and motorised rickshaws.*

▲   *Coping with the increase in car ownership means massive investment in infrastructure, as here in São Paulo, Brazil.*

trajectory of most rapidly growing third world cities is not sustainable: can technology, subsidy or road pricing, traffic management, and political will create a sustainable alternative?

Activity 12

Try to draw up a list comparing the early industrial cities of Europe in the nineteenth century with cities in the developed world and third world today. You can use the kind of criteria used in this chapter, such as air and water quality, transport and the scale of the city, global relations and impact etc.

4.12   *Summary*

●   We seem to be at a major threshold in the rate, scale and pattern of urban development.

●   The new international pattern of location of labour-intensive manufacturing has changed the role of cities in the more developed countries, leaving a legacy of contaminated land from the industrial era.

●   Cities in the developed world are becoming more low density, more dispersed and more reliant on the car.

●   Cities in the developed world consume resources and contribute to pollution globally, not just locally.

●   The use of motor vehicles now underpins the form of rich cities but also threatens their sustainability.

●   Global cities have emerged as command centres in the global economy, but with sharp internal social divisions.

- The major urban challenge is in the very rapidly growing third world cities.
- Rural-to-urban migration and rural poverty are key factors driving this urban growth, and though conditions are bad in the cities they are worse in the countryside.
- Most new housing in third world cities takes place in unauthorised areas where services are likely to be minimal at best, though these may later be regularised and services improved.
- There are major problems of water supply and water pollution, which are suffered mainly by the poor, and which will get worse with further growth unless remedial action is taken.
- Air pollution varies with degree of car use and industrial development, but again seems likely to get worse before it gets better.
- Transport arrangements in third world cities also threaten their sustainability.
- Large cities everywhere show signs of convergence – they are managed not planned, multi-ethnic with significant poverty populations living in hostile environments, and suffer traffic congestion.

5   Urban environments in the future

As we peer into the twenty-first century, what do we see? The aim of this final section is to summarise the lessons about urban environments from the preceding sections and to use these to speculate about what the future might hold. In addition it tries to highlight some initiatives that may point the way towards a more sustainable urban future.

5.1   Lessons from urban environments

What lessons emerge from this review of 5000 years of urban environments across the globe? As with all summaries and generalisations, the list is selective and subjective.

> *Activity 13*
>
> Try to make your own list of the lessons of urban development, before comparing it with our list of five points below.

1   Technology is important but it is only part of the story.

Throughout history, technology has helped to make urban living possible. The concentration of vast numbers of people in one place has many economic, social and cultural advantages, but it ultimately depends on technologies of transport and the provision of water and removal of waste and effluent. Building technologies made possible ever denser concentrations of people and activities, especially in the twentieth century through multi-storey construction techniques. Transport technologies made possible the spread of the city. Engineering offers ways to improve urban life still further, to make it more sustainable. For example, improved engine and exhaust designs have reduced vehicle emissions, and the telephone and the computer link have reduced the need to travel. Sewage treatment technology allows water to be recycled safely. Skills in building conservation and energy-efficient building design can make cities more sustainable places.

However, technology alone does not solve the problems. As you read this, children are dying in third world cities from water-borne diseases which we have had the ability to eradicate for over a century. To save the cities and their inhabitants we need the economic structures and political will to apply appropriate technologies.

2   Cities have relied on the exploitation of resources from far beyond their boundaries and on dispersal to dispose of their waste products.

From the earliest times cities have relied on a surplus of food being available from a surrounding area. Today they draw their water, energy, building materials and most other resources from across vast distances. They dispose of waste into the wider environment, too. It may still be possible to draw resources from afar, as long as these are renewable resources, but prevention or treatment at source will increasingly be necessary with respect to waste products and pollution. Cities invade local and global environments.

3   When trying to understand cities, we have to see them as part of one world.

Not only are cities part of extended networks for the appropriation of the resources they need and the disposal of their waste products, they are interconnected economically and politically. Cities have always had relationships with other cities; even the independent Greek city-states spawned their colonies and fought other cities. Colonialism created a pattern of global urban dependence that has become even more complex in the age of world trade, deregulated finance, new trading blocs and the new international division of labour. The New York commodity dealer in her Wall Street skyscraper is linked not just to the Filipino waiter who serves her lunch down by Battery Park, but to the family who leave their village in the Punjab and head to the *katchi abadi* in Faisalabad in the hope of getting work in the cotton mills. Locally and globally, cities are about domination of environments, people and other cities.

4   Cities are arenas of opportunity, culture, power and excitement.

Cities have flourished because people have wanted them. Certainly over the last two centuries rural people have voted with their feet and moved to the city in the hope of bettering their lives and those of their families. The opportunities were not just economic; the shackles of convention and conservatism from the village could be shaken. Urban environments have been fundamental to human liberation.

The city is potentially the highest expression of human achievement. It has the complexity of the computer, the beauty of a work of art; it brings together all sorts of different people and cultures and moulds them into something more than themselves. It is an environment which fosters learning and invention, creates wealth and culture. In many ways the urban legacy is itself a remarkable testimony to the achievement of sustainability – what wandering nomad five thousand years ago, eking out a subsistence existence, could possibly have believed that edifices could have been created to house millions of people and endure hundreds of years? It is little wonder that through time the city has held magical and symbolical meanings, celebrating the genius and power of those who lived in it, even if indirectly through their gods. From the palaces of Chang'an to the Government House of the colonial city, from the awe-inspiring mosques of Lahore to monuments to the proletariat in the Soviet bloc, and the cathedrals to capital that make the skyline of Manhattan so thrilling, the city has celebrated the power of human achievement.

5   *Good government and imaginative urban planning can move us towards more sustainable cities.*

What would a city celebrating sustainability look like? How might we move towards it? Section 2.1 stressed investment and collective responsibility as key factors in creating and managing urban environments. We can thus discern the basis for more sustainable cities. What is critical is a coherent development strategy to relate land use and transport, backed by a management approach which capitalises on the connections between ecology, safety, employment and the enjoyment of urban environments. To give one trivial example, solid waste collection and recycling create employment, are ecologically sensible and sustain healthy, attractive living environments. The task is to devise the flows of investment and the political structures to deliver such schemes. Urban parks and tree-planting can help air quality, reduce run-off, provide access to nature, visual delight and entertainment to people without barriers of cost, but to do this they must be adequately maintained and policed, activities which create employment.

Steps towards increasing the social sustainability begin with awareness of the needs of disadvantaged groups in the urban environment – those who are poor, young, old, disabled, female or members of minority ethnic groups. Barriers to their involvement in decisions about urban change have to be broken down. There are no easy answers, but there are some encouraging examples, as Section 5.4 will show.

Sustainability indicators, discussed in Section 2.2, have an important role to play. Despite the debates about which indicators should be chosen, the process of setting indicators and monitoring is a continuing public reminder of the need to move towards sustainability. Indicators show whether things are getting better or getting worse, and can be linked to targets for achieving improvements.

In short, urban planning has a critical role to play in moving urban environments towards sustainability. Such planning needs to be based on ecological principles and a commitment to equal opportunities. It needs to set a clear strategy, encourage demonstration projects that show the way to better environments, use action research to try out and assess new initiatives, and to back all these with monitoring of sustainability indicators.

Activity 14

You have time-travelled to a city in the year 2025. Write a letter to the folks back home telling them about it.

5.2   *A pessimistic future*

It is easy to be pessimistic about urban prospects for the next century. Many movies have painted a picture of the future city as a dystopia, the opposite to a Utopia. In the film *Bladerunner*, for example, executives commute to gigantic office towers in descendants of aeroplanes, while, in the streets below, the multi-ethnic workforce labours in back-street industries. There are abandoned houses occupied by bizarre urban drifters, as the infrastructure once shared by all city residents, but now needed only by the poor, decays:

> Given the distribution of incomes, given the foreseeable availability of resources – national, local, and worldwide – given present technology and given the present weakness of local government and the lack of interest of national governments in settlement problems, I don't see any solution for the Third World city.
>
> Third World cities are and they will increasingly become centres of competition for a plot to be invaded where you can build a shelter, for a room to rent, for a bed in a hospital, for a seat in a school or in a bus, essentially for the fewer stable adequately paid jobs, even for the space in a square or on a sidewalk where you can display and sell your merchandise, on which so many households depend.   (Hardoy, 1985, quoted in World Commission on Environment and Development, 1987, p. 239)

If we look more closely at trends in cities in the developed and less developed countries, and at the webs of global power into which they are locked, it is easy to project a scenario where there is so little that is sustainable that there are significant breakdowns. In the affluent world the city can already seem passé, the repository of the poor, and of those commercial and economic activities which inertia has not yet decentralised. The telecommunications technology privatises entertainment: why go downtown to the theatre or gallery when you can explore the world and beyond it in virtual reality in your home? Why go to lecture rooms when the university comes in texts, CD-ROMs, TV programmmes, computer-conferencing and over the telephone? The technology allows instant contact with people and access to files of data previously banked in buildings that resemble filing cabinets in their outward appearance. As capital finds ever more ingenious ways of replacing jobs by machine activities, and as jobs can be done from home or are not tied to any particular location, we could see the city as a residualised area, from which investment is withdrawn. If costs are less and returns better in greenfield locations favoured by higher spending, more reliable customers, utility companies may decide – on impeccably objective commercial criteria, and with only their shareholders' interests at heart – to cease to invest in renewing and upgrading obsolete urban infrastructure. The cities would become a kind of compound, a physical expression for social exclusion, no-go areas for those with jobs and income, because of fear of crime or the 'old' diseases; these, such as virulent and resistant forms of tuberculosis, are re-emerging as the public health provisions of the 'nanny state' are progressively withdrawn, to cut costs and allow for new investment in state-of-the-art fibre-optic technologies and tax inducements required by the wealth creators.

Meanwhile the mega-cities of the third world continue to swell as millions flock into them, and industry booms, drawn by the lax environmental controls – pity about the occasional mass fatalities consequent on leaks of poisonous substances, or the less visible pattern of increasing mortality rates due to long-term poisoning of water and air supplies. Water supplies are exhausted and become a focus for violent conflicts both internationally (as nations fight for their right to abstract water from transnational rivers) and locally as a community is accused of contaminating another's supply. Movement through the cities becomes impossible as the streets are permanently gridlocked, a situation made worse by abandoned vehicles as drivers collapse in the intense air pollution. Eventually people return to the fields to re-learn the skills of their ancestors in an attempt to secure a subsistence living from the land. As oceans rise due to global warming, mega-cities at the mouths of major rivers become flooded and abandoned.

5.3   *An optimisitic scenario*

Need it be so? Hopefully not, though if we are to avoid a dystopia we need to change the way our cities are at present. A more optimistic scenario might read something like this.

Back in the 1990s we all underestimated how quickly and dramatically attitudes could change, and despite all the lessons of history, we failed to appreciate just how dramatic could be the pace of technological change, and how adaptive and resilient capitalism was as a system. Of course, our urban environments are different to what they were in those days, but different for the better. Already more people in the wealthy world were living in small towns, but we did not appreciate just how liberating and environmentally friendly would be the impact of the new computer and telecommunications technologies. The rural revival brought young families and life to many remote regions that had for so long depended only on tourism and agricultural subsidies for income. But these were quite new types of settlements, each carefully self-sustaining, bartering environmental inputs and outputs with one another, and recycling all sorts of materials. The scale of these settlements is such that people can safely and conveniently walk or cycle around them, while the older people or the disabled have their special electrical buggies and no obstacles to exclude them from any part of the settlement. And all this has created so many new opportunities, not just in the local environmental management companies which look after each place, and in the new technologies of pollution control, but in education and caring for each other. Of course, incomes are not as high as before, but nor are our expenditures as we no longer need cars; we use less energy and waste less. Land and house prices are also cheaper in today's small towns than they used to be when most people had to live in a crowded city.

The old industrial cities have changed, too. They are so much greener now, thanks to the new urban parks programme. The rivers with their promenades look beautiful, and many tourists go on fishing holidays. The electric buses feed passengers into the light railway stations and the quality of public transport is so good that almost everybody uses it, so it is cheap to run. Of course, some people do still drive cars, but these are not like the old gas-guzzling polluters of the last century, and if people are prepared to pay the costs to use the road (two old 'pounds' a minute in central London), then I guess they really do need the use of the car. The magazine *The Big Issue* went out of business when government and local business started putting money into social housing instead of relying on people buying magazines to help what we used to call 'the homeless'.

But perhaps the greatest success story is in the third world. The massive technology transfer programme organised by the UN and the writing off of debt helped a lot, but in the end it was probably the fairer approach to trade that did most to take people out of poverty and thus to make it possible for them to afford not to abuse the environment. Rural consolidation and development programmes and literacy outreach initiatives have made rural life more attractive than could have been imagined in the days when we thought that cities might grow to sizes of 20 or 30 million people. As international companies insisted on level playing-fields and saw the need to care for the planet, environmental controls became uniform across the globe, with the rich countries making reparation payments to the poor for their previous depletion of world resources. Last but not least, the transition to sustainability was made possible by the attitudes of mutual support embedded in kinship networks in so many third world countries and by the leading role that women came to play in development.

5.4   Pointers for the future

These scenarios are fantasies, though they are not entirely removed from current debates about development and environmental policy. But are there concrete signs of progress that we can point to and build upon? There may be some in your town, or you may have read of local initiatives that link to the global agenda. Here are four to stimulate ideas, discussion and hope.

Box 7.3   Ecolonia: sustainable housing development as part of a National Environmental Policy Plan

In the Netherlands there are a number of energy-efficient demonstration housing developments which are being monitored and the results fed into housing policy and to the building industry. Ecolonia, near the town of Alphen an der Rijn, is a particularly interesting example. Part of the National Environmental Policy Plan, it is putting sustainable development into practice. It is supported by central government, but the Netherlands Municipalities Building Fund is responsible for the development. House prices start at around £80,000 and go up to £125,000, showing that the ideas are relevant to the mainstream housing market.

There are 101 'green' houses of varied design, constructed to test out different building technologies. Particular attention is paid to the way in which aspect can aid energy-saving. The lay-out does not exclude cars, but the paving, and the way houses or gardens open onto the routes, clearly indicate that this is a pedestrian environment. The centre piece of the design is an attractive reed-fringed lake. It is the reservoir for rainwater run-off from the houses, creating a closed water system, so the level varies with the seasons, and there is variety in flora and fauna.

Box 7.4   Český Krumlov: using culture, ecology and market forces for urban regeneration

The Southern Bohemian town of Český Krumlov has a population of 12 000, and, like many small Czech and Moravian towns, it has a marvellously intact historic core, with some 300 historic houses of Gothic, Renaissance and Baroque origin, a chateau (which includes an excellently preserved Baroque theatre), and a church which dominates the skyline. It has a spectacular setting on a deep meander on the River Vltava, and since 1992 has figured on UNESCO's list of 300 architectural sites of world importance.

A paper mill located six kilometres south of the town was a major employer under the communist regime, but was also a major source of pollution of both the river and the atmosphere because of its old technology and reliance on brown coal. More industry developed to the north of the town in the 1980s.

After 1989 the town council saw new possibilities for the town, and since then has tried to fashion a regeneration based on culture, ecology and public–private partnership. In February 1992 the town council founded the Český Krumlov Development Fund Limited, providing it with over 50 strategically important buildings in the town centre, including two major hotels, a brewery, shops and buildings used as offices. The Fund is an independent business, separate from – but owned by – the council. Its main aims are 'to manage, rent, lease, form joint-ventures or sell the properties received from the town to the benefit of the town and its citizens, in accordance with guidelines set by the Town Council'. Foundations have been set up linked to the Fund. The Social Foundation aims to help those with low incomes and disabilities, and one of its first actions has been to build a hostel for homeless people. The Cultural Foundation will support 'quality cultural activities that are less

commercially attractive'. The Sports Foundation is supporting sports teams for young people. The Schools Foundation will 'support ecological education of people, teaching of arts, with special attention to national heritage protection and the restoration of art works'. Industrial pollution has been tackled. The Council has also given the stigmatised gypsy minority jobs keeping the streets clean.

Box 7.5   The Orangi Pilot Project: community involvement to improve sanitation

In the early 1980s Akhter Hameed Khan, a community organiser, began working in the slums of Karachi. He asked the residents what problem he could help with, and was told 'the streets were filled with excreta and wastewater, making movement difficult and creating enormous health hazards'. They wanted a traditional sewerage system and asked Dr Khan to persuade the Karachi Development Authority (KDA) to provide it for free. When his petitions failed, he worked with the community to find alternatives.

The Orangi Pilot Project (OPP) was set up with a small amount of core external funding. The task was to reduce the provision of the sewerage facility to affordable levels and to establish an organisation capable of delivering and managing it. Thanks partly to the elimination of corruption, and to the provision of labour by community members, the cost for an in-house sanitary latrine and house sewer on the plot and underground sewers in the lanes and streets was less than $50 per household.

The OPP staff explained the benefits of sanitation, and provided technical help. The households paid their share of the costs, did their part of the construction and elected lane managers, typically representing groups of 15 households. Lane committees in turn elected members of neighbourhood committees (600 households) to manage the secondary sewers. As the powers of OPP grew through their achievements, they were able to persuade the municipality to provide funds for the construction of trunk sewers.

The OPP has helped to provide sewerage services for 600 000 poor people, and has established a model that is being replicated in other Pakistani cities, with residents being contracted to undertake the crucial tasks of maintenance of the system once it has been installed, a much more efficient practice than reliance on centralised and bureaucratic agencies.

Sullage streams in Orangi, 1985.

Enclosed drains being laid, Faisalabad.

Box 7.6   Curitiba: integrated planning of rapid urban growth

Curitiba, in the south of Brazil, is a city of over 1 million, growing rapidly through natural increase and migration from the surrounding countryside. Unlike so many of the cities in Brazil, it has managed to grow without environmental degradation. The reason is the town planning approach adopted, which has integrated land use and transport systems. Growth has been directed to high-density corridors with full public transport services. This has avoided the kind of traffic congestion and air pollution found in many other cities.

A view of the city.

The *favelas*, the shanty towns on the edge of the city, are still increasing, however. Curitiba gives poor residents of these areas a public transport voucher in exchange for a bag of litter and rubbish. Revenues saved through this system of solid waste collection are donated to community organisations to provide jobs in neighbourhood renewal and recycling. People's relations with their city are seen as important, too. Traditional buildings and streets are cherished, areas are designed for the pedestrian, more trees are being planted. Perhaps most important is the fact that Curitiba has set out clear priorities, to be an ecological city and one that is humane for all its people. That is the starting-point in creating and managing better urban environments.

The streets in Curitiba have separate car and bus lanes, and large 'bendy' buses to provide an efficient mass public transport system.

References

ABERLEY, D. (ed.) (1994) *Futures by Design: the practice of ecological planning*, Gabriola Island, British Columbia, New Society Publishers.

ASHTON, J. (ed.) (1992) *Healthy Cities*, Milton Keynes, Open University Press.

ASHWORTH, W. (1954) *The Genesis of Modern British Town Planning*, London, Routledge and Kegan Paul.

BATER, J. H. (1980) *The Soviet City: ideal and reality*, London, Edward Arnold.

BENEVOLO, L. (1980) *The History of the City*, London, Scolar Press.

BLOWERS, A. (ed.) (1993) *Planning for a Sustainable Environment: a report by the Town and Country Planning Association*, London, Earthscan Publications.

BLUNDEN, J. and REDDISH, A. (eds) (1996) *Energy, Resources and Environment*, London, Hodder and Stoughton/The Open University (second edition) (Book Three of this series).

BRIGGS, A. (1968) *Victorian Cities*, Harmondsworth, Pelican.

CENTRE FOR SCIENCE AND ENVIRONMENT (1983) *The State of India's Environment: a citizens' report*, New Delhi, Centre for Science and Environment.

CHURCH OF ENGLAND (1985) *Faith in the City: report of the Archbishop of Canterbury's Commission on Urban Priority Areas*, London, Christian Action.

CLAPP, B. W. (1994) *An Environmental History of Britain since the Industrial Revolution*, London, Longman.

COMMISSION OF THE EUROPEAN COMMUNITIES (1990) *Green Paper on the Urban Environment*, (EUR 12902 EN), Brussels, Commission of the European Communities.

EVELYN, J. (1661) *Fumifugium: Or the Inconvenience of the Aer and Smoake of London Dissipated together with some Remedies humbly proposed* (reprinted by The National Society for Clean Air, London, 1961).

FOSTER, J. (1977) *Class Struggle and the Industrial Revolution: early industrial capitalism in three English towns*, London, Methuen.

GILBERT, A. (1981)' Urban development in a world system', in Gilbert, A. and Gugler, J. (eds) *Cities, Poverty and Development: urbanization in the third world*, Oxford, Oxford University Press, pp. 11–26.

GIRARDET, H. (1992) *The Gaia Atlas of Cities: new directions for sustainable urban living*, London, Gaia Books.

GLOBAL TOMORROW COALITION (1986) 'Sustainable development and how to achieve it', WCED Public Hearing, Ottawa, 26–27 May.

GODWIN, G. (1859) *Town Swamps and Social Bridges*, London.

HAGUE, C. (1984) *The Development of Planning Thought: a critical perspective*, London, Hutchinson.

HAGUE, C. (1990) 'Planning and equity in Eastern Europe: raking through the rubble', *The Planner*, 30 March, pp. 19–21.

HALL, P. (1988) *Cities of Tomorrow: an intellectual history of urban planning and design in the twentieth century*, Oxford, Basil Blackwell.

HAMNETT, C. (1995) 'Controlling space: global cities', in Allen, J. and Hamnett, C. (eds) *A Shrinking World? Global unevenness and inequality*, Oxford, Oxford University Press/The Open University.

HARDOY, J., MITLIN, D. and SATTERTHWAITE, D. (1992) *Environmental Problems in Third World Cities*, London, Earthscan Publications.

HARDOY, J. (1985) International Institute for Environment and Development, WCED Public Hearing, São Paulo, 28–29 October.

HOWARD, E. (1970) *Garden Cities of To-Morrow*, London, Faber and Faber. (Originally published in 1902; earlier version entitled *To-morrow: a peaceful path to real reform* published in 1898.)

JACOBS, J. (1994) 'Natural and city ecosystems: thoughts on city ecology', in Aberley, D. (ed.), pp. 41–3.

KENWORTHY, J. R. and NEWMAN, P. W. G. (1990) 'Cities and transport energy: lessons from a global survey', *Ekistics*, No. 344–5, pp. 258–68.

KING, A. D. (1990) *Global Cities: post-imperialism and the internationalization of London*, London, Routledge.

LOCAL GOVERNMENT MANAGEMENT BOARD (1994) *Sustainability Indicators Research Project: report of Phase 1*, Luton, Local Government Management Board.

LOWDER, S. (1986) *Inside Third World Cities*, London, Routledge.

LOWE, M. D. (1992) 'Shaping cities', in Brown, L. R. *et al.*, *State of the World, 1992: a Worldwatch Institute report on progress toward a sustainable society*, London, Earthscan Publications, pp. 119–37.

MASSEY, D. and MCDOWELL, L. (1994) 'A woman's place?', in Massey, D. (ed.) *Space, Place and Gender*, Cambridge, Polity Press.

MERRETT, S. (1994) 'New age of planning', *Town and Country Planning*, June, pp. 164–5.

MOHOLY-NAGY, S. (1968) *Matrix of Man: an illustrated history of urban environments*, London, Pall Mall Press.

MUMFORD, L. (1961) *The City in History*, Harmondsworth, Penguin.

ORGANISATION FOR ECONOMIC CO-OPERATION AND DEVELOPMENT (1990) *Environmental Policies for Cities in the 1990s*, Paris, OECD.

REES, W. E. (1992) 'Ecological footprints and appropriated carrying capacity: what urban economics leaves out', *Environment and Urbanization*, Vol. 4, No. 2, pp. 121–30.

ROSELAND, M. (1992) *Toward Sustainable Communities: a resource book for municipal and local governments*, Ottawa, Canada, National Round Table on the Environment and the Economy.

ROSELAND, M. (1994) 'Ecological planning for sustainable communities', in Aberley, D. (ed.), pp. 70–8.

SARRE, P. and BROWN, S. (1996) 'Changing attitudes to Nature', Ch. 3 in Sarre, P. and Reddish, A. (eds).

SARRE, P. and REDDISH, A. (eds) (1996) *Environment and Society*, London, Hodder & Stoughton/The Open University (second edition) (Book One of this series).

SASSEN, S. (1991) *The Global City: New York, London, Tokyo*, Princeton, NJ, Princeton University Press.

SILVERTOWN, J. (1996) 'Ecosystems and populations', Ch. 7 in Sarre, P. and Reddish, A. (eds).

UNITED NATIONS (1991) *World Urbanization Prospects 1990: estimates and projections of urban and rural populations and of urban agglomerations*, New York, United Nations, ST/ESA/SER.A/121.

WILLIAMS, J. F. (1983) 'Cities of East Asia', in Brunn, S. D. and Williams, J. F. (eds) *Cities of the World: world regional urban development*, New York, Harper & Row, pp. 408–50.

WORLD BANK (1992) *World Development Report, 1992: development and the environment*, Oxford, Oxford University Press.

WORLD COMMISSION ON ENVIRONMENT AND DEVELOPMENT (1987) *Our Common Future*, (Brundtland Report), Oxford, Oxford University Press.

Further reading

BLOWERS, A. (1993) *Planning for a Sustainable Environment: a report by the Town and Country Planning Association*, London, Earthscan Publications.

HALL, P. (1988) *Cities of Tomorrow: an intellectual history of urban planning and design in the twentieth century*, Oxford, Basil Blackwell.

HARDOY, J., MITLIN, D. and SATTERTHWAITE, D. (1992) *Environmental Problems in Third World Cities*, London, Earthscan Publications.

Answers to Activities

Activity 4

Chang'an demonstrated typical features of an ancient city. It was walled, and had a central area enclosed within its own wall, where the key buildings housing the government and religious leaders were to be found. There was a clear plan to the city, which expressed the spiritual and cultural values of the society, while also representing the dominance of an elite within the built form of the settlement.

Activity 5

The Garden City sought a sustainable but flexible form of settlement, keeping everyone in reach of the countryside but also able to enjoy the benefits of city living. He saw what he called 'The Vanishing Point of Landlords' Rent' as a self-sustaining, self-regulating mechanism. The key to Howard's vision was local management and self-government, central themes in current thinking about social equity.

There are also ways in which they differ: Howard assumed the Earth's resources as 'abiding forever', showed no concern for the impact on nature, and no consideration for the depletion of energy and other natural resources (Merrett, 1994).

Activity 6

Points that might be considered are the extent to which the cities, already in a privileged location, widened the gap between themselves and those living in the countryside; though were the cities themselves neglected in favour of an extractive, rural-based colonial economy of primary products? Did colonialism bring the blessing of new sanitary technologies, or simply destroy existing urban communities? Health standards were appalling in the colonial cities, but they were not much better back in the industrial cities in the colonialist countries. What, too, about the structure of the city, its ethnic divides and also the institutions for managing urban development?

Acknowledgements

Grateful acknowledgement is made to the following sources for permission to reproduce material in this book:

Cover
Front cover, clockwise from top right: Stephen Best, Dr. David Snashall; Ed Buziak/Farming Information Centre, NFU; Farming Information Centre, NFU; Ed Buziak/Farming Information Centre, NFU; David Sims/ICCE; Roy Lawrance; Martin Bond/Environmental Picture Library; Alan Gilbert; *centre*: J. Goodman/Farming Information Centre, NFU; *back cover*: Information Service of the European Community.

Colour plate section
Plate 1: Patrick McClay/OXFAM; *Plate 2:* Sally and Richard Greenhill; *Plate 3a:* Steven C. Wilson/ENTHEOS; *Plates 3b and 4:* Mark Boulton/ICCE; *Plates 5 and 6:* Sally and Richard Greenhill; *Plates 7 and 8:* Tony Stone Photolibrary, London; *Plate 9:* Copyright © Geoffrey Sinclair/Environment Information Services; *Plates 10 and 11:* Peak District National Park; *Plate 12:* London Docklands Development Corporation; *Plate 13:* Cliff Hague; *Plate 14:* Copyright © John Goldblatt/Mexicolore; *Plate 15:* Copyright © Alicia Sanchez/Mexicolore.

Figures and text
Figures 1.1 and 1.8: Findlay, A. (1994) 'Population doubling times' and 'Life expectancy' in Unwin, T. (ed.) (1994) *Atlas of World Development*, John Wiley and Sons Ltd. Reprinted by permission of John Wiley and Sons Ltd; *Figure 1.2:* Barry, R. G. and Chorley, R. J. (1968) *Atmosphere, Weather and Climate*, Methuen & Co; *Figure 1.6:* Wood, A. and Clarke, J. *et al.* (eds) (1984) *Population and Development Projects in Africa*, Cambridge University Press; *Figures 1.7(a) and 1.11:* Findlay, A. and Findlay, A. (1987) *Population and Development in the Third World*, Methuen & Co.; *Figure 1.7(b):* Woods, R. (1987) *Theoretical Geography*, Methuen & Co.; *Figure 1.9:* Findlay, A. and White, P. (1986) *West European Population Change*, Croom Helm; *Figure 1.10:* Gibb, A. (1988) 'The demographic consequences of rapid industrial growth', *Occasional Papers*, Vol. 24, Department of Geography, University of Glasgow; *Figure 2.1:* Reproduced with permission from Professor P. D. Harvey, Literary executor of the late Professor C. D. Darlington; *Figure 2.3:* from *Physical-Geographical Atlas of the World*; *Figure 2.4:* Lydolph, P. E. (1985) *The Climate of the Earth*, Rowman and Littlefield; *Figure 2.5: The Times Atlas of the World*, copyright © 1968, Bartholomew; *Figure 3.3:* Redrawn from original illustration by Patricia J. Wynne from M. S. Swaminathan, 'Rice', in *Scientific American*, January 1984, International Edition; *Figure 3.5:* Redrawn from original illustration by Andrew Tomko from M. S. Swaminathan, 'Rice', in *Scientific American*, January 1984, International Edition; *Figure 4.1:* Erlichman, J., Consumer Affairs Correspondent, *The Guardian*, © James Erlichman, 1989; *Figure 4.4: Countryside Commission News*, Part 23 1986 and Part 29 1987, Countryside Commission; *Figure 4.6:* CAB International, Wallingford, Oxon; *Figure 4.12:* 'Countryside Conflicts' from 'In', 1986, Gower Publishing Ltd; *Figure 4.13:* Institute of Terrestrial Ecology, © NERC, 1987; *Figures 4.14 and 4.15:* Guardian News Services Ltd; *Figure 5.1:* Based upon the 1975 Ordnance Survey 1:1250 000 © Crown Copyright. Reproduced with the permission of the Controller of Her Majesty's Stationery Office; *Figure 5.3:* UK Centre for Economic and Environmental Development; *Figure 5.4:* Courtesy Terry Farrell & Partners; *Figure 5.5:*

United Kingdom Environmentally Sensitive Areas 1994, © Crown Copyright. Reproduced with the permission of the Controller of Her Majesty's Stationery Office; *Figure 5.6: The Sunday Times* © Times Newspapers Ltd; *Figure 6.1:* Source: US Department of Transportation (1981) *The US Automobile Industry, 1980*, Washington, DC, USGPO, p. 57. Photographed from Dicken, P. (1986) *Global Shift: industrial change in a turbulent world*, Harper & Row, p. 304; *Figure 6.2:* Mansell Collection; *Figure 6.3:* Delpeuch, B. (1992) *Seed and Surplus: an illustrated guide to the world food system*, translated by Finney, C. and Smith, A. (1994) Farmers' Link, English translation © Catholic Institute for International Relations (CIIR) 1994; *Figure 6.4:* Vidal, J. (1991) 'Global conservation threatened as Gatt declares war', *The Guardian*, 6 September 1991; *Figure 6.6:* The Ecologist (1993) *Whose Common Future?*, Earthscan Publications Ltd; *Figure 6.8:* 'Producer subsidy equivalents' from 'The Gatt negotiations on agriculture: what are the implications for developing countries?', *CAP Tales*, September 1993, Farmers' Link; *Figures 6.9, 6.10 and 6.11: Pocket Planet Earth Factfile*, (1995) © HarperCollins Cartographic 1995; *Figure 6.12:* Vidal, J. (1995) 'The power of the flower', *The Guardian*, 24 May 1995; *Figure 6.13:* Vidal, J. (1994) 'Trade marks', *The Guardian*, 17 June 1994; *Figure 7.1:* Adapted from Girardet, H. (1992) *The Gaia Atlas of Cities*, Gaia Books Limited; *Figure 7.2(a):* Howard, E. (1970) *Garden Cities of To-Morrow*, Faber and Faber; *Figure 7.3:* Hall, P. (1992) *Urban and Regional Planning*, pp. 194–5, Routledge; *Figure 7.4:* World Bank (1992) *World Development Report, 1992*, Figure 2.3, Oxford University Press, © 1992 The International Bank; *Box 6.1:* 'The power of the transnationals', *The Ecologist*, 1993, Earthscan Publications Ltd.

Photographs and cartoons
p.9: (left) Courtesy of the Royal Danish Ministry for Foreign Affairs/Foto: Lars-Kristian Crane, *(right)* Maggie Murray/Format; *p.19:* © D. Ross/ OXFAM 1984; *p.21:* Punchline by Christian, *New Internationalist*, © Christian 1988; *p.25: (both)* Maggie Murray/Format; *p.27:* © Andes Press Agency/Carlos Reyes; *p.30:* The Glasgow Room, The Mitchell Library, Glasgow; *p.31:* The Independent/Brian Harris; *p.32:* Alan Gilbert; *p.44:* Maggie Murray/Format; *p.47: (top)* Institute of Agricultural History and Museum of English Rural Life, University of Reading, *(bottom)* Farmers' *Weekly* Photo Library; *p.48:* Farmers' *Weekly* Photo Library/Peter Allen; *p.58:* Sharma Studios; *p.69:* © Andes Press Agency/Carlos Reyes; *p.73:* Courtesy of Professor R. B. Bryan, Soil Erosion Laboratory, Scarborough College, University of Toronto; *p.74:* Jeremy Hartley/ OXFAM; *p.76:* Christian Aid; *p.77:* Jeremy Hartley/OXFAM; *p.80:* © Patrick Sutherland; *p.81:* Punchline by Christian, *New Internationalist*, © Christian 1989; *p.100:* © Andes Press Agency/Julian Filokowski; *pp.102, 112, 119 (left):* Sally and Richard Greenhill; *p.119 (right):* Sally and Richard Greenhill/Society for Anglo-Chinese Understanding; *p.129:* Sheila Gray/Format; *p.133:* Farmers' *Weekly* Photo Library; *p.144:* Czech Embassy; *p.145:* Steven C. Wilson/ENTHEOS; *p.149: (both)* © Patrick Sutherland; *p.154: (top)* Farmers' *Weekly*/Keith Huggett; *p.154 (bottom) and p.155:* Farmers' *Weekly*/Charles Topham; *p.156:* Colin Molyneux/Farming Information Centre, NFU; *p.157:* © Thelwell; *p.163:* © Patrick Sutherland; *p.165: (both)* © Richard Denyer; *p.172:* © Thelwell; *p.178:* S. & O. Mathews Photography; *p.182:* Ramblers' Association; *p.185: (top)* Forestry Commission, *(bottom)* © Fay Godwin; *p.186:* Forestry Commission; *p.191: (top)* Mike England, *(bottom)* CEGB Photo Library/Nuclear Electric plc; *p.194:* © Thelwell; *p.195: (both)* Peak National Park; *p.199:* S. & O Mathews Photography; *p.220: (top)* Engraving by Mackenzie after J. Tassie, from Mansell Collection, *(bottom)* Mezzotint by

Index

intensive farming systems – *contd.*
  unit size 156
  veal 80, 155, 156
 modern pressures 234–5
intercropping 44, *44*, 62, 75
interest rates 35
International Monetary Fund (IMF) 36,
  214, 225
International Rice Research Institute
  (IRRI) 93, 110
international specialisation 217, 220,
  222, 236, 246
Intervention Board for Agricultural
  Produce 138
intervention boards 138
Iraq 31, 72, 268
Ireland 29–30, *209*, 276
 see also Northern Ireland
irrigation 40, 50, 58, 128, 146
 developing countries 72
 effects on soils 71–2
 Guanxian contour canals 96
 investment in 98, 100
 Khmer empire 96
 labour requirements 105
 organisational requirements 97, 98,
  105, 119
 rice cultivation 95–8, 100, 105, 118,
  119
 state intervention in development 100
 subak system (Bali) 97
 tank irrigation 96
irrigation, types 72, 96, 97, 100, 105
Italy 137, *137*, 279
Ivory Coast 35

Japan
 agriculture 63, 69, 105–10, *106*, 121
  green revolution 110, 111
  historical development 100, 101–4
  silk production 109
 control of world's capital 35
 diversification of rural employment 118
 world trade 213, 225, 247, 249
Java 57, 117

Kenya 9, *9*, 77
Kielder Reservoir (Northumberland,
  UK) 187
Kinder Scout mass trespass 194
Korea 4, 63, 100, *106*, 110, 111
Kuwait *4*, 31

labelling of source 245
labour 55–6, *56–7*, 62, 64–8, *65–8*, 216
 agriculture
  peripatetic harvesting contractors
  142
  rice cultivation 69, 103, 105–9, *108*,
  111–12, 117–18

temperate farming 128–9, *129*, 132,
  142, *143*, 173
child 27, *27*, 117
female 27, 129, *129*
fisheries 55
forestry 55
formal/informal sector 32, *32*
guest workers 35
migration 28–32, *30–1*
productivity 57, 64–8, 69, 117–18
reserve armies 35
smokestack industries 285
under capitalism 34
women and the household 278
Ladybower reservoir (Peak National
  Park, UK) 195
land 52, *53*
 capability maps 128
 capital input 129
 cash crops 81
 concentration in agriculture 137
 degradation 70, 79
  see also desertification; soil, erosion;
  soil, salinsation
 deindustrialisation 293
 ownership 16, 81, 103, 111, 215–16
 productivity 57, *57–64*, 58–64
  rice cultivation 106, *106*, 121–2
 reclamation 158–9
 tenure 52, 81, 103–4, 106, 110–11
 third-world prime land 81
landscape 79, 162, 167, 171, 175–7
 change in visual quality 172–3
 extensification 201
 land-use pressures (case study) 193–6
Latin America *4*, 59, 66–7, 68, 82, 241–2,
  243, *264*, *265*, *303*, *312*
 disease risks 114
 land ownership 81
 old world diseases 280
Le Corbusier 294
LEADER programme 206, *207*
least-developed countries, definition *4*
legume crops 43, 45, 88
leisure *see* recreation
Less Favoured Areas (EC Directive)
  159, 170, *171*, 177, 194
less-developed countries, definition *4*
Libya *4*
life expectancy 22, 25–6, 36
limestone extraction 196
livestock 126, 136
 animal feed 79, 129, 130, 150
 beef/veal cattle 42–3, 80, 136, 137,
  155–6
 dairying 79, 81, 133, 140, 153–5,
  154–5
 dependency on feed imports 81, 129
 diversification 203
 domestication 40, *40–1*, 45

energy to protein conversion 42–3,
  148, *148*, 156
 environmental impact 150–2, 153,
  156–7, 158–62
 nomadic herding 54
 pigs 40, 45, 126, 133, 137, 140, 155–6
 poultry 40–3, 45, 126, 137, 155–6
 ruminants/non-ruminants 42–3
 sheep 42, 136, 137, 154–5
 slurry 43, 45, 133, *133*, 153, *154*, 156–7
livestock grazing, moorland 151, 159,
  160, 161
living standards 22, 36
Longnor (Peak National Park, UK) 206
low input-output farming 167, 200

McSharry, Raymond 171, 199, 200
maize 45, 61, *62*, *78*, 136, 242
Malawi 241, 242–4
Malaysia 45, 110, 116, 231, 245
malnutrition 16–17, 80–1, 83
Malthus, Thomas 17–20, 21, 35
Malthusian checks 17, 18
Manchester Development Corporation
  295
manioc (cassava) 41, 42, 45, 129
Mao Zedung 37, 119
marketing boards 138
markets 224, 231
marshland, loss of 171
Marx, Karl 34–5
mass selection 94
maternal mortality 10
Maximum Admissible Concentrations
  164
meat, world trade 20–1, *21*
medical care 22, 25–6
Mediterranean farming 40, 146, 152, 206
methane (marsh gas) 91
Mexico 35, 45, 61, 222, 232, 249, 279
Mexico City 31
middle developing countries *4*
Middle East 6, 21, 45, 59, 67
 see also under country e.g. Iraq
migration 16, 23–6, 28–32, 67, 216
military power, urbanisation 266
military training, environmental effects
  189–90
milk
 quota scheme 155, 200
 world trade 20–1, *20*
millet 18, 50
mineral elements 86–9
 see also nutrient cycle
mineral extraction 188–9, 196
 world trade 213
mineralisation 89
Minerals Act (1951) 188
Minerals Exploration Act (1971) 188–9
Ministry of Defence (MoD) 190